Everyone's Guide to Planet Earth

Compiled by

Nicholle Rojas

Scribbles

Year of Publication 2018

ISBN : 9789352979516

Book Published by

Scribbles

(An Imprint of Alpha Editions)

email - alphaedis@gmail.com

Produced by: PediaPress GmbH
Limburg an der Lahn
Germany
http://pediapress.com/

Contents

Introduction

Earth

<indicator name="pp-default"> 🔒 </indicator>
<indicator name="featured-star"> ⭐ </indicator>

Earth

The Blue Marble photograph of Earth, taken during the Apollo 17 lunar mission in 1972

Orbital characteristics	
Epoch J2000	
Aphelion	152100000 km (94500000 mi; 1.017 AU)
Perihelion	147095000 km (91401000 mi; 0.98327 AU)
Semi-major axis	149598023 km (92955902 mi; 1.00000102 AU)
Eccentricity	0.0167086
Orbital period	365.256363004 d (1.00001742096 yr)

Average orbital speed	29.78 km/s (107200 km/h; 66600 mph)
Mean anomaly	358.617°
Inclination	• 7.155° to the Sun's equator; • 1.57869° to invariable plane; • 0.00005° to J2000 ecliptic
Longitude of ascending node	−11.26064° to J2000 ecliptic
Argument of perihelion	114.20783°
Satellites	• 1 natural satellite: the Moon • 5 quasi-satellites • >1 700 operational artificial satellites • >16 000 space debris
Physical characteristics	
Mean radius	6371.0 km (3958.8 mi)
Equatorial radius	6378.1 km (3963.2 mi)
Polar radius	6356.8 km (3949.9 mi)
Flattening	0.0033528 1/298.257222101 (ETRS89)
Circumference	• 40075.017 km equatorial (24901.461 mi)[1] • 40007.86 km meridional (24859.73 mi)[2]
Surface area	• 510072000 km^2 (196940000 sq mi) • 148940000 km^2 land (57510000 sq mi; 29.2%) • 361132000 km^2 water (139434000 sq mi; 70.8%)
Volume	1.08321 × 10^{12} km^3 (2.59876 × 10^{11} cu mi)
Mass	5.97237 × 10^{24} kg (1.31668 × 10^{25} lb) (3.0 × 10^{-6} $M_\odot$)
Mean density	5.514 g/cm^3 (0.1992 lb/cu in)
Surface gravity	9.807 m/s^2 (1 g; 32.18 ft/s^2)
Moment of inertia factor	0.3307
Escape velocity	11.186 km/s (40270 km/h; 25020 mph)
Sidereal rotation period	0.99726968 d (23h 56m 4.100s)
Equatorial rotation velocity	0.4651 km/s (1674.4 km/h; 1040.4 mph)
Axial tilt	23.4392811°
Albedo	• 0.367 geometric • 0.306 Bond

Surface temp.	min	mean	max
Kelvin	184 K	288 K	330 K
Celsius	−89.2 °C	14.9 °C	56.9 °C
Fahrenheit	−128.5 °F	58.7 °F	134.3 °F

Atmosphere	
Surface pressure	101.325 kPa (at MSL)
Composition by volume	• 78.08% nitrogen (N_2; dry air) • 20.95% oxygen (O_2) • 0.934% argon • 0.0408% carbon dioxide • ∼1% water vapor (climate variable)

Earth is the third planet from the Sun and the only astronomical object known to harbor life. According to radiometric dating and other sources of evidence, Earth formed over 4.5 billion years ago. Earth's gravity interacts with other objects in space, especially the Sun and the Moon, Earth's only natural satellite. Earth revolves around the Sun in 365.26 days, a period known as an Earth year. During this time, Earth rotates about its axis about 366.26 times.

Earth's axis of rotation is tilted with respect to its orbital plane, producing seasons on Earth. The gravitational interaction between Earth and the Moon causes ocean tides, stabilizes Earth's orientation on its axis, and gradually slows its rotation. Earth is the densest planet in the Solar System and the largest of the four terrestrial planets.

Earth's lithosphere is divided into several rigid tectonic plates that migrate across the surface over periods of many millions of years. About 71% of Earth's surface is covered with water, mostly by oceans. The remaining 29% is land consisting of continents and islands that together have many lakes, rivers and other sources of water that contribute to the hydrosphere. The majority of Earth's polar regions are covered in ice, including the Antarctic ice sheet and the sea ice of the Arctic ice pack. Earth's interior remains active with a solid iron inner core, a liquid outer core that generates the Earth's magnetic field, and a convecting mantle that drives plate tectonics.

Within the first billion years of Earth's history, life appeared in the oceans and began to affect the Earth's atmosphere and surface, leading to the proliferation of aerobic and anaerobic organisms. Some geological evidence indicates that life may have arisen as much as 4.1 billion years ago. Since then, the combination of Earth's distance from the Sun, physical properties, and geological history have allowed life to evolve and thrive.[3] In the history of the Earth, biodiversity has gone through long periods of expansion, occasionally punctuated by mass extinction events. Over 99% of all species that ever lived on Earth are

Figure 1: *An early mention of "eorðan" (earth) in Beowulf*

extinct. Estimates of the number of species on Earth today vary widely; most species have not been described. Over 7.6 billion humans live on Earth and depend on its biosphere and natural resources for their survival. Humans have developed diverse societies and cultures; politically, the world has about 200 sovereign states.

Name and etymology

The modern English word *Earth* developed from a wide variety of Middle English forms,[4] which derived from an Old English noun most often spelled *eorðe*.[5] It has cognates in every Germanic language, and their proto-Germanic root has been reconstructed as **erþō*. In its earliest appearances, *eorðe* was already being used to translate the many senses of Latin *terra* and Greek γῆ (*gē*): the ground,[6]</ref> its soil,[7]</ref> dry land,[8]
"And God called the dry land **Earth**; and the gathering together of the waters called he Seas."[9]</ref> the human world,[10]</ref> the surface of the world (including the sea),[11]
"Here first with mighty power the Everlasting Lord, the Helm of all created things, Almighty King, made **earth** and heaven, raised up the sky and founded the spacious land."[12]</ref> and the globe itself.[13]</ref> As with Terra and Gaia, Earth was a personified goddess in Germanic paganism: the Angles were listed by Tacitus as among the devotees of Nerthus,[14] and later Norse mythology included Jörð, a giantess often given as the mother of Thor.[15]

Originally, *earth* was written in lowercase, and from early Middle English, its definite sense as "the globe" was expressed as *the earth*. By Early Modern English, many nouns were capitalized, and *the earth* became (and often remained) *the Earth*, particularly when referenced along with other heavenly bodies. More recently, the name is sometimes simply given as *Earth*, by analogy with the names of the other planets. House styles now vary: Oxford spelling recognizes the lowercase form as the most common, with the capitalized form an acceptable variant. Another convention capitalizes "Earth" when appearing as a name (e.g. "Earth's atmosphere") but writes it in lowercase when preceded by *the* (e.g. "the atmosphere of the earth"). It almost

Figure 2: *Artist's impression of the early Solar System's planetary disk*

always appears in lowercase in colloquial expressions such as "what on earth are you doing?"[16]

Chronology

Formation

The oldest material found in the Solar System is dated to 4.5672±0.0006 billion years ago (Bya). By 4.54±0.04 Bya the primordial Earth had formed. The bodies in the Solar System formed and evolved with the Sun. In theory, a solar nebula partitions a volume out of a molecular cloud by gravitational collapse, which begins to spin and flatten into a circumstellar disk, and then the planets grow out of that disk with the Sun. A nebula contains gas, ice grains, and dust (including primordial nuclides). According to nebular theory, planetesimals formed by accretion, with the primordial Earth taking 10–20 million years (Mys) to form.

A subject of research is the formation of the Moon, some 4.53 Bya. A leading hypothesis is that it was formed by accretion from material loosed from Earth after a Mars-sized object, named Theia, hit Earth. In this view, the mass of Theia was approximately 10 percent of Earth, it hit Earth with a glancing blow and some of its mass merged with Earth. Between approximately 4.1 and 3.8 Bya, numerous asteroid impacts during the Late Heavy Bombardment caused significant changes to the greater surface environment of the Moon and, by inference, to that of Earth.

Figure 3. *Hoodoos at the Bryce Canyon National Park, Utah*

Geological history

Earth's atmosphere and oceans were formed by volcanic activity and outgassing. Water vapor from these sources condensed into the oceans, augmented by water and ice from asteroids, protoplanets, and comets. In this model, atmospheric "greenhouse gases" kept the oceans from freezing when the newly forming Sun had only 70% of its current luminosity. By 3.5 Bya, Earth's magnetic field was established, which helped prevent the atmosphere from being stripped away by the solar wind.

A crust formed when the molten outer layer of Earth cooled to form a solid. The two models that explain land mass propose either a steady growth to the present-day forms or, more likely, a rapid growth early in Earth history followed by a long-term steady continental area. Continents formed by plate tectonics, a process ultimately driven by the continuous loss of heat from Earth's interior. Over the period of hundreds of millions of years, the supercontinents have assembled and broken apart. Roughly 750 million years ago (Mya), one of the earliest known supercontinents, Rodinia, began to break apart. The continents later recombined to form Pannotia 600–540 Mya, then finally Pangaea, which also broke apart 180 Mya.

The present pattern of ice ages began about 40 Mya and then intensified during the Pleistocene about 3 Mya. High-latitude regions have since undergone repeated cycles of glaciation and thaw, repeating about every

40,000–100,000 years. The last continental glaciation ended 10,000 years ago.

Origin of life and evolution

Life timeline

θ —
500
1000
1500
2000
2500
3000
3500
4000
4500

Axis scale: million years

Also see: *Human timeline* and *Nature timeline*

Phylogenetic Tree of Life

Bacteria   Archaea   Eucarya

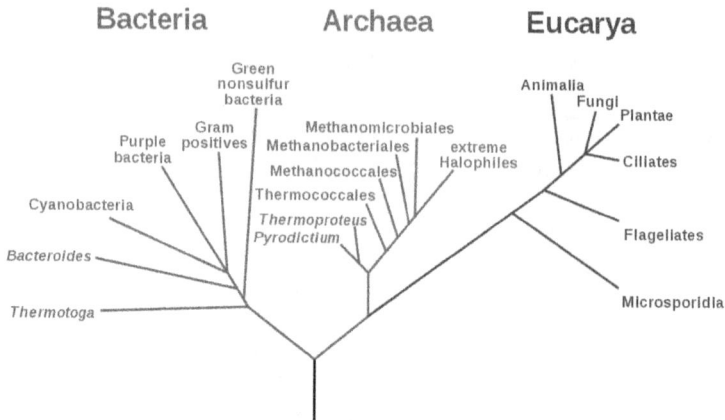

Figure 4: *Phylogenetic tree of life on Earth based on rRNA analysis*

Chemical reactions led to the first self-replicating molecules about four billion years ago. A half billion years later, the last common ancestor of all current life arose. The evolution of photosynthesis allowed the Sun's energy to be harvested directly by life forms. The resultant molecular oxygen (O$_2$) accumulated in the atmosphere and due to interaction with ultraviolet solar radiation, formed a protective ozone layer (O$_3$) in the upper atmosphere. The incorporation of smaller cells within larger ones resulted in the development of complex cells called eukaryotes. True multicellular organisms formed as cells within colonies became increasingly specialized. Aided by the absorption of harmful ultraviolet radiation by the ozone layer, life colonized Earth's surface. Among the earliest fossil evidence for life is microbial mat fossils found in 3.48 billion-year-old sandstone in Western Australia, biogenic graphite found in 3.7 billion-year-old metasedimentary rocks in Western Greenland, and remains of biotic material found in 4.1 billion-year-old rocks in Western Australia. The earliest direct evidence of life on Earth is contained in 3.45 billion-year-old Australian rocks showing fossils of microorganisms.

During the Neoproterozoic, 750 to 580 Mya, much of Earth might have been covered in ice. This hypothesis has been termed "Snowball Earth", and it is of particular interest because it preceded the Cambrian explosion, when multicellular life forms significantly increased in complexity. Following the Cambrian explosion, 535 Mya, there have been five mass extinctions. The most recent such event was 66 Mya, when an asteroid impact triggered the extinction of the non-avian dinosaurs and other large reptiles, but spared some small animals such as mammals, which at the time resembled shrews. Mammalian life has diversified over the past 66 Mys, and several million years ago an African ape-like animal such as *Orrorin tugenensis* gained the ability to stand upright. This facilitated tool use and encouraged communication that provided the nutrition and stimulation needed for a larger brain, which led to the evolution of humans. The development of agriculture, and then civilization, led to humans having an influence on Earth and the nature and quantity of other life forms that continues to this day.

Future

Earth's expected long-term future is tied to that of the Sun. Over the next 1.1 Bys, solar luminosity will increase by 10%, and over the next 3.5 Bys by 40%. The Earth's increasing surface temperature will accelerate the inorganic carbon cycle, reducing CO_2 concentration to levels lethally low for plants (10 ppm for C4 photosynthesis) in approximately 500–900 Mys. The lack of vegetation will result in the loss of oxygen in the atmosphere, making animal life impossible. After another billion years all surface water will have disappeared and the mean global temperature will reach 70 °C (158 °F). From that point, the Earth is expected to be habitable for another 500 Ma, possibly up to 2.3 Ga if nitrogen is removed from the atmosphere. Even if the Sun were eternal and stable, 27% of the water in the modern oceans will descend to the mantle in one billion years, due to reduced steam venting from mid-ocean ridges.

The Sun will evolve to become a red giant in about 5 Bys. Models predict that the Sun will expand to roughly 1 AU (150 million km; 93 million mi), about 250 times its present radius. Earth's fate is less clear. As a red giant, the Sun will lose roughly 30% of its mass, so, without tidal effects, Earth will move to an orbit 1.7 AU (250 million km; 160 million mi) from the Sun when the star reaches its maximum radius. Most, if not all, remaining life will be destroyed by the Sun's increased luminosity (peaking at about 5,000 times its present level). A 2008 simulation indicates that Earth's orbit will eventually decay due to tidal effects and drag, causing it to enter the Sun's atmosphere and be vaporized.

Shape of the Earth
distances of relief points to the geocentre

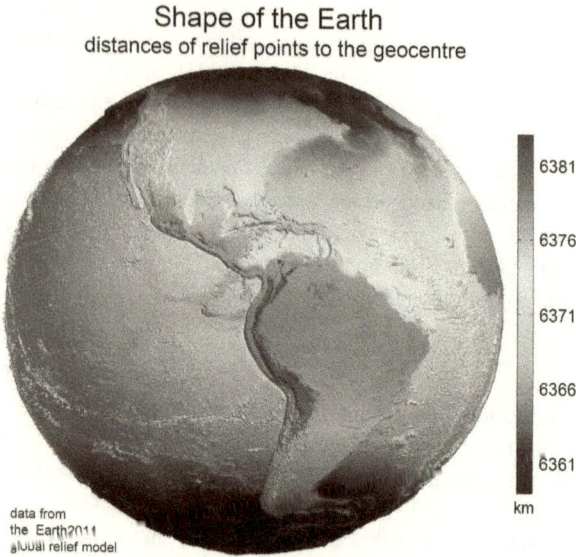

```
6381
6376
6371
6366
6361
```

data from
the Earth2011
global relief model

km

Figure 5: *Shape of planet Earth. Shown are distances between surface relief and the geocentre. The South American Andes summits are visible as elevated areas. Data from the Earth2014 global relief model.*

Physical characteristics

Shape

The shape of Earth is approximately oblate spheroidal. Due to rotation, the Earth is flattened at the poles and bulging around the equator. The diameter of the Earth at the equator is 43 kilometres (27 mi) larger than the pole-to-pole diameter. Thus the point on the surface farthest from Earth's center of mass is the summit of the equatorial Chimborazo volcano in Ecuador. The average diameter of the reference spheroid is 12,742 kilometres (7,918 mi). Local topography deviates from this idealized spheroid, although on a global scale these deviations are small compared to Earth's radius: The maximum deviation of only 0.17% is at the Mariana Trench (10,911 metres (35,797 ft) below local sea level), whereas Mount Everest (8,848 metres (29,029 ft) above local sea level) represents a deviation of 0.14%.[17]</ref>

In geodesy, the exact shape that Earth's oceans would adopt in the absence of land and perturbations such as tides and winds is called the geoid. More precisely, the geoid is the surface of gravitational equipotential at mean sea level.

Chemical composition

Chemical composition of the crust

Compound	Formula	Composition Continental	Composition Oceanic
silica	SiO_2	60.6%	48.6%
alumina	Al_2O_3	15.9%	16.5%
lime	CaO	6.41%	12.3%
magnesia	MgO	4.66%	6.8%
iron oxide	FeO_T	6.71%	6.2%
sodium oxide	Na_2O	3.07%	2.6%
potassium oxide	K_2O	1.81%	0.4%
titanium dioxide	TiO_2	0.72%	1.4%
phosphorus pentoxide	P_2O_5	0.13%	0.3%
manganese oxide	MnO	0.10%	1.4%
Total		**100.1%**	**99.9%**

Earth's mass is approximately 5.97×10^{24} kg (5,970 Yg). It is composed mostly of iron (32.1%), oxygen (30.1%), silicon (15.1%), magnesium (13.9%), sulfur (2.9%), nickel (1.8%), calcium (1.5%), and aluminium (1.4%), with the remaining 1.2% consisting of trace amounts of other elements. Due to mass segregation, the core region is estimated to be primarily composed of iron (88.8%), with smaller amounts of nickel (5.8%), sulfur (4.5%), and less than 1% trace elements.

The most common rock constituents of the crust are nearly all oxides: chlorine, sulfur, and fluorine are the important exceptions to this and their total amount in any rock is usually much less than 1%. Over 99% of the crust is composed of 11 oxides, principally silica, alumina, iron oxides, lime, magnesia, potash, and soda.

Internal structure

Earth's interior, like that of the other terrestrial planets, is divided into layers by their chemical or physical (rheological) properties. The outer layer is a chemically distinct silicate solid crust, which is underlain by a highly viscous solid mantle. The crust is separated from the mantle by the Mohorovičić discontinuity. The thickness of the crust varies from about 6 kilometres (3.7 mi) under the oceans to 30–50 km (19–31 mi) for the continents. The crust and the cold, rigid, top of the upper mantle are collectively known as the lithosphere, and it is of the lithosphere that the tectonic plates are composed. Beneath the lithosphere is the asthenosphere, a relatively low-viscosity layer on which the lithosphere rides. Important changes in crystal structure within the mantle occur at 410 and 660 km (250 and 410 mi) below the surface, spanning a transition zone that separates the upper and lower mantle. Beneath the mantle, an extremely low viscosity liquid outer core lies above a solid inner core. The Earth's inner core might rotate at a slightly higher angular velocity than the remainder of the planet, advancing by 0.1–0.5° per year. The radius of the inner core is about one fifth of that of Earth.

Geologic layers of Earth

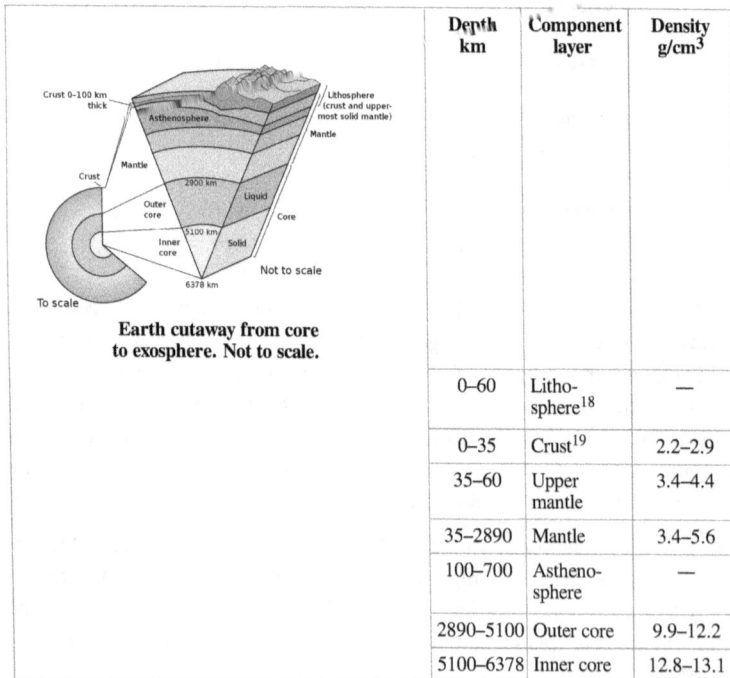

Earth cutaway from core to exosphere. Not to scale.

Depth km	Component layer	Density g/cm^3
0–60	Litho-sphere[18]	—
0–35	Crust[19]	2.2–2.9
35–60	Upper mantle	3.4–4.4
35–2890	Mantle	3.4–5.6
100–700	Astheno-sphere	—
2890–5100	Outer core	9.9–12.2
5100–6378	Inner core	12.8–13.1

Heat

Earth's internal heat comes from a combination of residual heat from planetary accretion (about 20%) and heat produced through radioactive decay (80%). The major heat-producing isotopes within Earth are potassium-40, uranium-238, and thorium-232. At the center, the temperature may be up to 6,000 °C

(10,830 °F), and the pressure could reach 360 GPa (52 million psi). Because much of the heat is provided by radioactive decay, scientists postulate that early in Earth's history, before isotopes with short half-lives were depleted, Earth's heat production was much higher. At approximately 3 Ga, twice the present-day heat would have been produced, increasing the rates of mantle convection and plate tectonics, and allowing the production of uncommon igneous rocks such as komatiites that are rarely formed today.

Present-day major heat-producing isotopes

Iso-tope	Heat release W/kg isotope	Half-life years	Mean mantle concentration kg isotope/kg mantle	Heat release W/kg mantle
238U	94.6×10^{-6}	4.47×10^9	30.8×10^{-9}	2.91×10^{-12}
235U	569×10^{-6}	0.704×10^9	0.22×10^{-9}	0.125×10^{-12}
232Th	26.4×10^{-6}	14.0×10^9	124×10^{-9}	3.27×10^{-12}
40K	29.2×10^{-6}	1.25×10^9	36.9×10^{-9}	1.08×10^{-12}

The mean heat loss from Earth is 87 mW m^{-2}, for a global heat loss of 4.42×10^{13} W. A portion of the core's thermal energy is transported toward the crust by mantle plumes, a form of convection consisting of upwellings of higher-temperature rock. These plumes can produce hotspots and flood basalts. More of the heat in Earth is lost through plate tectonics, by mantle upwelling associated with mid-ocean ridges. The final major mode of heat loss is through conduction through the lithosphere, the majority of which occurs under the oceans because the crust there is much thinner than that of the continents.

Tectonic plates

Earth's major plates

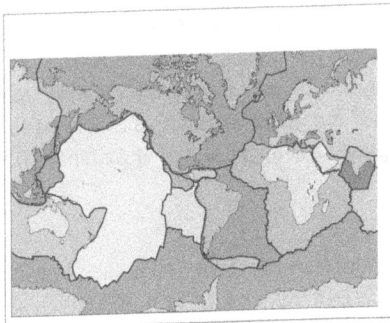

Plate name	Area 10^6 km²
Pacific Plate	103.3
African Plate	78.0
North American Plate	75.9
Eurasian Plate	67.8
Antarctic Plate	60.9
Indo-Australian Plate	47.2
South American Plate	43.6

Earth's mechanically rigid outer layer, the lithosphere, is divided into tectonic plates. These plates are rigid segments that move relative to each other at one of three boundaries types: At convergent boundaries, two plates come together; at divergent boundaries, two plates are pulled apart; and at transform boundaries, two plates slide past one another laterally. Along these plate boundaries, earthquakes, volcanic activity, mountain-building, and oceanic trench formation can occur. The tectonic plates ride on top of the asthenosphere, the solid but less-viscous part of the upper mantle that can flow and move along with the plates.

As the tectonic plates migrate, oceanic crust is subducted under the leading edges of the plates at convergent boundaries. At the same time, the upwelling of mantle material at divergent boundaries creates mid-ocean ridges. The combination of these processes recycles the oceanic crust back into the mantle. Due to this recycling, most of the ocean floor is less than 100 Ma old. The oldest oceanic crust is located in the Western Pacific and is estimated to be 200 Ma old. By comparison, the oldest dated continental crust is 4,030 Ma.

The seven major plates are the Pacific, North American, Eurasian, African, Antarctic, Indo-Australian, and South American. Other notable plates include the Arabian Plate, the Caribbean Plate, the Nazca Plate off the west coast of South America and the Scotia Plate in the southern Atlantic Ocean. The Australian Plate fused with the Indian Plate between 50 and 55 Mya. The fastest-moving plates are the oceanic plates, with the Cocos Plate advancing at a rate of 75 mm/a (3.0 in/year) and the Pacific Plate moving 52–69 mm/a (2.0–2.7 in/year). At the other extreme, the slowest-moving plate is the Eurasian Plate, progressing at a typical rate of 21 mm/a (0.83 in/year).

Surface

The total surface area of Earth is about 510 million km² (197 million sq mi). Of this, 70.8%, or 361.13 million km² (139.43 million sq mi), is below sea level and covered by ocean water. Below the ocean's surface are much of the

Figure 6: *Mountains build up when tectonic plates move toward each other, forcing rock up. The highest mountain on Earth above sea level is Mount Everest.*

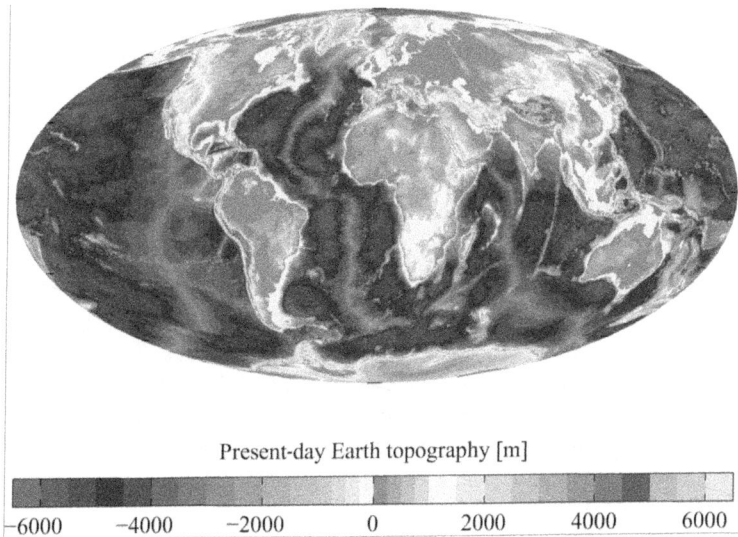

Present-day Earth topography [m]

-6000	-4000	-2000	0	2000	4000	6000

Figure 7: *Present-day Earth altimetry and bathymetry.*
Data from the National Geophysical Data Center.

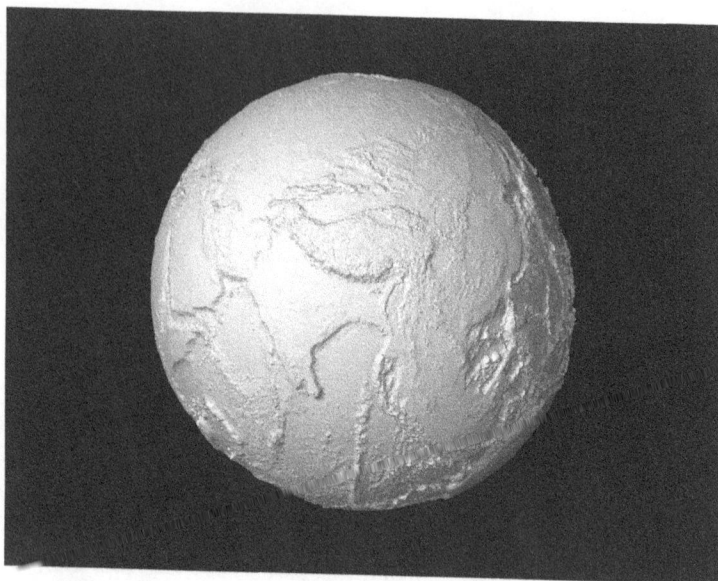

Figure 8: *Current Earth without water, elevation exaggerated 20 times (click/enlarge to "spin" 3D-globe).*

continental shelf, mountains, volcanoes, oceanic trenches, submarine canyons, oceanic plateaus, abyssal plains, and a globe-spanning mid-ocean ridge system. The remaining 29.2%, or 148.94 million km^2 (57.51 million sq mi), not covered by water has terrain that varies greatly from place to place and consists of mountains, deserts, plains, plateaus, and other landforms. Tectonics and erosion, volcanic eruptions, flooding, weathering, glaciation, the growth of coral reefs, and meteorite impacts are among the processes that constantly reshape the Earth's surface over geological time.

The continental crust consists of lower density material such as the igneous rocks granite and andesite. Less common is basalt, a denser volcanic rock that is the primary constituent of the ocean floors. Sedimentary rock is formed from the accumulation of sediment that becomes buried and compacted together. Nearly 75% of the continental surfaces are covered by sedimentary rocks, although they form about 5% of the crust. The third form of rock material found on Earth is metamorphic rock, which is created from the transformation of pre-existing rock types through high pressures, high temperatures, or both. The most abundant silicate minerals on Earth's surface include quartz, feldspars, amphibole, mica, pyroxene and olivine. Common carbonate minerals include calcite (found in limestone) and dolomite.

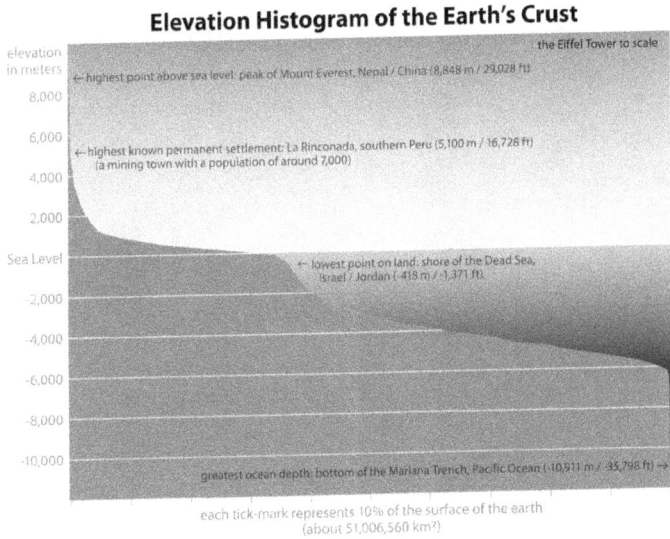

Figure 9: *Elevation histogram of Earth's surface*

The elevation of the land surface varies from the low point of −418 m (−1,371 ft) at the Dead Sea, to a maximum altitude of 8,848 m (29,029 ft) at the top of Mount Everest. The mean height of land above sea level is about 797 m (2,615 ft).

The pedosphere is the outermost layer of Earth's continental surface and is composed of soil and subject to soil formation processes. The total arable land is 10.9% of the land surface, with 1.3% being permanent cropland. Close to 40% of Earth's land surface is used for agriculture, or an estimated 16.7 million km^2 (6.4 million sq mi) of cropland and 33.5 million km^2 (12.9 million sq mi) of pastureland.

Hydrosphere

The abundance of water on Earth's surface is a unique feature that distinguishes the "Blue Planet" from other planets in the Solar System. Earth's hydrosphere consists chiefly of the oceans, but technically includes all water surfaces in the world, including inland seas, lakes, rivers, and underground waters down to a depth of 2,000 m (6,600 ft). The deepest underwater location is Challenger Deep of the Mariana Trench in the Pacific Ocean with a depth of 10,911.4 m (35,799 ft).

Figure 10: *Satellite image of Earth cloud cover using NASA's Moderate-Resolution Imaging Spectroradiometer*

The mass of the oceans is approximately 1.35×10^{18} metric tons or about 1/4400 of Earth's total mass. The oceans cover an area of 361.8 million km² (139.7 million sq mi) with a mean depth of 3,682 m (12,080 ft), resulting in an estimated volume of 1.332 billion km³ (320 million cu mi). If all of Earth's crustal surface were at the same elevation as a smooth sphere, the depth of the resulting world ocean would be 2.7 to 2.8 km (1.68 to 1.74 mi).

About 97.5% of the water is saline; the remaining 2.5% is fresh water. Most fresh water, about 68.7%, is present as ice in ice caps and glaciers.

The average salinity of Earth's oceans is about 35 grams of salt per kilogram of sea water (3.5% salt). Most of this salt was released from volcanic activity or extracted from cool igneous rocks. The oceans are also a reservoir of dissolved atmospheric gases, which are essential for the survival of many aquatic life forms. Sea water has an important influence on the world's climate, with the oceans acting as a large heat reservoir. Shifts in the oceanic temperature distribution can cause significant weather shifts, such as the El Niño–Southern Oscillation.

Atmosphere

The atmospheric pressure at Earth's sea level averages 101.325 kPa (14.696 psi), with a scale height of about 8.5 km (5.3 mi). A dry atmosphere is composed of 78.084% nitrogen, 20.946% oxygen, 0.934% argon, and trace amounts of carbon dioxide and other gaseous molecules. Water vapor content varies between 0.01% and 4% but averages about 1%. The height of the troposphere varies with latitude, ranging between 8 km (5 mi) at the poles to

17 km (11 mi) at the equator, with some variation resulting from weather and seasonal factors.

Earth's biosphere has significantly altered its atmosphere. Oxygenic photosynthesis evolved 2.7 Gya, forming the primarily nitrogen–oxygen atmosphere of today. This change enabled the proliferation of aerobic organisms and, indirectly, the formation of the ozone layer due to the subsequent conversion of atmospheric O

2 into O

3. The ozone layer blocks ultraviolet solar radiation, permitting life on land. Other atmospheric functions important to life include transporting water vapor, providing useful gases, causing small meteors to burn up before they strike the surface, and moderating temperature. This last phenomenon is known as the greenhouse effect: trace molecules within the atmosphere serve to capture thermal energy emitted from the ground, thereby raising the average temperature. Water vapor, carbon dioxide, methane, nitrous oxide, and ozone are the primary greenhouse gases in the atmosphere. Without this heat-retention effect, the average surface temperature would be –18 °C (0 °F), in contrast to the current +15 °C (59 °F), and life on Earth probably would not exist in its current form. In May 2017, glints of light, seen as twinkling from an orbiting satellite a million miles away, were found to be reflected light from ice crystals in the atmosphere.

Weather and climate

<templatestyles src="Multiple_image/styles.css" />

Hurricane Felix seen from low Earth orbit, September 2007

Lenticular cloud over an ice pressure ridge near Mount Discovery, Antarctica, November 2013

Massive clouds above the Mojave Desert, February 2016

Earth's atmosphere has no definite boundary, slowly becoming thinner and fading into outer space. Three-quarters of the atmosphere's mass is contained within the first 11 km (6.8 mi) of the surface. This lowest layer is called the troposphere. Energy from the Sun heats this layer, and the surface below, causing expansion of the air. This lower-density air then rises and is replaced by cooler, higher-density air. The result is atmospheric circulation that drives the weather and climate through redistribution of thermal energy.

The primary atmospheric circulation bands consist of the trade winds in the equatorial region below 30° latitude and the westerlies in the mid-latitudes between 30° and 60°. Ocean currents are also important factors in determining climate, particularly the thermohaline circulation that distributes thermal energy from the equatorial oceans to the polar regions.

Water vapor generated through surface evaporation is transported by circulatory patterns in the atmosphere. When atmospheric conditions permit an uplift of warm, humid air, this water condenses and falls to the surface as precipitation. Most of the water is then transported to lower elevations by river systems and usually returned to the oceans or deposited into lakes. This water cycle is a vital mechanism for supporting life on land and is a primary factor in the erosion of surface features over geological periods. Precipitation patterns vary widely, ranging from several meters of water per year to less than a millimeter. Atmospheric circulation, topographic features, and temperature differences determine the average precipitation that falls in each region.

The amount of solar energy reaching Earth's surface decreases with increasing latitude. At higher latitudes, the sunlight reaches the surface at lower angles, and it must pass through thicker columns of the atmosphere. As a result, the mean annual air temperature at sea level decreases by about 0.4 °C (0.7 °F) per degree of latitude from the equator. Earth's surface can be subdivided into specific latitudinal belts of approximately homogeneous climate. Ranging from the equator to the polar regions, these are the tropical (or equatorial), subtropical, temperate and polar climates.

This latitudinal rule has several anomalies:

- Proximity to oceans moderates the climate. For example, the Scandinavian Peninsula has more moderate climate than similarly northern latitudes of northern Canada.
- The wind enables this moderating effect. The windward side of a land mass experiences more moderation than the leeward side. In the Northern Hemisphere, the prevailing wind is west-to-east, and western coasts tend to be milder than eastern coasts. This is seen in Eastern North America and Western Europe, where rough continental climates appear on the east coast on parallels with mild climates on the other side of the ocean. In the Southern Hemisphere, the prevailing wind is east-to-west, and the eastern coasts are milder.
- The distance from the Earth to the Sun varies. The Earth is closest to the Sun (at perihelion) in January, which is summer in the Southern Hemisphere. It is furthest away (at aphelion) in July, which is summer in the Northern Hemisphere, and only 93.55% of the solar radiation from the Sun falls on a given square area of land than at perihelion. Despite this, there are larger land masses in the Northern Hemisphere, which are easier to heat than the seas. Consequently, summers are 2.3 °C (4 °F) warmer in the Northern Hemisphere than in the Southern Hemisphere under similar conditions.
- The climate is colder at high altitudes than at sea level because of the decreased air density.

The commonly used Köppen climate classification system has five broad groups (humid tropics, arid, humid middle latitudes, continental and cold polar), which are further divided into more specific subtypes. The Köppen system rates regions of terrain based on observed temperature and precipitation.

The highest air temperature ever measured on Earth was 56.7 °C (134.1 °F) in Furnace Creek, California, in Death Valley, in 1913. The lowest air temperature ever directly measured on Earth was –89.2 °C (–128.6 °F) at Vostok Station in 1983, but satellites have used remote sensing to measure temperatures as low as –94.7 °C (–138.5 °F) in East Antarctica. These temperature records are only measurements made with modern instruments from the 20th century onwards and likely do not reflect the full range of temperature on Earth.

Upper atmosphere

Above the troposphere, the atmosphere is usually divided into the stratosphere, mesosphere, and thermosphere. Each layer has a different lapse rate, defining the rate of change in temperature with height. Beyond these, the exosphere thins out into the magnetosphere, where the geomagnetic fields interact with the solar wind. Within the stratosphere is the ozone layer, a component that partially shields the surface from ultraviolet light and thus is important for life

Figure 11: *This view from orbit shows the full moon partially obscured by Earth's atmosphere.*

on Earth. The Kármán line, defined as 100 km above Earth's surface, is a working definition for the boundary between the atmosphere and outer space.

Thermal energy causes some of the molecules at the outer edge of the atmosphere to increase their velocity to the point where they can escape from Earth's gravity. This causes a slow but steady loss of the atmosphere into space. Because unfixed hydrogen has a low molecular mass, it can achieve escape velocity more readily, and it leaks into outer space at a greater rate than other gases. The leakage of hydrogen into space contributes to the shifting of Earth's atmosphere and surface from an initially reducing state to its current oxidizing one. Photosynthesis provided a source of free oxygen, but the loss of reducing agents such as hydrogen is thought to have been a necessary precondition for the widespread accumulation of oxygen in the atmosphere. Hence the ability of hydrogen to escape from the atmosphere may have influenced the nature of life that developed on Earth. In the current, oxygen-rich atmosphere most hydrogen is converted into water before it has an opportunity to escape. Instead, most of the hydrogen loss comes from the destruction of methane in the upper atmosphere.

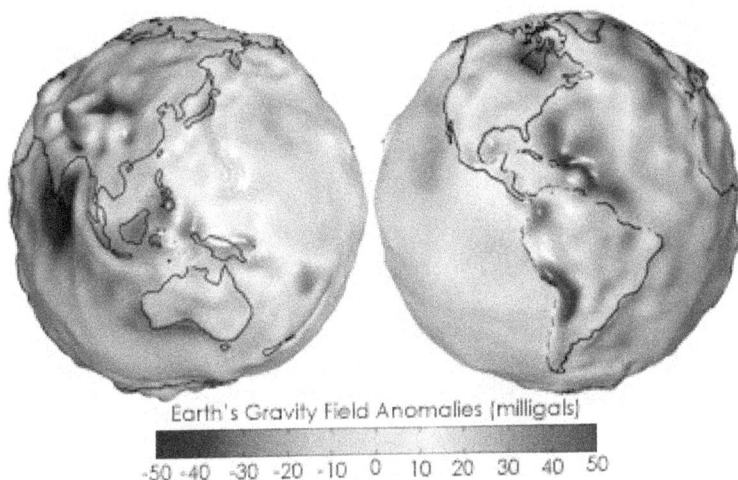

Figure 12: *Earth's gravity measured by NASA's GRACE mission, showing deviations from the theoretical gravity. Red shows where gravity is stronger than the smooth, standard value, and blue shows where it is weaker.*

Gravitational field

The gravity of Earth is the acceleration that is imparted to objects due to the distribution of mass within the Earth. Near the Earth's surface, gravitational acceleration is approximately 9.8 m/s^2 (32 ft/s^2). Local differences in topography, geology, and deeper tectonic structure cause local and broad, regional differences in the Earth's gravitational field, known as gravitational anomalies.

Magnetic field

The main part of Earth's magnetic field is generated in the core, the site of a dynamo process that converts the kinetic energy of thermally and compositionally driven convection into electrical and magnetic field energy. The field extends outwards from the core, through the mantle, and up to Earth's surface, where it is, approximately, a dipole. The poles of the dipole are located close to Earth's geographic poles. At the equator of the magnetic field, the magnetic-field strength at the surface is 3.05 $\times$ 10^{-5} T, with global magnetic dipole moment of 7.91 $\times$ 10^{15} T m^3. The convection movements in the core are chaotic; the magnetic poles drift and periodically change alignment. This causes secular variation of the main field and field reversals at irregular intervals averaging a few times every million years. The most recent reversal occurred approximately 700,000 years ago.

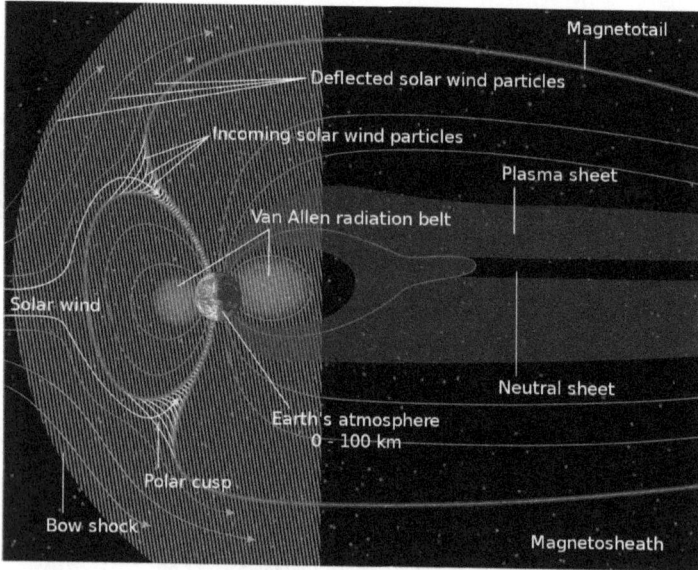

Figure 13: *Schematic of Earth's magnetosphere. The solar wind flows from left to right*

Magnetosphere

The extent of Earth's magnetic field in space defines the magnetosphere. Ions
and electrons of the solar wind are deflected by the magnetosphere; solar wind
pressure compresses the dayside of the magnetosphere, to about 10 Earth radii,
and extends the nightside magnetosphere into a long tail. Because the velocity
of the solar wind is greater than the speed at which waves propagate through the
solar wind, a supersonic bowshock precedes the dayside magnetosphere within
the solar wind. Charged particles are contained within the magnetosphere; the
plasmasphere is defined by low-energy particles that essentially follow magnetic field lines as Earth rotates; the ring current is defined by medium-energy
particles that drift relative to the geomagnetic field, but with paths that are still
dominated by the magnetic field, and the Van Allen radiation belt are formed
by high-energy particles whose motion is essentially random, but otherwise
contained by the magnetosphere.

During magnetic storms and substorms, charged particles can be deflected
from the outer magnetosphere and especially the magnetotail, directed along
field lines into Earth's ionosphere, where atmospheric atoms can be excited
and ionized, causing the aurora.

Figure 14: *Earth's rotation imaged by DSCOVR EPIC on 29 May 2016, a few weeks before the solstice.*

Orbit and rotation

Rotation

Earth's rotation period relative to the Sun—its mean solar day—is 86,400 seconds of mean solar time (86,400.0025 SI seconds). Because Earth's solar day is now slightly longer than it was during the 19th century due to tidal deceleration, each day varies between 0 and 2 SI ms longer.

Earth's rotation period relative to the fixed stars, called its *stellar day* by the International Earth Rotation and Reference Systems Service (IERS), is 86,164.0989 seconds of mean solar time (UT1), or 23h 56m 4.0989s. Earth's rotation period relative to the precessing or moving mean vernal equinox, misnamed its *sidereal day*, is 86,164.0905 seconds of mean solar time (UT1) (23h 56m 4.0905s). Thus the sidereal day is shorter than the stellar day by about 8.4 ms. The length of the mean solar day in SI seconds is available from the IERS for the periods 1623–2005 and 1962–2005.

Apart from meteors within the atmosphere and low-orbiting satellites, the main apparent motion of celestial bodies in Earth's sky is to the west at a rate of 15°/h = 15'/min. For bodies near the celestial equator, this is equivalent to an apparent diameter of the Sun or the Moon every two minutes; from Earth's

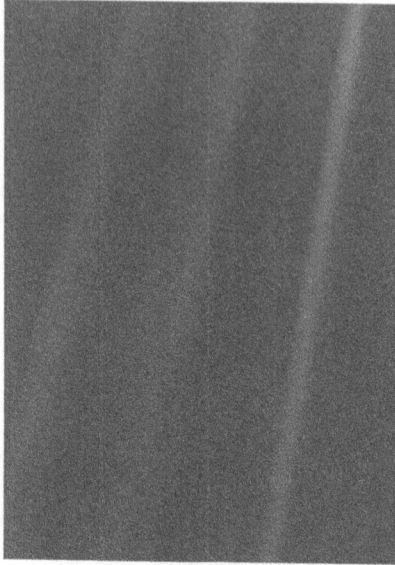

Figure 15: *The Pale Blue Dot photo taken in 1990 by the Voyager 1 spacecraft showing Earth (center right) from nearly 6.4 billion km (4 billion mi) away*

surface, the apparent sizes of the Sun and the Moon are approximately the same.

Orbit

Earth orbits the Sun at an average distance of about 150 million km (93 million mi) every 365.2564 mean solar days, or one sidereal year. This gives an apparent movement of the Sun eastward with respect to the stars at a rate of about 1°/day, which is one apparent Sun or Moon diameter every 12 hours. Due to this motion, on average it takes 24 hours—a solar day—for Earth to complete a full rotation about its axis so that the Sun returns to the meridian. The orbital speed of Earth averages about 29.78 km/s (107,200 km/h; 66,600 mph), which is fast enough to travel a distance equal to Earth's diameter, about 12,742 km (7,918 mi), in seven minutes, and the distance to the Moon, 384,000 km (239,000 mi), in about 3.5 hours.

The Moon and Earth orbit a common barycenter every 27.32 days relative to the background stars. When combined with the Earth–Moon system's common orbit around the Sun, the period of the synodic month, from new moon to new moon, is 29.53 days. Viewed from the celestial north pole, the motion of Earth, the Moon, and their axial rotations are all counterclockwise. Viewed

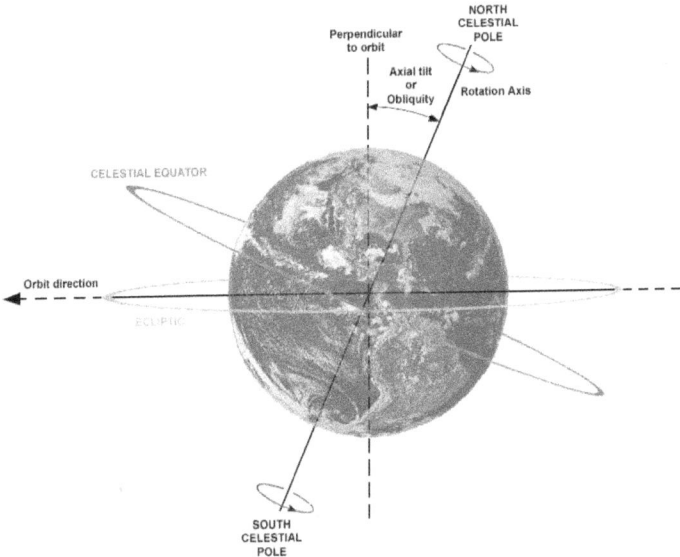

Figure 16: *Earth's axial tilt (or obliquity) and
its relation to the rotation axis and plane of orbit*

from a vantage point above the north poles of both the Sun and Earth, Earth orbits in a counterclockwise direction about the Sun. The orbital and axial planes
are not precisely aligned: Earth's axis is tilted some 23.44 degrees from the
perpendicular to the Earth–Sun plane (the ecliptic), and the Earth–Moon plane
is tilted up to ±5.1 degrees against the Earth–Sun plane. Without this tilt, there
would be an eclipse every two weeks, alternating between lunar eclipses and
solar eclipses.

The Hill sphere, or the sphere of gravitational influence, of the Earth is about
1.5 million km (930,000 mi) in radius. This is the maximum distance at which
the Earth's gravitational influence is stronger than the more distant Sun and
planets. Objects must orbit the Earth within this radius, or they can become
unbound by the gravitational perturbation of the Sun.

Earth, along with the Solar System, is situated in the Milky Way and orbits
about 28,000 light-years from its center. It is about 20 light-years above the
galactic plane in the Orion Arm.

Axial tilt and seasons

The axial tilt of the Earth is approximately 23.439281° with the axis of its
orbit plane, always pointing towards the Celestial Poles. Due to Earth's axial

tilt, the amount of sunlight reaching any given point on the surface varies over the course of the year. This causes the seasonal change in climate, with summer in the Northern Hemisphere occurring when the Tropic of Cancer is facing the Sun, and winter taking place when the Tropic of Capricorn in the Southern Hemisphere faces the Sun. During the summer, the day lasts longer, and the Sun climbs higher in the sky. In winter, the climate becomes cooler and the days shorter. In northern temperate latitudes, the Sun rises north of true east during the summer solstice, and sets north of true west, reversing in the winter. The Sun rises south of true east in the summer for the southern temperate zone and sets south of true west.

Above the Arctic Circle, an extreme case is reached where there is no daylight at all for part of the year, up to six months at the North Pole itself, a polar night. In the Southern Hemisphere, the situation is exactly reversed, with the South Pole oriented opposite the direction of the North Pole. Six months later, this pole will experience a midnight sun, a day of 24 hours, again reversing with the South Pole.

By astronomical convention, the four seasons can be determined by the solstices—the points in the orbit of maximum axial tilt toward or away from the Sun—and the equinoxes, when the direction of the tilt and the direction to the Sun are perpendicular. In the Northern Hemisphere, winter solstice currently occurs around 21 December; summer solstice is near 21 June, spring equinox is around 20 March and autumnal equinox is about 22 or 23 September. In the Southern Hemisphere, the situation is reversed, with the summer and winter solstices exchanged and the spring and autumnal equinox dates swapped.

The angle of Earth's axial tilt is relatively stable over long periods of time. Its axial tilt does undergo nutation; a slight, irregular motion with a main period of 18.6 years. The orientation (rather than the angle) of Earth's axis also changes over time, precessing around in a complete circle over each 25,800 year cycle; this precession is the reason for the difference between a sidereal year and a tropical year. Both of these motions are caused by the varying attraction of the Sun and the Moon on Earth's equatorial bulge. The poles also migrate a few meters across Earth's surface. This polar motion has multiple, cyclical components, which collectively are termed quasiperiodic motion. In addition to an annual component to this motion, there is a 14-month cycle called the Chandler wobble. Earth's rotational velocity also varies in a phenomenon known as length-of-day variation.

In modern times, Earth's perihelion occurs around 3 January, and its aphelion around 4 July. These dates change over time due to precession and other orbital factors, which follow cyclical patterns known as Milankovitch cycles. The changing Earth–Sun distance causes an increase of about 6.9% in solar energy

Figure 17: *The Rocky Mountains in Canada overlook Moraine Lake.*

reaching Earth at perihelion relative to aphelion. Because the Southern Hemisphere is tilted toward the Sun at about the same time that Earth reaches the closest approach to the Sun, the Southern Hemisphere receives slightly more energy from the Sun than does the northern over the course of a year. This effect is much less significant than the total energy change due to the axial tilt, and most of the excess energy is absorbed by the higher proportion of water in the Southern Hemisphere.

A study from 2016 suggested that Planet Nine tilted all Solar System planets, including Earth's, by about six degrees.

Habitability

A planet that can sustain life is termed habitable, even if life did not originate there. Earth provides liquid water—an environment where complex organic molecules can assemble and interact, and sufficient energy to sustain metabolism. The distance of Earth from the Sun, as well as its orbital eccentricity, rate of rotation, axial tilt, geological history, sustaining atmosphere, and magnetic field all contribute to the current climatic conditions at the surface.

Biosphere

A planet's life forms inhabit ecosystems, whose total is sometimes said to form a "biosphere". Earth's biosphere is thought to have begun evolving about 3.5 Gya. The biosphere is divided into a number of biomes, inhabited by broadly similar plants and animals. On land, biomes are separated primarily by differences in latitude, height above sea level and humidity. Terrestrial biomes lying within the Arctic or Antarctic Circles, at high altitudes or in extremely arid areas are relatively barren of plant and animal life; species diversity reaches a peak in humid lowlands at equatorial latitudes.

In July 2016, scientists reported identifying a set of 355 genes from the last universal common ancestor (LUCA) of all organisms living on Earth.

Natural resources and land use

Estimated human land use, 2000

Land use	Mha
Cropland	1,510–1,611
Pastures	2,500–3,410
Natural forests	3,143–3,871
Planted forests	126–215
Urban areas	66–351
Unused, productive land	356–445

Earth has resources that have been exploited by humans. Those termed non-renewable resources, such as fossil fuels, only renew over geological timescales.

Large deposits of fossil fuels are obtained from Earth's crust, consisting of coal, petroleum, and natural gas. These deposits are used by humans both for energy production and as feedstock for chemical production. Mineral ore bodies have also been formed within the crust through a process of ore genesis, resulting from actions of magmatism, erosion, and plate tectonics. These bodies form concentrated sources for many metals and other useful elements.

Earth's biosphere produces many useful biological products for humans, including food, wood, pharmaceuticals, oxygen, and the recycling of many organic wastes. The land-based ecosystem depends upon topsoil and fresh water, and the oceanic ecosystem depends upon dissolved nutrients washed down from the land. In 1980, 50.53 million km^2 (19.51 million sq mi) of Earth's

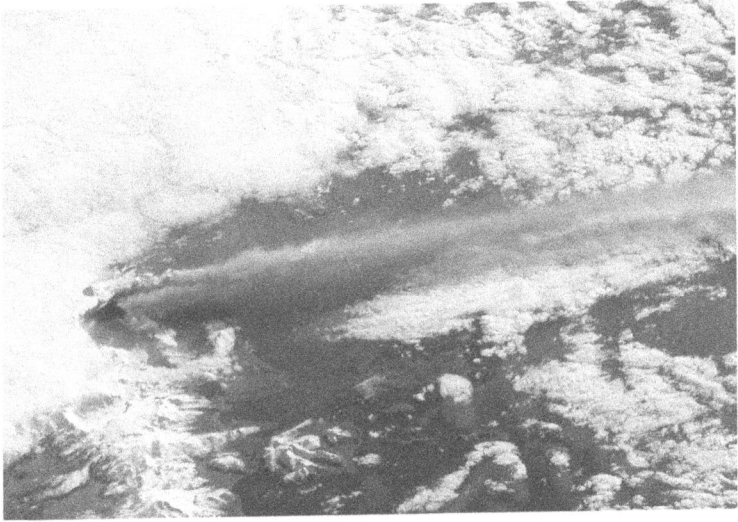

Figure 18: *A volcano injecting hot ash into the atmosphere*

land surface consisted of forest and woodlands, 67.88 million km^2 (26.21 million sq mi) was grasslands and pasture, and 15.01 million km^2 (5.80 million sq mi) was cultivated as croplands. The estimated amount of irrigated land in 1993 was 2,481,250 km^2 (958,020 sq mi). Humans also live on the land by using building materials to construct shelters.

Natural and environmental hazards

Large areas of Earth's surface are subject to extreme weather such as tropical cyclones, hurricanes, or typhoons that dominate life in those areas. From 1980 to 2000, these events caused an average of 11,800 human deaths per year. Many places are subject to earthquakes, landslides, tsunamis, volcanic eruptions, tornadoes, sinkholes, blizzards, floods, droughts, wildfires, and other calamities and disasters.

Many localized areas are subject to human-made pollution of the air and water, acid rain and toxic substances, loss of vegetation (overgrazing, deforestation, desertification), loss of wildlife, species extinction, soil degradation, soil depletion and erosion.

There is a scientific consensus linking human activities to global warming due to industrial carbon dioxide emissions. This is predicted to produce changes such as the melting of glaciers and ice sheets, more extreme temperature ranges, significant changes in weather and a global rise in average sea levels.

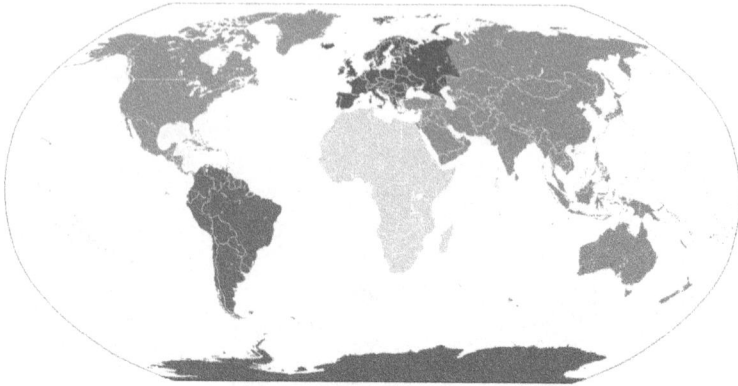

Figure 19:
The seven continents of Earth:[20]

Human geography

Cartography, the study and practice of map-making, and geography, the study of the lands, features, inhabitants and phenomena on Earth, have historically been the disciplines devoted to depicting Earth. Surveying, the determination of locations and distances, and to a lesser extent navigation, the determination of position and direction, have developed alongside cartography and geography, providing and suitably quantifying the requisite information.

Earth's human population reached approximately seven billion on 31 October 2011. Projections indicate that the world's human population will reach 9.2 billion in 2050. Most of the growth is expected to take place in developing nations. Human population density varies widely around the world, but a majority live in Asia. By 2020, 60% of the world's population is expected to be living in urban, rather than rural, areas.

It is estimated that one-eighth of Earth's surface is suitable for humans to live on – three-quarters of Earth's surface is covered by oceans, leaving one-quarter as land. Half of that land area is desert (14%), high mountains (27%), or other unsuitable terrains. The northernmost permanent settlement in the world is Alert, on Ellesmere Island in Nunavut, Canada. (82°28′N) The southernmost is the Amundsen–Scott South Pole Station, in Antarctica, almost exactly at the South Pole. (90°S)

Independent sovereign nations claim the planet's entire land surface, except for some parts of Antarctica, a few land parcels along the Danube river's western bank, and the unclaimed area of Bir Tawil between Egypt and Sudan. As of

2015[21], there are 193 sovereign states that are member states of the United Nations, plus two observer states and 72 dependent territories and states with limited recognition. Earth has never had a sovereign government with authority over the entire globe, although some nation-states have striven for world domination and failed.

The United Nations is a worldwide intergovernmental organization that was created with the goal of intervening in the disputes between nations, thereby avoiding armed conflict. The U.N. serves primarily as a forum for international diplomacy and international law. When the consensus of the membership permits, it provides a mechanism for armed intervention.

The first human to orbit Earth was Yuri Gagarin on 12 April 1961. In total, about 487 people have visited outer space and reached orbit as of 30 July 2010[21], and, of these, twelve have walked on the Moon. Normally, the only humans in space are those on the International Space Station. The station's crew, made up of six people, is usually replaced every six months. The farthest that humans have traveled from Earth is 400,171 km (248,655 mi), achieved during the Apollo 13 mission in 1970.

Moon

Characteristics

Diameter	3,474.8 km
Mass	7.349×10^{22} kg
Semi-major axis	384,400 km
Orbital period	27^d 7^h 43.7^m

The Moon is a relatively large, terrestrial, planet-like natural satellite, with a diameter about one-quarter of Earth's. It is the largest moon in the Solar System relative to the size of its planet, although Charon is larger relative to the

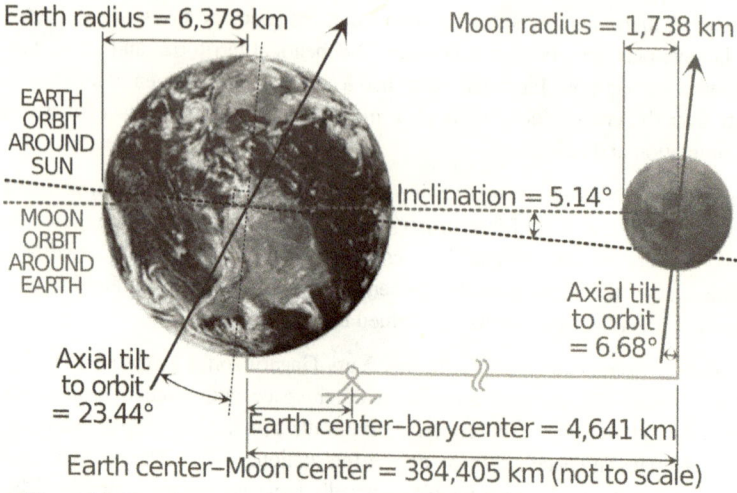

Figure 20: *Details of the Earth–Moon system, showing the radius of each object and the Earth–Moon barycenter. The Moon's axis is located by Cassini's third law.*

dwarf planet Pluto. The natural satellites of other planets are also referred to as "moons", after Earth's.

The gravitational attraction between Earth and the Moon causes tides on Earth. The same effect on the Moon has led to its tidal locking: its rotation period is the same as the time it takes to orbit Earth. As a result, it always presents the same face to the planet. As the Moon orbits Earth, different parts of its face are illuminated by the Sun, leading to the lunar phases; the dark part of the face is separated from the light part by the solar terminator.

Due to their tidal interaction, the Moon recedes from Earth at the rate of approximately 38 mm/a (1.5 in/year). Over millions of years, these tiny modifications—and the lengthening of Earth's day by about 23 μs/yr—add up to significant changes. During the Devonian period, for example, (approximately 410 Mya) there were 400 days in a year, with each day lasting 21.8 hours.

The Moon may have dramatically affected the development of life by moderating the planet's climate. Paleontological evidence and computer simulations show that Earth's axial tilt is stabilized by tidal interactions with the Moon. Some theorists think that without this stabilization against the torques applied by the Sun and planets to Earth's equatorial bulge, the rotational axis might be chaotically unstable, exhibiting chaotic changes over millions of years, as appears to be the case for Mars.

Viewed from Earth, the Moon is just far enough away to have almost the same apparent-sized disk as the Sun. The angular size (or solid angle) of these two

Figure 21: *Tracy Caldwell Dyson viewing Earth from the ISS Cupola, 2010*

bodies match because, although the Sun's diameter is about 400 times as large as the Moon's, it is also 400 times more distant. This allows total and annular solar eclipses to occur on Earth.

The most widely accepted theory of the Moon's origin, the giant-impact hypothesis, states that it formed from the collision of a Mars-size protoplanet called Theia with the early Earth. This hypothesis explains (among other things) the Moon's relative lack of iron and volatile elements and the fact that its composition is nearly identical to that of Earth's crust.

Asteroids and artificial satellites

Earth has at least five co-orbital asteroids, including 3753 Cruithne and 2002 AA29. A trojan asteroid companion, 2010 TK7, is librating around the leading Lagrange triangular point, L4, in the Earth's orbit around the Sun.

The tiny near-Earth asteroid 2006 RH120 makes close approaches to the Earth–Moon system roughly every twenty years. During these approaches, it can orbit Earth for brief periods of time.

As of August 2017[21], there were 1,738 operational, human-made satellites orbiting Earth. There are also inoperative satellites, including Vanguard 1, the oldest satellite currently in orbit, and over 16,000 pieces of tracked space debris. Earth's largest artificial satellite is the International Space Station.

Figure 22: *Earthrise, taken by astronauts on board Apollo 8*

Cultural and historical viewpoint

The standard astronomical symbol of Earth consists of a cross circumscribed by a circle, $\oplus$, representing the four corners of the world.

Human cultures have developed many views of the planet. Earth is sometimes personified as a deity. In many cultures it is a mother goddess that is also the primary fertility deity, and by the mid-20th century, the Gaia Principle compared Earth's environments and life as a single self-regulating organism leading to broad stabilization of the conditions of habitability.[22] Creation myths in many religions involve the creation of Earth by a supernatural deity or deities.

Scientific investigation has resulted in several culturally transformative shifts in people's view of the planet. Initial belief in a flat Earth was gradually displaced in the Greek colonies of southern Italy during the late 6th century BC by the idea of spherical Earth, which was attributed to both the philosophers Pythagoras and Parmenides. By the end of the 5th century BC, the sphericity of Earth was universally accepted among Greek intellectuals. Earth was generally believed to be the center of the universe until the 16th century, when scientists first conclusively demonstrated that it was a moving object, comparable to the other planets in the Solar System. Due to the efforts of influential Christian scholars and clerics such as James Ussher, who sought to determine the age of Earth through analysis of genealogies in Scripture, Westerners before the 19th century generally believed Earth to be a few thousand years old at most. It was only during the 19th century that geologists realized Earth's age was at least many millions of years.

Lord Kelvin used thermodynamics to estimate the age of Earth to be between 20 million and 400 million years in 1864, sparking a vigorous debate on the subject; it was only when radioactivity and radioactive dating were discovered in the late 19th and early 20th centuries that a reliable mechanism for determining Earth's age was established, proving the planet to be billions of years old. The perception of Earth shifted again in the 20th century when humans first viewed it from orbit, and especially with photographs of Earth returned by the Apollo program.

<templatestyles src="Multiple_image/styles.css" />

Clickable

Life on Earth

(view · discuss)

0.2 Mya

Humans

140 Mya

Flowers

200 Mya

Mammals

240 Mya

Dinosaurs

3500 Mya

Oxygen

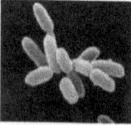

4280 Mya

Microorganisms

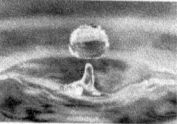

4410 Mya

Water

4540 Mya

Earth

<templatestyles src="Multiple_image/styles.css" />

Clickable

Location of the Earth in the Universe

(view · discuss)

Earth

Solar System

Gould Belt

Orion Arm

Milky Way

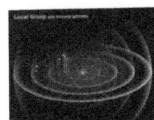

Local Group

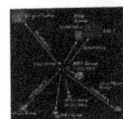

Virgo SCl

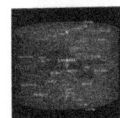

Laniakea SCl

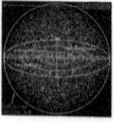

Our Universe

References

Further reading

* Comins, Neil F. (2001). *Discovering the Essential Universe*[23] (2nd ed.). New York: W. H. Freeman. Bibcode: 2003deu..book.....C[24]. ISBN 0-7167-5804-0. OCLC 52082611[25].

External links

<indicator name="spoken-icon"> ⊛⟫ </indicator>

* *National Geographic* encyclopedic entry about Earth[26]
* Earth – Profile[27] – Solar System Exploration[28] – NASA
* Earth – Climate Changes Cause Shape to Change[29] – NASA
* Earth – Astronaut Photography Gateway[30] – NASA
* Earth Observatory[31] – NASA
* Earth – Audio (29:28) – Cain/Gay – Astronomy Cast (2007)[32]
* Earth – Videos – International Space Station:
  * Video (01:02)[33] – Earth (time-lapse)
  * Video (00:27)[34] – Earth and auroras (time-lapse)
* United States Geological Survey[35] – USGS
* Google Earth 3D[36], interactive map

History of Earth

History of Earth

The **history of Earth** concerns the development of planet Earth from its formation to the present day. Nearly all branches of natural science have contributed to understanding of the main events of Earth's past, characterized by constant geological change and biological evolution.

The geological time scale (GTS), as defined by international convention,[37] depicts the large spans of time from the beginning of the Earth to the present, and its divisions chronicle some definitive events of Earth history. (In the graphic: Ga means "billion years ago"; Ma, "million years ago".) Earth formed around 4.54 billion years ago, approximately one-third the age of the universe, by accretion from the solar nebula. Volcanic outgassing probably created the primordial atmosphere and then the ocean, but the early atmosphere contained almost no oxygen. Much of the Earth was molten because of frequent collisions with other bodies which led to extreme volcanism. While Earth was in its earliest stage (Early Earth), a giant impact collision with a planet-sized body named Theia is thought to have formed the Moon. Over time, the Earth cooled, causing the formation of a solid crust, and allowing liquid water on the surface.

The Hadean eon represents the time before a reliable (fossil) record of life; it began with the formation of the planet and ended 4.0 billion years ago. The following Archean and Proterozoic eons produced the beginnings of life on Earth and its earliest evolution. The succeeding eon is the Phanerozoic, divided into three eras: the Palaeozoic, an era of arthropods, fishes, and the first life on land; the Mesozoic, which spanned the rise, reign, and climactic extinction of the non-avian dinosaurs; and the Cenozoic, which saw the rise of mammals. Recognizable humans emerged at most 2 million years ago, a vanishingly small period on the geological scale.

2 Ma:
First Hominins
230-66 Ma: 4550 Ma:
Non-avian dinosaurs   **Formation of the Earth**

c. 380 Ma: Hominins
First vertebrate land animals   Mammals
                                Land plants
c. 530 Ma: Animals
Cambrian explosion   Multicellular life
                                Eukaryotes   4527 Ma:
750-635 Ma: Prokaryotes   Formation of the Moon
Two Snowball Earths

c. 4000 Ma: End of the
Late Heavy Bombardment;
first life

Hadean

Paleozoic   Mesozoic

1 Ga

c. 3200 Ma:
Earliest start
of Photosynthesis

Proterozoic

Archean

.5 Ga

2 Ga

c. 2300 Ma:
Atmosphere becomes oxygen-rich;
first Snowball Earth

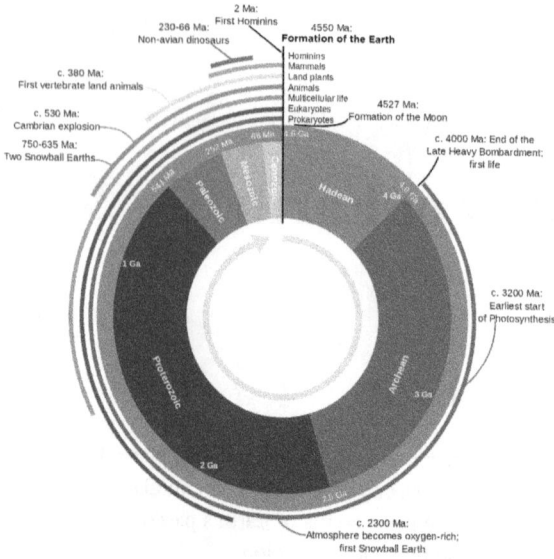

The earliest undisputed evidence of life on Earth dates at least from 3.5 billion years ago, during the Eoarchean Era, after a geological crust started to solidify following the earlier molten Hadean Eon. There are microbial mat fossils such as stromatolites found in 3.48 billion-year-old sandstone discovered in Western Australia. Other early physical evidence of a biogenic substance is graphite in 3.7 billion-year-old metasedimentary rocks discovered in southwestern Greenland as well as "remains of biotic life" found in 4.1 billion-year-old rocks in Western Australia.[38] According to one of the researchers, "If life arose relatively quickly on Earth ... then it could be common in the universe."

Photosynthetic organisms appeared between 3.2 and 2.4 billion years ago and began enriching the atmosphere with oxygen. Life remained mostly small and microscopic until about 580 million years ago, when complex multicellular life arose, developed over time, and culminated in the Cambrian Explosion about 541 million years ago. This sudden diversification of life forms produced most of the major phyla known today, and divided the Proterozoic Eon from the Cambrian Period of the Paleozoic Era. It is estimated that 99 percent of all species that ever lived on Earth, over five billion, have gone extinct. Estimates on the number of Earth's current species range from 10 million to 14 million, of which about 1.2 million are documented, but over 86 percent have not been described. However, it was recently claimed that 1 trillion species currently live on Earth, with only one-thousandth of one percent described.

The Earth's crust has constantly changed since its formation, as has life has since its first appearance. Species continue to evolve, taking on new forms, splitting into daughter species, or going extinct in the face of ever-changing physical environments. The process of plate tectonics continues to shape the Earth's continents and oceans and the life they harbor. Human activity is now a dominant force affecting global change, harming the biosphere, the Earth's surface, hydrosphere, and atmosphere with the loss of wild lands, over-exploitation of the oceans, production of greenhouse gases, degradation of the ozone layer, and general degradation of soil, air, and water quality.

Eons

In geochronology, time is generally measured in mya (megayears or million years), each unit representing the period of approximately 1,000,000 years in the past. The history of Earth is divided into four great eons, starting 4,540 mya with the formation of the planet. Each eon saw the most significant changes in Earth's composition, climate and life. Each eon is subsequently divided into eras, which in turn are divided into periods, which are further divided into epochs.

Eon	Time (mya)	Description
Hadean	4,540–4,000	The Earth is formed out of debris around the solar protoplanetary disk. There is no life. Temperatures are extremely hot, with frequent volcanic activity and hellish environments. The atmosphere is nebular. Possible early oceans or bodies of liquid water. The moon is formed around this time, probably due to a protoplanet's collision into Earth.
Archean	4,000–2,500	Prokaryote life, the first form of life, emerges at the very beginning of this eon, in a process known as abiogenesis. The continents of Ur, Vaalbara and Kenorland may have been formed around this time. The atmosphere is composed of volcanic and greenhouse gases.
Protero-zoic	2,500–541	Eukaryotes, a more complex form of life, emerge, including some forms of multicellular organisms. Bacteria begin producing oxygen, shaping the third and current of Earth's atmospheres. Plants, later animals and possibly earlier forms of fungi form around this time. The early and late phases of this eon may have undergone "Snowball Earth" periods, in which all of the planet suffered below-zero temperatures. The early continents of Columbia, Rodinia and Pannotia may have formed around this time, in that order.
Phanero-zoic	541–present	Complex life, including vertebrates, begin to dominate the Earth's ocean in a process known as the Cambrian explosion. Pangaea forms and later dissolves into Laurasia and Gondwana. Gradually, life expands to land and all familiar forms of plants, animals and fungi begin appearing, including annelids, insects and reptiles. Several mass extinctions occur, among which birds, the descendants of dinosaurs, and more recently mammals emerge. Modern animals—including humans—evolve at the most recent phases of this eon.

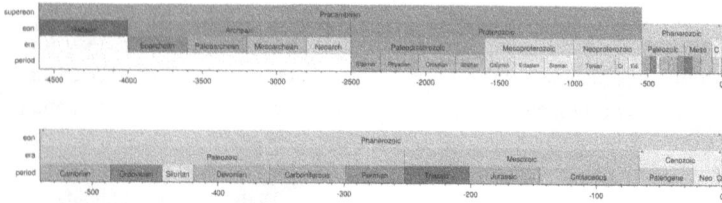

Geologic time scale

The history of the Earth can be organized chronologically according to the geologic time scale, which is split into intervals based on stratigraphic analysis. The following four timelines show the geologic time scale. The first shows the entire time from the formation of the Earth to the present, but this gives little space for the most recent eon. Therefore, the second timeline shows an expanded view of the most recent eon. In a similar way, the most recent era is expanded in the third timeline, and the most recent period is expanded in the fourth timeline.

Millions of Years

Solar System formation

The standard model for the formation of the Solar System (including the Earth) is the solar nebula hypothesis. In this model, the Solar System formed from a large, rotating cloud of interstellar dust and gas called the solar nebula. It was composed of hydrogen and helium created shortly after the Big Bang 13.8 Ga (billion years ago) and heavier elements ejected by supernovae. About 4.5 Ga, the nebula began a contraction that may have been triggered by the shock wave from a nearby supernova. A shock wave would have also made the nebula rotate. As the cloud began to accelerate, its angular momentum, gravity, and inertia flattened it into a protoplanetary disk perpendicular to its axis of rotation. Small perturbations due to collisions and the angular momentum of other large debris created the means by which kilometer-sized protoplanets began to form, orbiting the nebular center.

The center of the nebula, not having much angular momentum, collapsed rapidly, the compression heating it until nuclear fusion of hydrogen into helium began. After more contraction, a T Tauri star ignited and evolved into

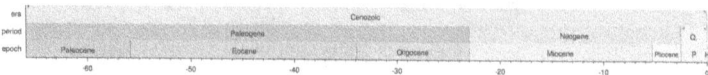

period		Quaternary					
epoch			Pleistocene				Holocen
age		Gelasian	Calabrian		Middle		Late
		-2		-1			0

Figure 23: *An artist's rendering of a protoplanetary disk*

the Sun. Meanwhile, in the outer part of the nebula gravity caused matter to condense around density perturbations and dust particles, and the rest of the protoplanetary disk began separating into rings. In a process known as runaway accretion, successively larger fragments of dust and debris clumped together to form planets. Earth formed in this manner about 4.54 billion years ago (with an uncertainty of 1%) and was largely completed within 10–20 million years. The solar wind of the newly formed T Tauri star cleared out most of the material in the disk that had not already condensed into larger bodies. The same process is expected to produce accretion disks around virtually all newly forming stars in the universe, some of which yield planets.

The proto-Earth grew by accretion until its interior was hot enough to melt the heavy, siderophile metals. Having higher densities than the silicates, these metals sank. This so-called *iron catastrophe* resulted in the separation of a primitive mantle and a (metallic) core only 10 million years after the Earth began to form, producing the layered structure of Earth and setting up the formation of Earth's magnetic field.[39] J. A. Jacobs was the first to suggest that the inner core—a solid center distinct from the liquid outer core—is freezing and growing out of the liquid outer core due to the gradual cooling of Earth's interior (about 100 degrees Celsius per billion years).

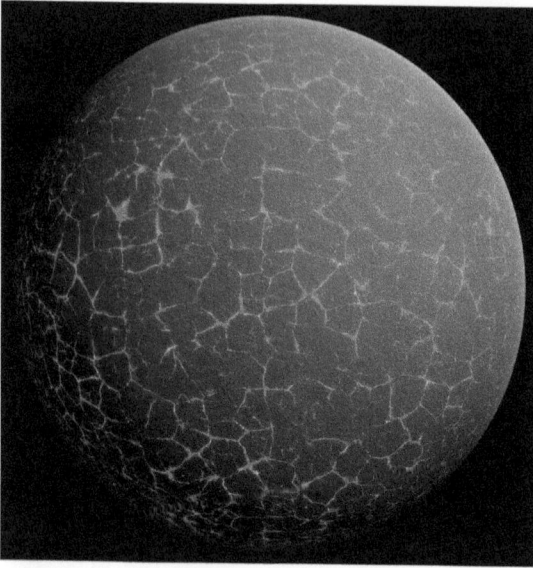

Figure 24: *Artist's conception of Hadean Eon Earth, when it was much hotter and inhospitable to all forms of life.*

Hadean and Archean Eons

The first eon in Earth's history, the *Hadean*, begins with the Earth's formation and is followed by the *Archean* eon at 3.8 Ga.:[145] The oldest rocks found on Earth date to about 4.0 Ga, and the oldest detrital zircon crystals in rocks to about 4.4 Ga, soon after the formation of the Earth's crust and the Earth itself. The giant impact hypothesis for the Moon's formation states that shortly after formation of an initial crust, the proto-Earth was impacted by a smaller protoplanet, which ejected part of the mantle and crust into space and created the Moon.

From crater counts on other celestial bodies, it is inferred that a period of intense meteorite impacts, called the *Late Heavy Bombardment*, began about 4.1 Ga, and concluded around 3.8 Ga, at the end of the Hadean. In addition, volcanism was severe due to the large heat flow and geothermal gradient. Nevertheless, detrital zircon crystals dated to 4.4 Ga show evidence of having undergone contact with liquid water, suggesting that the Earth already had oceans or seas at that time.

By the beginning of the Archean, the Earth had cooled significantly. Present life forms could not have survived at Earth's surface, because the Archean atmosphere lacked oxygen hence had no ozone layer to block ultraviolet light.

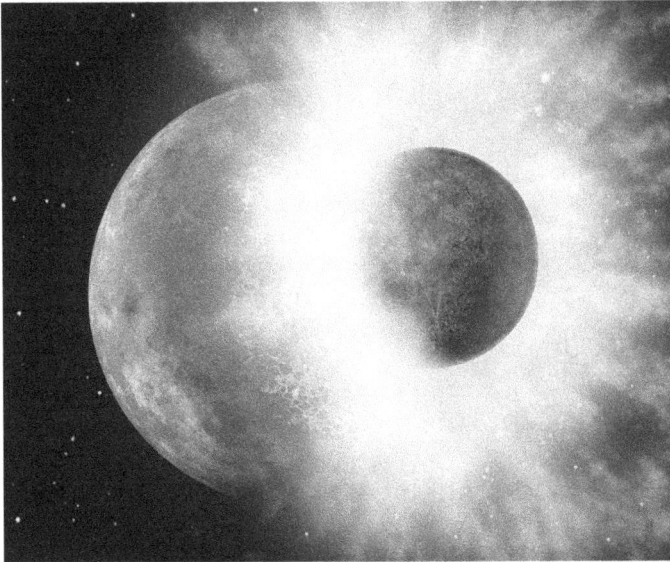

Figure 25: *Artist's impression of the enormous collision that probably formed the Moon*

Nevertheless, it is believed that primordial life began to evolve by the early Archean, with candidate fossils dated to around 3.5 Ga. Some scientists even speculate that life could have begun during the early Hadean, as far back as 4.4 Ga, surviving the possible Late Heavy Bombardment period in hydrothermal vents below the Earth's surface.

Formation of the Moon

Earth's only natural satellite, the Moon, is larger relative to its planet than any other satellite in the solar system.[40] but Pluto is defined as a dwarf planet.</ref> During the Apollo program, rocks from the Moon's surface were brought to Earth. Radiometric dating of these rocks shows that the Moon is 4.53 ± 0.01 billion years old, formed at least 30 million years after the solar system. New evidence suggests the Moon formed even later, 4.48 ± 0.02 Ga, or 70–110 million years after the start of the Solar System.

Theories for the formation of the Moon must explain its late formation as well as the following facts. First, the Moon has a low density (3.3 times that of water, compared to 5.5 for the earth) and a small metallic core. Second, there is virtually no water or other volatiles on the moon. Third, the Earth and Moon

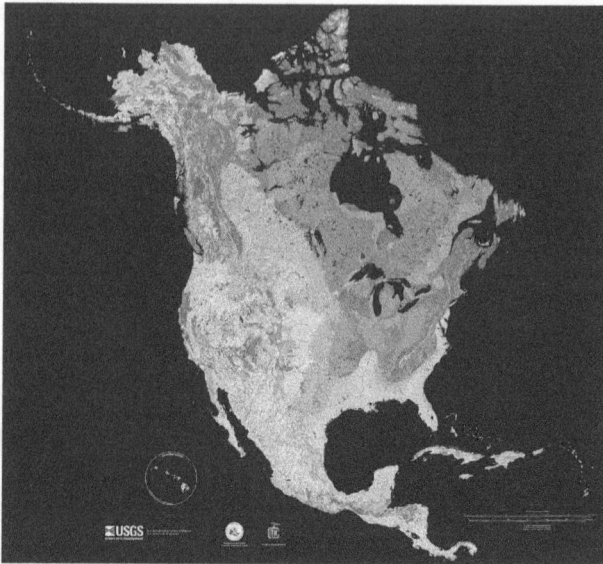

Figure 26: *Geologic map of North America, color-coded by age. The reds and pinks indicate rock from the Archean.*

have the same oxygen isotopic signature (relative abundance of the oxygen isotopes). Of the theories proposed to account for these phenomena, one is widely accepted: The *giant impact hypothesis* proposes that the Moon originated after a body the size of Mars (sometimes named Theia) struck the proto-Earth a glancing blow.[256]

The collision released about 100 million times more energy than the more recent Chicxulub impact that is believed to have caused the extinction of the dinosaurs. It was enough to vaporize some of the Earth's outer layers and melt both bodies.[256] A portion of the mantle material was ejected into orbit around the Earth. The giant impact hypothesis predicts that the Moon was depleted of metallic material, explaining its abnormal composition. The ejecta in orbit around the Earth could have condensed into a single body within a couple of weeks. Under the influence of its own gravity, the ejected material became a more spherical body: the Moon.

First continents

Mantle convection, the process that drives plate tectonics, is a result of heat flow from the Earth's interior to the Earth's surface.[2] It involves the creation of

rigid tectonic plates at mid-oceanic ridges. These plates are destroyed by subduction into the mantle at subduction zones. During the early Archean (about
3.0 Ga) the mantle was much hotter than today, probably around 1,600 °C
(2,910 °F),[82] so convection in the mantle was faster. Although a process similar to present-day plate tectonics did occur, this would have gone faster too.
It is likely that during the Hadean and Archean, subduction zones were more
common, and therefore tectonic plates were smaller.[258]

The initial crust, formed when the Earth's surface first solidified, totally disappeared from a combination of this fast Hadean plate tectonics and the intense
impacts of the Late Heavy Bombardment. However, it is thought that it was
basaltic in composition, like today's oceanic crust, because little crustal differentiation had yet taken place.[258] The first larger pieces of continental crust,
which is a product of differentiation of lighter elements during partial melting
in the lower crust, appeared at the end of the Hadean, about 4.0 Ga. What
is left of these first small continents are called cratons. These pieces of late
Hadean and early Archean crust form the cores around which today's continents grew.

The oldest rocks on Earth are found in the North American craton of Canada.
They are tonalites from about 4.0 Ga. They show traces of metamorphism
by high temperature, but also sedimentary grains that have been rounded by
erosion during transport by water, showing that rivers and seas existed then.
Cratons consist primarily of two alternating types of terranes. The first are
so-called greenstone belts, consisting of low-grade metamorphosed sedimentary rocks. These "greenstones" are similar to the sediments today found in
oceanic trenches, above subduction zones. For this reason, greenstones are
sometimes seen as evidence for subduction during the Archean. The second
type is a complex of felsic magmatic rocks. These rocks are mostly tonalite,
trondhjemite or granodiorite, types of rock similar in composition to granite
(hence such terranes are called TTG-terranes). TTG-complexes are seen as the
relics of the first continental crust, formed by partial melting in basalt.[Chapter 5]

Oceans and atmosphere

Earth is often described as having had three atmospheres. The first atmosphere, captured from the solar nebula, was composed of light (atmophile)
elements from the solar nebula, mostly hydrogen and helium. A combination
of the solar wind and Earth's heat would have driven off this atmosphere, as
a result of which the atmosphere is now depleted of these elements compared
to cosmic abundances. After the impact which created the moon, the molten
Earth released volatile gases; and later more gases were released by volcanoes,
completing a second atmosphere rich in greenhouse gases but poor in oxygen.

Stages

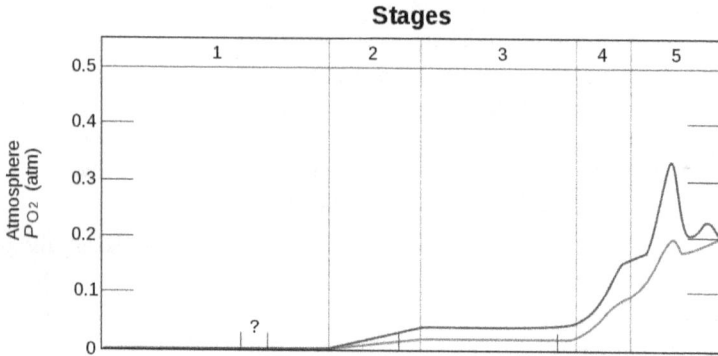

Figure 27: *Graph showing range of estimated partial pressure of atmospheric oxygen through geologic time*

:256 Finally, the third atmosphere, rich in oxygen, emerged when bacteria began to produce oxygen about 2.8 Ga.:[83–84,116–117]

In early models for the formation of the atmosphere and ocean, the second atmosphere was formed by outgassing of volatiles from the Earth's interior. Now it is considered likely that many of the volatiles were delivered during accretion by a process known as *impact degassing* in which incoming bodies vaporize on impact. The ocean and atmosphere would, therefore, have started to form even as the Earth formed. The new atmosphere probably contained water vapor, carbon dioxide, nitrogen, and smaller amounts of other gases.

Planetesimals at a distance of 1 astronomical unit (AU), the distance of the Earth from the Sun, probably did not contribute any water to the Earth because the solar nebula was too hot for ice to form and the hydration of rocks by water vapor would have taken too long. The water must have been supplied by meteorites from the outer asteroid belt and some large planetary embryos from beyond 2.5 AU. Comets may also have contributed. Though most comets are today in orbits farther away from the Sun than Neptune, computer simulations show that they were originally far more common in the inner parts of the solar system.:[130–132]

As the Earth cooled, clouds formed. Rain created the oceans. Recent evidence suggests the oceans may have begun forming as early as 4.4 Ga. By the start of the Archean eon, they already covered much of the Earth. This early formation has been difficult to explain because of a problem known as the faint young Sun paradox. Stars are known to get brighter as they age, and at the time of its formation the Sun would have been emitting only 70% of its current power. Thus, the Sun has become 30% brighter in the last 4.5 billion years.[41] Many models indicate that the Earth would have been covered in ice. A likely solution is that there was enough carbon dioxide and methane to produce a greenhouse effect. The carbon dioxide would have been produced by volcanoes and the methane by early microbes. Another greenhouse gas, ammonia, would have been ejected by volcanos but quickly destroyed by ultraviolet radiation.[83]

Origin of life

Life timeline

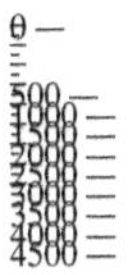

Axis scale: million years

Also see: *Human timeline* and *Nature timeline*

One of the reasons for interest in the early atmosphere and ocean is that they form the conditions under which life first arose. There are many models, but little consensus, on how life emerged from non-living chemicals; chemical systems created in the laboratory fall well short of the minimum complexity for a living organism.

The first step in the emergence of life may have been chemical reactions that produced many of the simpler organic compounds, including nucleobases and amino acids, that are the building blocks of life. An experiment in 1953 by Stanley Miller and Harold Urey showed that such molecules could form in an atmosphere of water, methane, ammonia and hydrogen with the aid of sparks to mimic the effect of lightning. Although atmospheric composition was probably different from that used by Miller and Urey, later experiments with more realistic compositions also managed to synthesize organic molecules. Computer simulations show that extraterrestrial organic molecules could have formed in the protoplanetary disk before the formation of the Earth.

Additional complexity could have been reached from at least three possible starting points: self-replication, an organism's ability to produce offspring that are similar to itself; metabolism, its ability to feed and repair itself; and external cell membranes, which allow food to enter and waste products to leave, but exclude unwanted substances.

Replication first: RNA world

Even the simplest members of the three modern domains of life use DNA to record their "recipes" and a complex array of RNA and protein molecules to "read" these instructions and use them for growth, maintenance, and self-replication.

The discovery that a kind of RNA molecule called a ribozyme can catalyze both its own replication and the construction of proteins led to the hypothesis that earlier life-forms were based entirely on RNA. They could have formed an RNA world in which there were individuals but no species, as mutations and horizontal gene transfers would have meant that the offspring in each generation were quite likely to have different genomes from those that their parents started with. RNA would later have been replaced by DNA, which is more stable and therefore can build longer genomes, expanding the range of capabilities a single organism can have. Ribozymes remain as the main components of ribosomes, the "protein factories" of modern cells.

Although short, self-replicating RNA molecules have been artificially produced in laboratories, doubts have been raised about whether natural non-biological

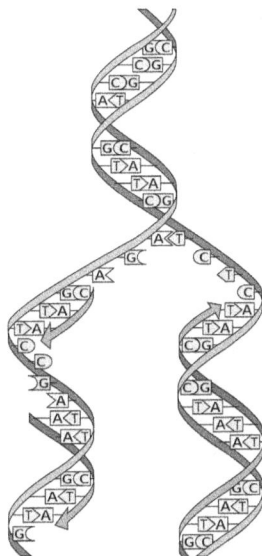

Figure 28: *The replicator in virtually all known life is de-oxyribonucleic acid. DNA is far more complex than the original replicator and its replication systems are highly elaborate.*

synthesis of RNA is possible. The earliest ribozymes may have been formed of simpler nucleic acids such as PNA, TNA or GNA, which would have been replaced later by RNA. Other pre-RNA replicators have been posited, including crystals:[150] and even quantum systems.

In 2003 it was proposed that porous metal sulfide precipitates would assist RNA synthesis at about 100 °C (212 °F) and at ocean-bottom pressures near hydrothermal vents. In this hypothesis, the proto-cells would be confined in the pores of the metal substrate until the later development of lipid membranes.

Metabolism first: iron–sulfur world

Another long-standing hypothesis is that the first life was composed of protein molecules. Amino acids, the building blocks of proteins, are easily synthesized in plausible prebiotic conditions, as are small peptides (polymers of amino acids) that make good catalysts.:[295–297] A series of experiments starting in 1997 showed that amino acids and peptides could form in the presence of carbon monoxide and hydrogen sulfide with iron sulfide and nickel sulfide as catalysts. Most of the steps in their assembly required temperatures of about 100 °C (212 °F) and moderate pressures, although one stage required 250 °C (482 °F) and a pressure equivalent to that found under 7 kilometers (4.3 mi) of rock. Hence, self-sustaining synthesis of proteins could have occurred near hydrothermal vents.

A difficulty with the metabolism-first scenario is finding a way for organisms to evolve. Without the ability to replicate as individuals, aggregates of molecules would have "compositional genomes" (counts of molecular species in the aggregate) as the target of natural selection. However, a recent model shows that such a system is unable to evolve in response to natural selection.

Membranes first: Lipid world

It has been suggested that double-walled "bubbles" of lipids like those that form the external membranes of cells may have been an essential first step. Experiments that simulated the conditions of the early Earth have reported the formation of lipids, and these can spontaneously form liposomes, double-walled "bubbles", and then reproduce themselves. Although they are not intrinsically information-carriers as nucleic acids are, they would be subject to natural selection for longevity and reproduction. Nucleic acids such as RNA might then have formed more easily within the liposomes than they would have outside.

The clay theory

Some clays, notably montmorillonite, have properties that make them plausible accelerators for the emergence of an RNA world: they grow by self-replication of their crystalline pattern, are subject to an analog of natural selection (as the clay "species" that grows fastest in a particular environment rapidly becomes dominant), and can catalyze the formation of RNA molecules. Although this idea has not become the scientific consensus, it still has active supporters.:[150-158]

Research in 2003 reported that montmorillonite could also accelerate the conversion of fatty acids into "bubbles", and that the bubbles could encapsulate RNA attached to the clay. Bubbles can then grow by absorbing additional lipids and dividing. The formation of the earliest cells may have been aided by similar processes.

A similar hypothesis presents self-replicating iron-rich clays as the progenitors of nucleotides, lipids and amino acids.

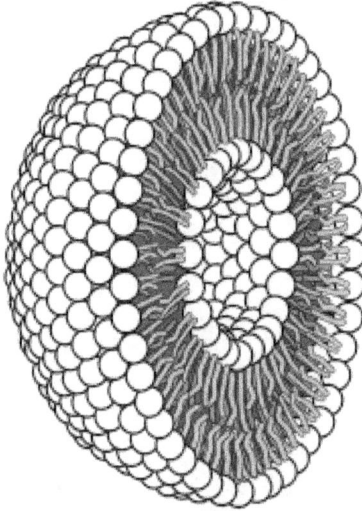

Figure 29: *Cross-section through a liposome*

Last universal ancestor

It is believed that of this multiplicity of protocells, only one line survived. Current phylogenetic evidence suggests that the last universal ancestor (LUA) lived during the early Archean eon, perhaps 3.5 Ga or earlier. This LUA cell is the ancestor of all life on Earth today. It was probably a prokaryote, possessing a cell membrane and probably ribosomes, but lacking a nucleus or membrane-bound organelles such as mitochondria or chloroplasts. Like modern cells, it used DNA as its genetic code, RNA for information transfer and protein synthesis, and enzymes to catalyze reactions. Some scientists believe that instead of a single organism being the last universal common ancestor, there were populations of organisms exchanging genes by lateral gene transfer.

Proterozoic Eon

The Proterozoic eon lasted from 2.5 Ga to 542 Ma (million years) ago.[130] In this time span, cratons grew into continents with modern sizes. The change to an oxygen-rich atmosphere was a crucial development. Life developed from prokaryotes into eukaryotes and multicellular forms. The Proterozoic saw a couple of severe ice ages called snowball Earths. After the last Snowball Earth about 600 Ma, the evolution of life on Earth accelerated. About

Figure 30: *Lithified stromatolites on the shores of Lake Thetis, Western Australia. Archean stromatolites are the first direct fossil traces of life on Earth.*

580 Ma, the Ediacaran biota formed the prelude for the Cambrian Explosion.Wikipedia:Citation needed

Oxygen revolution

The earliest cells absorbed energy and food from the surrounding environment. They used fermentation, the breakdown of more complex compounds into less complex compounds with less energy, and used the energy so liberated to grow and reproduce. Fermentation can only occur in an *anaerobic* (oxygen-free) environment. The evolution of photosynthesis made it possible for cells to derive energy from the Sun.[:377]

Most of the life that covers the surface of the Earth depends directly or indirectly on photosynthesis. The most common form, oxygenic photosynthesis, turns carbon dioxide, water, and sunlight into food. It captures the energy of sunlight in energy-rich molecules such as ATP, which then provide the energy to make sugars. To supply the electrons in the circuit, hydrogen is stripped from water, leaving oxygen as a waste product. Some organisms, including purple bacteria and green sulfur bacteria, use an anoxygenic form of photosynthesis that uses alternatives to hydrogen stripped from water as electron donors; examples are hydrogen sulfide, sulfur and iron. Such extremophile

Figure 31: *A banded iron formation from the 3.15 Ga Moories Group, Barberton Greenstone Belt, South Africa. Red layers represent the times when oxygen was available; gray layers were formed in anoxic circumstances.*

organisms are restricted to otherwise inhospitable environments such as hot springs and hydrothermal vents.[379–382]

The simpler anoxygenic form arose about 3.8 Ga, not long after the appearance of life. The timing of oxygenic photosynthesis is more controversial; it had certainly appeared by about 2.4 Ga, but some researchers put it back as far as 3.2 Ga. The latter "probably increased global productivity by at least two or three orders of magnitude". Among the oldest remnants of oxygen-producing lifeforms are fossil stromatolites.

At first, the released oxygen was bound up with limestone, iron, and other minerals. The oxidized iron appears as red layers in geological strata called banded iron formations that formed in abundance during the Siderian period (between 2500 Ma and 2300 Ma).[133] When most of the exposed readily reacting minerals were oxidized, oxygen finally began to accumulate in the atmosphere. Though each cell only produced a minute amount of oxygen, the combined metabolism of many cells over a vast time transformed Earth's atmosphere to its current state. This was Earth's third atmosphere.[50–51;83–84,116–117]

Some oxygen was stimulated by solar ultraviolet radiation to form ozone, which collected in a layer near the upper part of the atmosphere. The ozone layer absorbed, and still absorbs, a significant amount of the ultraviolet radiation that once had passed through the atmosphere. It allowed cells to colonize

the surface of the ocean and eventually the land: without the ozone layer, ultraviolet radiation bombarding land and sea would have caused unsustainable
levels of mutation in exposed cells.[219-220]

Photosynthesis had another major impact. Oxygen was toxic; much life on
Earth probably died out as its levels rose in what is known as the *oxygen catastrophe*. Resistant forms survived and thrived, and some developed the ability
to use oxygen to increase their metabolism and obtain more energy from the
same food.

Snowball Earth

The natural evolution of the Sun made it progressively more luminous during
the Archean and Proterozoic eons; the Sun's luminosity increases 6% every
billion years.[165] As a result, the Earth began to receive more heat from the
Sun in the Proterozoic eon. However, the Earth did not get warmer. Instead,
the geological record suggests it cooled dramatically during the early Proterozoic. Glacial deposits found in South Africa date back to 2.2 Ga, at which
time, based on paleomagnetic evidence, they must have been located near the
equator. Thus, this glaciation, known as the Huronian glaciation, may have
been global. Some scientists suggest this was so severe that the Earth was
frozen over from the poles to the equator, a hypothesis called Snowball Earth.

The Huronian ice age might have been caused by the increased oxygen concentration in the atmosphere, which caused the decrease of methane (CH_4)
in the atmosphere. Methane is a strong greenhouse gas, but with oxygen it
reacts to form CO_2, a less effective greenhouse gas.[172] When free oxygen
became available in the atmosphere, the concentration of methane could have
decreased dramatically, enough to counter the effect of the increasing heat flow
from the Sun.

However, the term Snowball Earth is more commonly used to describe later
extreme ice ages during the Cryogenian period. There were four periods, each
lasting about 10 million years, between 750 and 580 million years ago, when
the earth is thought to have been covered with ice apart from the highest mountains, and average temperatures were about –50 °C (–58 °F). The snowball
may have been partly due to the location of the supercontinent Rodinia strad-
dling the Equator. Carbon dioxide combines with rain to weather rocks to
form carbonic acid, which is then washed out to sea, thus extracting the greenhouse gas from the atmosphere. When the continents are near the poles, the
advance of ice covers the rocks, slowing the reduction in carbon dioxide, but
in the Cryogienian the weathering of Rodinia was able to continue unchecked
until the ice advanced to the tropics. The process may have finally been reversed by the emission of carbon dioxide from volcanoes or the destabilization

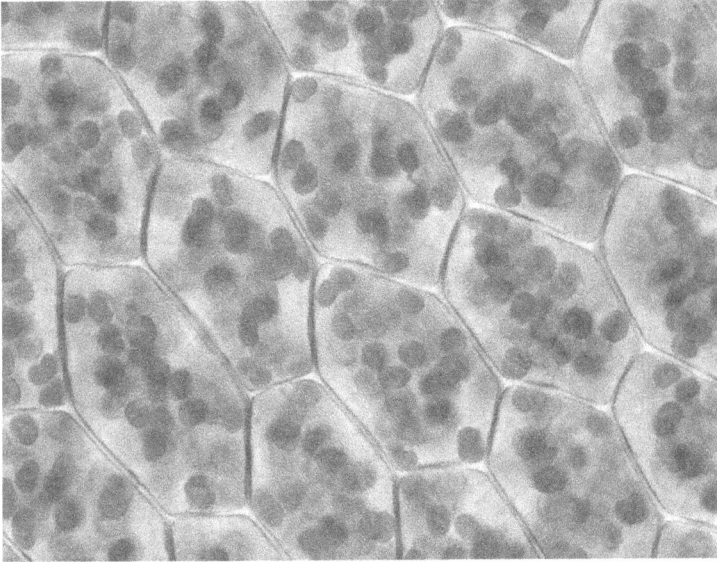

Figure 32: *Chloroplasts in the cells of a moss*

of methane gas hydrates. According to the alternative Slushball Earth theory, even at the height of the ice ages there was still open water at the Equator.

Emergence of eukaryotes

Modern taxonomy classifies life into three domains. The time of their origin is uncertain. The Bacteria domain probably first split off from the other forms of life (sometimes called Neomura), but this supposition is controversial. Soon after this, by 2 Ga, the Neomura split into the Archaea and the Eukarya. Eukaryotic cells (Eukarya) are larger and more complex than prokaryotic cells (Bacteria and Archaea), and the origin of that complexity is only now becoming known. Wikipedia:Citation needed

Around this time, the first proto-mitochondrion was formed. A bacterial cell related to today's *Rickettsia*, which had evolved to metabolize oxygen, entered a larger prokaryotic cell, which lacked that capability. Perhaps the large cell attempted to digest the smaller one but failed (possibly due to the evolution of prey defenses). The smaller cell may have tried to parasitize the larger one. In any case, the smaller cell survived inside the larger cell. Using oxygen, it metabolized the larger cell's waste products and derived more energy. Part of this excess energy was returned to the host. The smaller cell replicated inside the larger one. Soon, a stable symbiosis developed between the large cell and the smaller cells inside it. Over time, the host cell acquired some genes from the smaller cells, and the two kinds became dependent on each other: the larger cell could not survive without the energy produced by the smaller ones, and these, in turn, could not survive without the raw materials provided by the larger cell. The whole cell is now considered a single organism, and the smaller cells are classified as organelles called mitochondria.

A similar event occurred with photosynthetic cyanobacteria entering large heterotrophic cells and becoming chloroplasts.[:60–61:536–539] Probably as a result of these changes, a line of cells capable of photosynthesis split off from the other eukaryotes more than 1 billion years ago. There were probably several such inclusion events. Besides the well-established endosymbiotic theory of the cellular origin of mitochondria and chloroplasts, there are theories that cells led to peroxisomes, spirochetes led to cilia and flagella, and that perhaps a DNA virus led to the cell nucleus, though none of them are widely accepted.

Archaeans, bacteria, and eukaryotes continued to diversify and to become more complex and better adapted to their environments. Each domain repeatedly split into multiple lineages, although little is known about the history of the archaea and bacteria. Around 1.1 Ga, the supercontinent Rodinia was assembling. The plant, animal, and fungi lines had split, though they still existed as solitary cells. Some of these lived in colonies, and gradually a division of labor began to take place; for instance, cells on the periphery might have started to assume different roles from those in the interior. Although the division between a colony with specialized cells and a multicellular organism is not always clear, around 1 billion years ago, the first multicellular plants emerged, probably green algae. Possibly by around 900 Ma[:488] true multicellularity had also evolved in animals.Wikipedia:Citation needed

At first, it probably resembled today's sponges, which have totipotent cells that allow a disrupted organism to reassemble itself.[:483–487] As the division of labor was completed in all lines of multicellular organisms, cells became more specialized and more dependent on each other; isolated cells would die.Wikipedia:Citation needed

Supercontinents in the Proterozoic

Reconstructions of tectonic plate movement in the past 250 million years (the Cenozoic and Mesozoic eras) can be made reliably using fitting of continental margins, ocean floor magnetic anomalies and paleomagnetic poles. No ocean crust dates back further than that, so earlier reconstructions are more difficult. Paleomagnetic poles are supplemented by geologic evidence such as orogenic belts, which mark the edges of ancient plates, and past distributions of flora and fauna. The further back in time, the scarcer and harder to interpret the data get and the more uncertain the reconstructions.[:370]

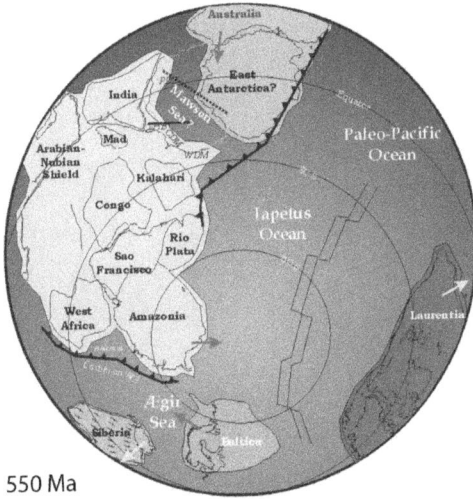

Figure 33: *A reconstruction of Pannotia (550 Ma).*

Throughout the history of the Earth, there have been times when continents collided and formed a supercontinent, which later broke up into new continents. About 1000 to 830 Ma, most continental mass was united in the supercontinent Rodinia.[370] Rodinia may have been preceded by Early-Middle Proterozoic continents called Nuna and Columbia.[374]

After the break-up of Rodinia about 800 Ma, the continents may have formed another short-lived supercontinent around 550 Ma. The hypothetical supercontinent is sometimes referred to as Pannotia or Vendia.[321-322] The evidence for it is a phase of continental collision known as the Pan-African orogeny, which joined the continental masses of current-day Africa, South America, Antarctica and Australia. The existence of Pannotia depends on the timing of the rifting between Gondwana (which included most of the landmass now in the Southern Hemisphere, as well as the Arabian Peninsula and the Indian subcontinent) and Laurentia (roughly equivalent to current-day North America).[374] It is at least certain that by the end of the Proterozoic eon, most of the continental mass lay united in a position around the south pole.

Figure 34: *A 580 million year old fossil of Spriggina floundensi, an animal from the Ediacaran period. Such life forms could have been ancestors to the many new forms that originated in the Cambrian Explosion.*

Late Proterozoic climate and life

The end of the Proterozoic saw at least two Snowball Earths, so severe that the surface of the oceans may have been completely frozen. This happened about 716.5 and 635 Ma, in the Cryogenian period. The intensity and mechanism of both glaciations are still under investigation and harder to explain than the early Proterozoic Snowball Earth. Most paleoclimatologists think the cold episodes were linked to the formation of the supercontinent Rodinia. Because Rodinia was centered on the equator, rates of chemical weathering increased and carbon dioxide (CO_2) was taken from the atmosphere. Because CO_2 is an important greenhouse gas, climates cooled globally.Wikipedia:Citation needed In the same way, during the Snowball Earths most of the continental surface was covered with permafrost, which decreased chemical weathering again, leading to the end of the glaciations. An alternative hypothesis is that enough carbon dioxide escaped through volcanic outgassing that the resulting greenhouse effect raised global temperatures. Increased volcanic activity resulted from the break-up of Rodinia at about the same time.Wikipedia:Citation needed

The Cryogenian period was followed by the Ediacaran period, which was characterized by a rapid development of new multicellular lifeforms. Whether there is a connection between the end of the severe ice ages and the increase in diversity of life is not clear, but it does not seem coincidental. The new forms of life, called Ediacara biota, were larger and more diverse than ever. Though the taxonomy of most Ediacaran life forms is unclear, some were ancestors of groups of modern life. Important developments were the origin of muscular and neural cells. None of the Ediacaran fossils had hard body parts like skeletons. These first appear after the boundary between the Proterozoic and Phanerozoic eons or Ediacaran and Cambrian periods.Wikipedia:Citation needed

Phanerozoic Eon

The Phanerozoic is the current eon on Earth, which started approximately 542 million years ago. It consists of three eras: The Paleozoic, Mesozoic, and Cenozoic, and is the time when multi-cellular life greatly diversified into almost all the organisms known today.

The Paleozoic ("old life") era was the first and longest era of the Phanerozoic eon, lasting from 542 to 251 Ma. During the Paleozoic, many modern groups of life came into existence. Life colonized the land, first plants, then animals. Two major extinctions occurred. The continents formed at the break-up of Pannotia and Rodinia at the end of the Proterozoic slowly moved together again, forming the supercontinent Pangaea in the late Paleozoic.Wikipedia:Citation needed

The Mesozoic ("middle life") era lasted from 251 Ma to 66 Ma. It is subdivided into the Triassic, Jurassic, and Cretaceous periods. The era began with the Permian–Triassic extinction event, the most severe extinction event in the fossil record; 95% of the species on Earth died out. It ended with the Cretaceous–Paleogene extinction event that wiped out the dinosaurs.Wikipedia:Citation needed.

The Cenozoic ("new life") era began at 66 Ma, and is subdivided into the Paleogene, Neogene, and Quaternary periods. These three periods are further split into seven sub-divisions, with the Paleogene composed of The Paleocene, Eocene, and Oligocene, the Neogene divided into the Miocene, Pliocene, and the Quaternary composed of the Pleistocene, and Holocene. Mammals, birds, amphibians, crocodilians, turtles, and lepidosaurs survived the Cretaceous–Paleogene extinction event that killed off the non-avian dinosaurs and many other forms of life, and this is the era during which they diversified into their modern forms.Wikipedia:Citation needed

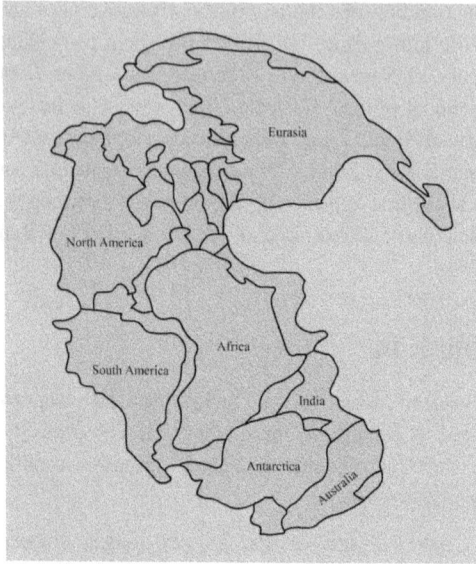

Figure 35: *Pangaea was a supercontinent that existed from about 300 to 180 Ma. The outlines of the modern continents and other landmasses are indicated on this map.*

Tectonics, paleogeography and climate

At the end of the Proterozoic, the supercontinent Pannotia had broken apart into the smaller continents Laurentia, Baltica, Siberia and Gondwana. During periods when continents move apart, more oceanic crust is formed by volcanic activity. Because young volcanic crust is relatively hotter and less dense than old oceanic crust, the ocean floors rise during such periods. This causes the sea level to rise. Therefore, in the first half of the Paleozoic, large areas of the continents were below sea level.Wikipedia:Citation needed

Early Paleozoic climates were warmer than today, but the end of the Ordovician saw a short ice age during which glaciers covered the south pole, where the huge continent Gondwana was situated. Traces of glaciation from this period are only found on former Gondwana. During the Late Ordovician ice age, a few mass extinctions took place, in which many brachiopods, trilobites, Bryozoa and corals disappeared. These marine species could probably not contend with the decreasing temperature of the sea water.

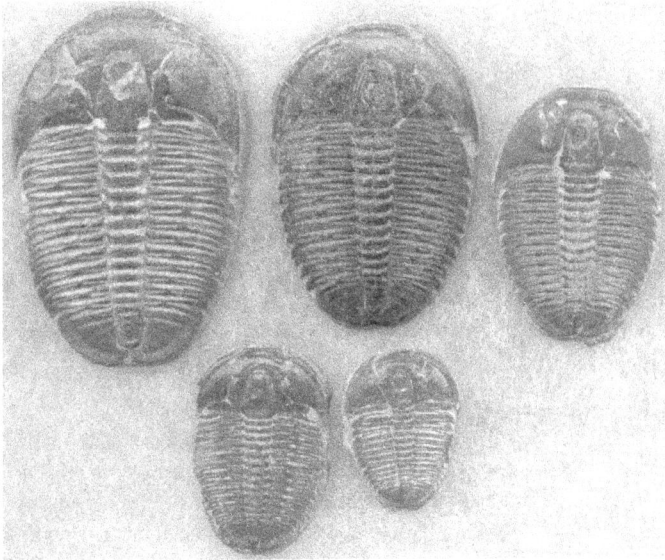

Figure 36: *Trilobites first appeared during the Cambrian period and were among the most widespread and diverse groups of Paleozoic organisms.*

The continents Laurentia and Baltica collided between 450 and 400 Ma, during the Caledonian Orogeny, to form Laurussia (also known as Euramerica). Traces of the mountain belt this collision caused can be found in Scandinavia, Scotland, and the northern Appalachians. In the Devonian period (416–359 Ma) Gondwana and Siberia began to move towards Laurussia. The collision of Siberia with Laurussia caused the Uralian Orogeny, the collision of Gondwana with Laurussia is called the Variscan or Hercynian Orogeny in Europe or the Alleghenian Orogeny in North America. The latter phase took place during the Carboniferous period (359–299 Ma) and resulted in the formation of the last supercontinent, Pangaea.

By 180 Ma, Pangaea broke up into Laurasia and Gondwana.Wikipedia:Citation needed

Cambrian explosion

The rate of the evolution of life as recorded by fossils accelerated in the Cambrian period (542–488 Ma). The sudden emergence of many new species, phyla, and forms in this period is called the Cambrian Explosion. The biological fomenting in the Cambrian Explosion was unpreceded before and since that time.[229] Whereas the Ediacaran life forms appear yet primitive and not easy to put in any modern group, at the end of the Cambrian most modern phyla were already present. The development of hard body parts such as shells, skeletons or exoskeletons in animals like molluscs, echinoderms, crinoids and arthropods (a well-known group of arthropods from the lower Paleozoic are the trilobites) made the preservation and fossilization of such life forms easier than those of their Proterozoic ancestors. For this reason, much more is known about life in and after the Cambrian than about that of older periods. Some of these Cambrian groups appear complex but are seemingly quite different from modern life; examples are *Anomalocaris* and *Haikouichthys*. More recently, however, these seem to have found a place in modern classification. Wikipedia:Citation needed

During the Cambrian, the first vertebrate animals, among them the first fishes, had appeared.[357] A creature that could have been the ancestor of the fishes, or was probably closely related to it, was *Pikaia*. It had a primitive notochord, a structure that could have developed into a vertebral column later. The first fishes with jaws (Gnathostomata) appeared during the next geological period, the Ordovician. The colonisation of new niches resulted in massive body sizes. In this way, fishes with increasing sizes evolved during the early Paleozoic, such as the titanic placoderm *Dunkleosteus*, which could grow 7 meters (23 ft) long.Wikipedia:Citation needed

The diversity of life forms did not increase greatly because of a series of mass extinctions that define widespread biostratigraphic units called *biomeres*. After each extinction pulse, the continental shelf regions were repopulated by similar life forms that may have been evolving slowly elsewhere. By the late Cambrian, the trilobites had reached their greatest diversity and dominated nearly all fossil assemblages.[34]

Colonization of land

Oxygen accumulation from photosynthesis resulted in the formation of an ozone layer that absorbed much of the Sun's ultraviolet radiation, meaning unicellular organisms that reached land were less likely to die, and prokaryotes began to multiply and become better adapted to survival out of the water. Prokaryote lineages had probably colonized the land as early as 2.6 Ga even before the origin of the eukaryotes. For a long time, the land remained barren of multicellular organisms. The supercontinent Pannotia formed around 600 Ma and then broke apart a short 50 million years later. Fish, the earliest vertebrates, evolved in the oceans around 530 Ma.[354] A major extinction event occurred near the end of the Cambrian period, which ended 488 Ma.

Several hundred million years ago, plants (probably resembling algae) and fungi started growing at the edges of the water, and then out of it.[138–140] The

Figure 37: *Artist's conception of Devonian flora*

oldest fossils of land fungi and plants date to 480–460 Ma, though molecular evidence suggests the fungi may have colonized the land as early as 1000 Ma and the plants 700 Ma. Initially remaining close to the water's edge, mutations and variations resulted in further colonization of this new environment. The timing of the first animals to leave the oceans is not precisely known: the oldest clear evidence is of arthropods on land around 450 Ma, perhaps thriving and becoming better adapted due to the vast food source provided by the terrestrial plants. There is also unconfirmed evidence that arthropods may have appeared on land as early as 530 Ma.

Evolution of tetrapods

At the end of the Ordovician period, 443 Ma, additional extinction events occurred, perhaps due to a concurrent ice age. Around 380 to 375 Ma, the first tetrapods evolved from fish. Fins evolved to become limbs that the first tetrapods used to lift their heads out of the water to breathe air. This would let them live in oxygen-poor water, or pursue small prey in shallow water. They may have later ventured on land for brief periods. Eventually, some of them became so well adapted to terrestrial life that they spent their adult lives on land, although they hatched in the water and returned to lay their eggs. This was the origin of the amphibians. About 365 Ma, another period of extinction occurred, perhaps as a result of global cooling. Plants evolved seeds, which

Figure 38: *Tiktaalik, a fish with limb-like fins and a predecessor of tetrapods. Reconstruction from fossils about 375 million years old.*

dramatically accelerated their spread on land, around this time (by approximately 360 Ma).

About 20 million years later (340 Ma:293–296), the amniotic egg evolved, which could be laid on land, giving a survival advantage to tetrapod embryos. This resulted in the divergence of amniotes from amphibians. Another 30 million years (310 Ma:254–256) saw the divergence of the synapsids (including mammals) from the sauropsids (including birds and reptiles). Other groups of organisms continued to evolve, and lines diverged—in fish, insects, bacteria, and so on—but less is known of the details.Wikipedia:Citation needed

After yet another, the most severe extinction of the period (251∼250 Ma), around 230 Ma, dinosaurs split off from their reptilian ancestors. The Triassic–Jurassic extinction event at 200 Ma spared many of the dinosaurs, and they soon became dominant among the vertebrates. Though some mammalian lines began to separate during this period, existing mammals were probably small animals resembling shrews.:169

The boundary between avian and non-avian dinosaurs is not clear, but *Archaeopteryx*, traditionally considered one of the first birds, lived around 150 Ma.

The earliest evidence for the angiosperms evolving flowers is during the Cretaceous period, some 20 million years later (132 Ma).

Extinctions

The first of five great mass extinctions was the Ordovician-Silurian extinction. Its possible cause was the intense glaciation of Gondwana, which eventually led to a snowball earth. 60% of marine invertebrates became extinct and 25% of all families.Wikipedia:Citation needed

The second mass extinction was the Late Devonian extinction, probably caused by the evolution of trees, which could have led to the depletion of greenhouse

Figure 39: *Dinosaurs were the dominant terrestrial vertebrates throughout most of the Mesozoic*

gases (like CO2) or the eutrophication of water. 70% of all species became extinct.Wikipedia:Citation needed

The third mass extinction was the Permian-Triassic, or the Great Dying, event was possibly caused by some combination of the Siberian Traps volcanic event, an asteroid impact, methane hydrate gasification, sea level fluctuations, and a major anoxic event. Either the proposed Wilkes Land crater in Antarctica or Bedout structure off the northwest coast of Australia may indicate an impact connection with the Permian-Triassic extinction. But it remains uncertain whether either these or other proposed Permian-Triassic boundary craters are either real impact craters or even contemporaneous with the Permian-Triassic extinction event. This was by far the deadliest extinction ever, with about 57% of all families and 83% of all genera killed.

The fourth mass extinction was the Triassic-Jurassic extinction event in which almost all synapsids and archosaurs became extinct, probably due to new competition from dinosaurs.Wikipedia:Citation needed

The fifth and most recent mass extinction was the K-T extinction. In 66 Ma, a 10-kilometer (6.2 mi) asteroid struck Earth just off the Yucatán Peninsula – somewhere in the south western tip of then Laurasia – where the Chicxulub crater is today. This ejected vast quantities of particulate matter and vapor into the air that occluded sunlight, inhibiting photosynthesis. 75% of all life, including the non-avian dinosaurs, became extinct, marking the end of the Cretaceous period and Mesozoic era.Wikipedia:Citation needed

Diversification of mammals

The first true mammals evolved in the shadows of dinosaurs and other large archosaurs that filled the world by the late Triassic. The first mammals were very small, and were probably nocturnal to escape predation. Mammal diversification truly began only after the Cretaceous-Paleogene extinction event. By the early Paleocene the earth recovered from the extinction, and mammalian diversity increased. Creatures like *Ambulocetus* took to the oceans to eventually evolve into whales, whereas some creatures, like primates, took to the trees. This all changed during the mid to late Eocene when the circum-Antarctic current formed between Antarctica and Australia which disrupted weather patterns on a global scale. Grassless savannas began to predominate much of the landscape, and mammals such as *Andrewsarchus* rose up to become the largest known terrestrial predatory mammal ever, and early whales like *Basilosaurus* took control of the seas. Wikipedia:Citation needed

The evolution of grass brought a remarkable change to the Earth's landscape, and the new open spaces created pushed mammals to get bigger and bigger. Grass started to expand in the Miocene, and the Miocene is where many modern-day mammals first appeared. Giant ungulates like *Paraceratherium* and *Deinotherium* evolved to rule the grasslands. The evolution of grass also brought primates down from the trees, and started human evolution. The first big cats evolved during this time as well. The Tethys Sea was closed off by the collision of Africa and Europe.

The formation of Panama was perhaps the most important geological event to occur in the last 60 million years. Atlantic and Pacific currents were closed off from each other, which caused the formation of the Gulf Stream, which made Europe warmer. The land bridge allowed the isolated creatures of South America to migrate over to North America, and vice versa. Various species migrated south, leading to the presence in South America of llamas, the spectacled bear, kinkajous and jaguars.Wikipedia:Citation needed

Three million years ago saw the start of the Pleistocene epoch, which featured dramatic climactic changes due to the ice ages. The ice ages led to the evolution of modern man in Saharan Africa and expansion. The mega-fauna that dominated fed on grasslands that, by now, had taken over much of the subtropical world. The large amounts of water held in the ice allowed for various bodies of water to shrink and sometimes disappear such as the North Sea and the Bering Strait. It is believed by many that a huge migration took place along

Beringia which is why, today, there are camels (which evolved and became extinct in North America), horses (which evolved and became extinct in North America), and Native Americans. The ending of the last ice age coincided with the expansion of man, along with a massive die out of ice age megafauna. This extinction, nicknamed "the Sixth Extinction", has been going ever since.Wikipedia:Citation needed

Human evolution

Hominin timeline

Axis scale: million years

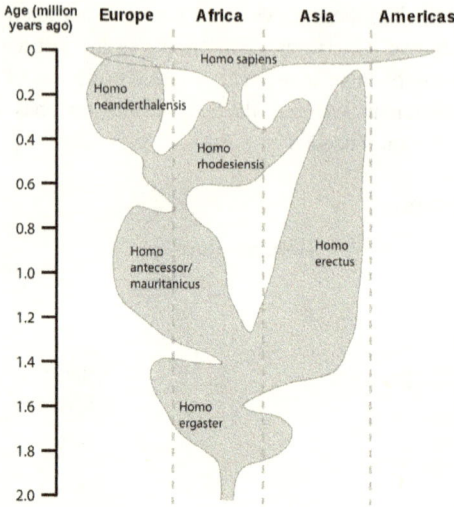

Figure 40: *A reconstruction of human history based on fossil data.*

Also see: *Life timeline* and *Nature timeline*

A small African ape living around 6 Ma was the last animal whose descendants would include both modern humans and their closest relatives, the chimpanzees.[100–101] Only two branches of its family tree have surviving descendants. Very soon after the split, for reasons that are still unclear, apes in one branch developed the ability to walk upright.[95–99] Brain size increased rapidly, and by 2 Ma, the first animals classified in the genus *Homo* had appeared.[300] Of course, the line between different species or even genera is somewhat arbitrary as organisms continuously change over generations. Around the same time, the other branch split into the ancestors of the common chimpanzee and the ancestors of the bonobo as evolution continued simultaneously in all life forms.[100–101]

The ability to control fire probably began in *Homo erectus* (or *Homo ergaster*), probably at least 790,000 years ago but perhaps as early as 1.5 Ma.[67] The use and discovery of controlled fire may even predate *Homo erectus*. Fire was possibly used by the early Lower Paleolithic (Oldowan) hominid *Homo habilis* or strong australopithecines such as *Paranthropus*.

It is more difficult to establish the origin of language; it is unclear whether *Homo erectus* could speak or if that capability had not begun until *Homo sapiens*.[67] As brain size increased, babies were born earlier, before their heads grew too large to pass through the pelvis. As a result, they exhibited more plasticity, and thus possessed an increased capacity to learn and required a longer period of dependence. Social skills became more complex, language became more sophisticated, and tools became more elaborate. This contributed to further cooperation and intellectual development.[7] Modern humans (*Homo sapiens*) are believed to have originated around 200,000 years ago or earlier in Africa; the oldest fossils date back to around 160,000 years ago.

The first humans to show signs of spirituality are the Neanderthals (usually classified as a separate species with no surviving descendants); they buried their dead, often with no sign of food or tools.[17] However, evidence of more sophisticated beliefs, such as the early Cro-Magnon cave paintings (probably with magical or religious significance)[17-19] did not appear until 32,000 years ago. Cro-Magnons also left behind stone figurines such as Venus of Willendorf, probably also signifying religious belief.[17-19] By 11,000 years ago, *Homo sapiens* had reached the southern tip of South America, the last of the uninhabited continents (except for Antarctica, which remained undiscovered until 1820 AD). Tool use and communication continued to improve, and interpersonal relationships became more intricate.Wikipedia:Citation needed

Civilization

Throughout more than 90% of its history, *Homo sapiens* lived in small bands as nomadic hunter-gatherers.[8] As language became more complex, the ability to remember and communicate information resulted, according to a theory proposed by Richard Dawkins, in a new replicator: the meme. Ideas could be exchanged quickly and passed down the generations. Cultural evolution quickly outpaced biological evolution, and history proper began. Between 8500 and 7000 BC, humans in the Fertile Crescent in the Middle East began the systematic husbandry of plants and animals: agriculture. This spread to neighboring regions, and developed independently elsewhere, until most *Homo sapiens* lived sedentary lives in permanent settlements as farmers. Not all societies abandoned nomadism, especially those in isolated areas of the globe poor in domesticable plant species, such as Australia. However, among those civilizations that did adopt agriculture, the relative stability and increased productivity provided by farming allowed the population to expand.Wikipedia:Citation needed

Agriculture had a major impact; humans began to affect the environment as never before. Surplus food allowed a priestly or governing class to arise, followed by increasing division of labor. This led to Earth's first civilization at

Figure 41: *Vitruvian Man by Leonardo da Vinci epitomizes the advances in art and science seen during the Renaissance.*

Sumer in the Middle East, between 4000 and 3000 BC.[15] Additional civilizations quickly arose in ancient Egypt, at the Indus River valley and in China. The invention of writing enabled complex societies to arise: record-keeping and libraries served as a storehouse of knowledge and increased the cultural transmission of information. Humans no longer had to spend all their time working for survival, enabling the first specialized occupations (e.g. craftsmen, merchants, priests, etc...). Curiosity and education drove the pursuit of knowledge and wisdom, and various disciplines, including science (in a primitive form), arose. This in turn led to the emergence of increasingly larger and more complex civilizations, such as the first empires, which at times traded with one another, or fought for territory and resources.

By around 500 BC, there were advanced civilizations in the Middle East, Iran, India, China, and Greece, at times expanding, at times entering into decline.[3] In 221 BC, China became a single polity that would grow to spread its culture throughout East Asia, and it has remained the most populous nation in the world. The fundamentals of Western civilization were largely shaped in Ancient Greece, with the world's first democratic government and major advances in philosophy, science, and mathematics, and in Ancient Rome in law, government, and engineering. The Roman Empire was Christianized by Emperor Constantine in the early 4th century and declined by the end of the 5th.

Beginning with the 7th century, Christianization of Europe began. In 610, Islam was founded and quickly became the dominant religion in Western Asia. The House of Wisdom was established in Abbasid-era Baghdad, Iraq. It is considered to have been a major intellectual center during the Islamic Golden Age, where Muslim scholars in Baghdad and Cairo flourished from the ninth to the thirteenth centuries until the Mongol sack of Baghdad in 1258 AD. In 1054 AD the Great Schism between the Roman Catholic Church and the Eastern Orthodox Church led to the prominent cultural differences between Western and Eastern Europe.Wikipedia:Citation needed

In the 14th century, the Renaissance began in Italy with advances in religion, art, and science.:317–319 At that time the Christian Church as a political entity lost much of its power. In 1492, Christopher Columbus reached the Americas, initiating great changes to the new world. European civilization began to change beginning in 1500, leading to the scientific and industrial revolutions. That continent began to exert political and cultural dominance over human societies around the world, a time known as the Colonial era (also see Age of Discovery).:295–299 In the 18th century a cultural movement known as the Age of Enlightenment further shaped the mentality of Europe and contributed to its secularization. From 1914 to 1918 and 1939 to 1945, nations around the world were embroiled in world wars. Established following World War I, the League of Nations was a first step in establishing international institutions to settle disputes peacefully. After failing to prevent World War II, mankind's bloodiest conflict, it was replaced by the United Nations. After the war, many new states were formed, declaring or being granted independence in a period of decolonization. The United States and Soviet Union became the world's dominant superpowers for a time, and they held an often-violent rivalry known as the Cold War until the dissolution of the latter. In 1992, several European nations joined in the European Union. As transportation and communication improved, the economies and political affairs of nations around the world have become increasingly intertwined. This globalization has often produced both conflict and cooperation.Wikipedia:Citation needed

Recent events

Change has continued at a rapid pace from the mid-1940s to today. Technological developments include nuclear weapons, computers, genetic engineering, and nanotechnology. Economic globalization, spurred by advances in communication and transportation technology, has influenced everyday life in many parts of the world. Cultural and institutional forms such as democracy, capitalism, and environmentalism have increased influence. Major concerns and problems such as disease, war, poverty, violent radicalism, and recently, human-caused climate change have risen as the world population increases.Wikipedia:Citation needed

Figure 42: *Astronaut Bruce McCandless II outside of the space shuttle Challenger in 1984*

In 1957, the Soviet Union launched the first artificial satellite into orbit and,
soon afterward, Yuri Gagarin became the first human in space. Neil Armstrong, an American, was the first to set foot on another astronomical object,
the Moon. Unmanned probes have been sent to all the known planets in the
solar system, with some (such as Voyager) having left the solar system. Five
space agencies, representing over fifteen countries, have worked together to
build the International Space Station. Aboard it, there has been a continuous
human presence in space since 2000. The World Wide Web became a part of
everyday life in the 1990s, and since then has become an indispensable source
of information in the developed world.Wikipedia:Citation needed

References

Further reading

<templatestyles src="Template:Refbegin/styles.css" />

- Dalrymple, G. B. (1991). *The Age of the Earth*. California: Stanford
  University Press. ISBN 978-0-8047-1569-0.

- Dalrymple, G. Brent (2001). "The age of the Earth in the twentieth century: a problem (mostly) solved"[42]. *Geological Society, London, Special Publications*. **190** (1): 205–221. Bibcode: 2001GSLSP.190..205D[43]. doi: 10.1144/GSL.SP.2001.190.01.14[44]. Retrieved 2012-04-13.
- Dawkins, Richard (2004). *The Ancestor's Tale: A Pilgrimage to the Dawn of Life*. Boston: Houghton Mifflin Company. ISBN 978-0-618-00583-3.
- Gradstein, F. M.; Ogg, James George; Smith, Alan Gilbert, eds. (2004). *A Geological Time Scale 2004*. Reprinted with corrections 2006. Cambridge University Press. ISBN 978-0-521-78673-7.
- Gradstein, Felix M.; Ogg, James G.; van Kranendonk, Martin (2008). On the Geological Time Scale 2008[45] (PDF) (Report). International Commission on Stratigraphy. Fig. 2. Archived from the original[46] (PDF) on 28 October 2012. Retrieved 20 April 2012.
- Levin, H. L. (2009). *The Earth through time* (9th ed.). Saunders College Publishing. ISBN 978-0-470-38774-0.
- Lunine, J. I. (1999). *Earth: evolution of a habitable world*. United Kingdom: Cambridge University Press. ISBN 978-0-521-64423-5.
- McNeill, Willam H. (1999) [1967]. *A World History* (4th ed.). New York: Oxford University Press. ISBN 978-0-19-511615-1.
- Melosh, H. J.; Vickery, A. M. & Tonks, W. B. (1993). *Impacts and the early environment and evolution of the terrestrial planets*, in Levy, H.J. & Lunine, J.I. (eds.): *Protostars and Planets III*, University of Arizona Press, Tucson, pp. 1339–1370.
- Stanley, Steven M. (2005). *Earth system history* (2nd ed.). New York: Freeman. ISBN 978-0-7167-3907-4.
- Stern, T. W.; Bleeker, W. (1998). "Age of the world's oldest rocks refined using Canada's SHRIMP: The Acasta Gneiss Complex, Northwest Territories, Canada". *Geoscience Canada*. **25**: 27–31.
- Wetherill, G. W. (1991). "Occurrence of Earth-Like Bodies in Planetary Systems". *Science*. **253** (5019): 535–538. Bibcode: 1991Sci...253..535W[47]. doi: 10.1126/science.253.5019.535[48]. PMID 17745185[49].

External links

<indicator name="spoken-icon"> ◑⟩⟩ </indicator>

- Davies, Paul. " Quantum leap of life[50]". *The Guardian*. 2005 December 20. – discusses speculation on the role of quantum systems in the origin of life

- Evolution timeline[51] (uses Shockwave). Animated story of life shows everything from the big bang to the formation of the earth and the development of bacteria and other organisms to the ascent of man.
- 25 biggest turning points in earth History[52] BBC
- Evolution of the Earth[53]. Timeline of the most important events in the evolution of the Earth.
- The Earth's Origins[54] on *In Our Time* at the BBC
- Ageing the Earth[55], BBC Radio 4 discussion with Richard Corfield, Hazel Rymer & Henry Gee (*In Our Time*, Nov. 20, 2003)

Geological history

Geological history of Earth

The **geological history of Earth** follows the major events in Earth's past based on the geologic time scale, a system of chronological measurement based on the study of the planet's rock layers (stratigraphy). Earth formed about 4.54 billion years ago by accretion from the solar nebula, a disk-shaped mass of dust and gas left over from the formation of the Sun, which also created the rest of the Solar System.

Earth was initially molten due to extreme volcanism and frequent collisions with other bodies. Eventually, the outer layer of the planet cooled to form a solid crust when water began accumulating in the atmosphere. The Moon formed soon afterwards, possibly as a result of the impact of a planetoid with the Earth. Outgassing and volcanic activity produced the primordial atmosphere. Condensing water vapor, augmented by ice delivered from comets, produced the oceans.

As the surface continually reshaped itself over hundreds of millions of years, continents formed and broke apart. They migrated across the surface, occasionally combining to form a supercontinent. Roughly 750[56] million years ago, the earliest-known supercontinent Rodinia, began to break apart. The continents later recombined to form Pannotia, 600 to 540[57] million years ago, then finally Pangaea, which broke apart 200[58] million years ago.

The present pattern of ice ages began about 40[59] million years ago, then intensified at the end of the Pliocene. The polar regions have since undergone repeated cycles of glaciation and thaw, repeating every 40,000–100,000 years. The last glacial period of the current ice age ended about 10,000 years ago.

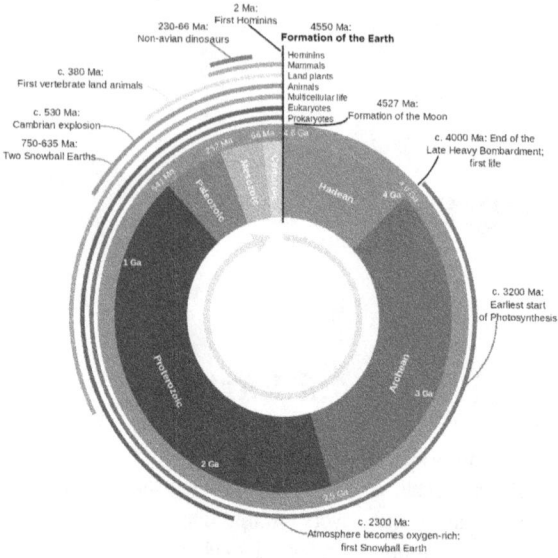

Figure 43: *Geologic time represented in a diagram called a geological clock, showing the relative lengths of the eons of Earth's history and noting major events*

Precambrian

Life timeline

Figure 44: *Artist's conception of a protoplanetary disc*

Axis scale: million years

Also see: *Human timeline* and *Nature timeline*

The Precambrian includes approximately 90% of geologic time. It extends from 4.6 billion years ago to the beginning of the Cambrian Period (about 541 Ma). It includes three eons, the Hadean, Archean, and Proterozoic.

Major volcanic events altering the Earth's environment and causing extinctions may have occurred 10 times in the past 3 billion years.

Hadean Eon

During Hadean time (4.6–4 Ga), the Solar System was forming, probably within a large cloud of gas and dust around the sun, called an accretion disc from which Earth formed 4,500[60] million years ago. The Hadean Eon is not formally recognized, but it essentially marks the era before we have adequate record of significant solid rocks. The oldest dated zircons date from about 4,400[61] million years ago.

Earth was initially molten due to extreme volcanism and frequent collisions with other bodies. Eventually, the outer layer of the planet cooled to form a solid crust when water began accumulating in the atmosphere. The Moon formed soon afterwards, possibly as a result of the impact of a large planetoid with the Earth. Some of this object's mass merged with the Earth, significantly altering its internal composition, and a portion was ejected into space. Some of the material survived to form an orbiting moon. More recent potassium isotopic studies suggest that the Moon was formed by a smaller, high-energy, high-angular-momentum giant impact cleaving off a significant portion of the Earth. Outgassing and volcanic activity produced the primordial atmosphere. Condensing water vapor, augmented by ice delivered from comets, produced the oceans.

During the Hadean the Late Heavy Bombardment occurred (approximately 4,100 to 3,800[62] million years ago) during which a large number of impact craters are believed to have formed on the Moon, and by inference on Earth, Mercury, Venus and Mars as well.

Archean Eon

The Earth of the early Archean (4,000 to 2,500[63] million years ago) may have had a different tectonic style. During this time, the Earth's crust cooled enough that rocks and continental plates began to form. Some scientists think because the Earth was hotter, that plate tectonic activity was more vigorous than it is today, resulting in a much greater rate of recycling of crustal material. This may have prevented cratonisation and continent formation until the mantle cooled and convection slowed down. Others argue that the subcontinental lithospheric mantle is too buoyant to subduct and that the lack of Archean rocks is a function of erosion and subsequent tectonic events.

In contrast to the Proterozoic, Archean rocks are often heavily metamorphized deep-water sediments, such as graywackes, mudstones, volcanic sediments and banded iron formations. Greenstone belts are typical Archean formations, consisting of alternating high- and low-grade metamorphic rocks. The high-grade rocks were derived from volcanic island arcs, while the low-grade metamorphic rocks represent deep-sea sediments eroded from the neighboring island rocks and deposited in a forearc basin. In short, greenstone belts represent sutured protocontinents.

The Earth's magnetic field was established 3.5 billion years ago. The solar wind flux was about 100 times the value of the modern Sun, so the presence of the magnetic field helped prevent the planet's atmosphere from being stripped away, which is what probably happened to the atmosphere of Mars. However, the field strength was lower than at present and the magnetosphere was about half the modern radius.

Proterozoic Eon

The geologic record of the Proterozoic (2,500 to 541[64] million years ago) is more complete than that for the preceding Archean. In contrast to the deep-water deposits of the Archean, the Proterozoic features many strata that were laid down in extensive shallow epicontinental seas; furthermore, many of these rocks are less metamorphosed than Archean-age ones, and plenty are unaltered. Study of these rocks show that the eon featured massive, rapid continental accretion (unique to the Proterozoic), supercontinent cycles, and wholly modern orogenic activity. Roughly 750[56] million years ago,[65] the earliest-known supercontinent Rodinia, began to break apart. The continents later recombined to form Pannotia, 600–540 Ma.

The first-known glaciations occurred during the Proterozoic, one began shortly after the beginning of the eon, while there were at least four during the Neoproterozoic, climaxing with the Snowball Earth of the Varangian glaciation.

Phanerozoic Eon

The **Phanerozoic** Eon is the current eon in the geologic timescale. It covers roughly 541 million years. During this period continents drifted about, eventually collected into a single landmass known as Pangea and then split up into the current continental landmasses.

The Phanerozoic is divided into three eras – the Paleozoic, the Mesozoic and the Cenozoic.

Most of biological evolution occurred during this time period.

Paleozoic Era

The **Paleozoic** spanned from roughly 541 to 252[66] million years ago (Ma) and is subdivided into six geologic periods; from oldest to youngest they are the Cambrian, Ordovician, Silurian, Devonian, Carboniferous and Permian. Geologically, the Paleozoic starts shortly after the breakup of a supercontinent called Pannotia and at the end of a global ice age. Throughout the early Paleozoic, the Earth's landmass was broken up into a substantial number of relatively small continents. Toward the end of the era the continents gathered together into a supercontinent called Pangaea, which included most of the Earth's land area.

Cambrian Period

The **Cambrian** is a major division of the geologic timescale that begins about 541.0 ± 1.0 Ma. Cambrian continents are thought to have resulted from the breakup of a Neoproterozoic supercontinent called Pannotia. The waters of the Cambrian period appear to have been widespread and shallow. Continental drift rates may have been anomalously high. Laurentia, Baltica and Siberia remained independent continents following the break-up of the supercontinent of Pannotia. Gondwana started to drift toward the South Pole. Panthalassa covered most of the southern hemisphere, and minor oceans included the Proto-Tethys Ocean, Iapetus Ocean and Khanty Ocean.

Ordovician period

The **Ordovician** period started at a major extinction event called the Cambrian–Ordovician extinction event some time about 485.4 ± 1.9 Ma. During the Ordovician the southern continents were collected into a single continent called Gondwana. Gondwana started the period in the equatorial latitudes and, as the period progressed, drifted toward the South Pole. Early in the Ordovician the continents Laurentia, Siberia and Baltica were still independent continents (since the break-up of the supercontinent Pannotia earlier), but Baltica began to move toward Laurentia later in the period, causing the Iapetus Ocean to shrink between them. Also, Avalonia broke free from Gondwana and began to head north toward Laurentia. The Rheic Ocean was formed as a result of this. By the end of the period, Gondwana had neared or approached the pole and was largely glaciated.

The Ordovician came to a close in a series of extinction events that, taken together, comprise the second-largest of the five major extinction events in Earth's history in terms of percentage of genera that became extinct. The only larger one was the Permian-Triassic extinction event. The extinctions occurred approximately 447 to 444[67] million years ago and mark the boundary between the Ordovician and the following Silurian Period.

The most-commonly accepted theory is that these events were triggered by the onset of an ice age, in the Hirnantian faunal stage that ended the long, stable greenhouse conditions typical of the Ordovician. The ice age was probably not as long-lasting as once thought; study of oxygen isotopes in fossil brachiopods shows that it was probably no longer than 0.5 to 1.5 million years. The event was preceded by a fall in atmospheric carbon dioxide (from 7000ppm to 4400ppm) which selectively affected the shallow seas where most organisms lived. As the southern supercontinent Gondwana drifted over the South Pole, ice caps formed on it. Evidence of these ice caps have been detected in Upper Ordovician rock strata of North Africa and then-adjacent northeastern South America, which were south-polar locations at the time.

Silurian Period

The **Silurian** is a major division of the geologic timescale that started about 443.8 ± 1.5 Ma. During the Silurian, Gondwana continued a slow southward drift to high southern latitudes, but there is evidence that the Silurian ice caps were less extensive than those of the late Ordovician glaciation. The melting of ice caps and glaciers contributed to a rise in sea levels, recognizable from the fact that Silurian sediments overlie eroded Ordovician sediments, forming an unconformity. Other cratons and continent fragments drifted together near the equator, starting the formation of a second supercontinent known as Euramerica. The vast ocean of Panthalassa covered most of the northern hemisphere.

Other minor oceans include Proto-Tethys, Paleo-Tethys, Rheic Ocean, a seaway of Iapetus Ocean (now in between Avalonia and Laurentia), and newly formed Ural Ocean.

Devonian Period

The **Devonian** spanned roughly from 419 to 359 Ma. The period was a time of great tectonic activity, as Laurasia and Gondwana drew closer together. The continent Euramerica (or Laurussia) was created in the early Devonian by the collision of Laurentia and Baltica, which rotated into the natural dry zone along the Tropic of Capricorn. In these near-deserts, the Old Red Sandstone sedimentary beds formed, made red by the oxidized iron (hematite) characteristic of drought conditions. Near the equator Pangaea began to consolidate from the plates containing North America and Europe, further raising the northern Appalachian Mountains and forming the Caledonian Mountains in Great Britain and Scandinavia. The southern continents remained tied together in the supercontinent of Gondwana. The remainder of modern Eurasia lay in the Northern Hemisphere. Sea levels were high worldwide, and much of the land lay submerged under shallow seas. The deep, enormous Panthalassa (the "universal ocean") covered the rest of the planet. Other minor oceans were Paleo-Tethys, Proto-Tethys, Rheic Ocean and Ural Ocean (which was closed during the collision with Siberia and Baltica).

Carboniferous Period

The **Carboniferous** extends from about 358.9 ± 0.4 to about 298.9 ± 0.15 Ma.

A global drop in sea level at the end of the Devonian reversed early in the Carboniferous; this created the widespread epicontinental seas and carbonate deposition of the Mississippian. There was also a drop in south polar temperatures; southern Gondwana was glaciated throughout the period, though it is uncertain if the ice sheets were a holdover from the Devonian or not. These conditions apparently had little effect in the deep tropics, where lush coal swamps flourished within 30 degrees of the northernmost glaciers. A mid-Carboniferous drop in sea-level precipitated a major marine extinction, one that hit crinoids and ammonites especially hard. This sea-level drop and the associated unconformity in North America separate the Mississippian Period from the Pennsylvanian period.

The Carboniferous was a time of active mountain building, as the supercontinent Pangea came together. The southern continents remained tied together in the supercontinent Gondwana, which collided with North America-Europe (Laurussia) along the present line of eastern North America. This continental collision resulted in the Hercynian orogeny in Europe, and the Alleghenian orogeny in North America; it also extended the newly uplifted Appalachians

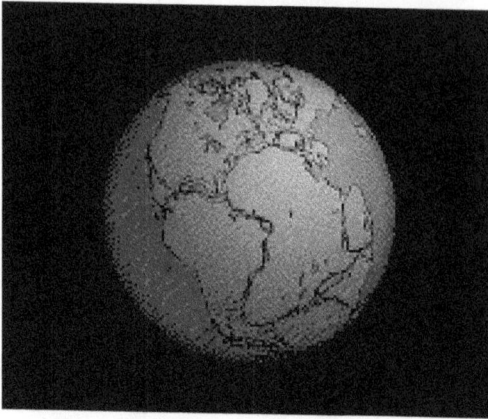

Figure 45: *Pangaea separation animation*

southwestward as the Ouachita Mountains. In the same time frame, much of present eastern Eurasian plate welded itself to Europe along the line of the Ural mountains. There were two major oceans in the Carboniferous the Panthalassa and Paleo-Tethys. Other minor oceans were shrinking and eventually closed the Rheic Ocean (closed by the assembly of South and North America), the small, shallow Ural Ocean (which was closed by the collision of Baltica, and Siberia continents, creating the Ural Mountains) and Proto-Tethys Ocean.

Permian Period

The **Permian** extends from about 298.9 ± 0.15 to 252.17 ± 0.06 Ma.

During the Permian all the Earth's major land masses, except portions of East Asia, were collected into a single supercontinent known as Pangaea. Pangaea straddled the equator and extended toward the poles, with a corresponding effect on ocean currents in the single great ocean (*Panthalassa*, the *universal sea*), and the Paleo-Tethys Ocean, a large ocean that was between Asia and Gondwana. The Cimmeria continent rifted away from Gondwana and drifted north to Laurasia, causing the Paleo-Tethys to shrink. A new ocean was growing on its southern end, the Tethys Ocean, an ocean that would dominate much of the Mesozoic Era. Large continental landmasses create climates with extreme variations of heat and cold ("continental climate") and monsoon conditions with highly seasonal rainfall patterns. Deserts seem to have been widespread on Pangaea.

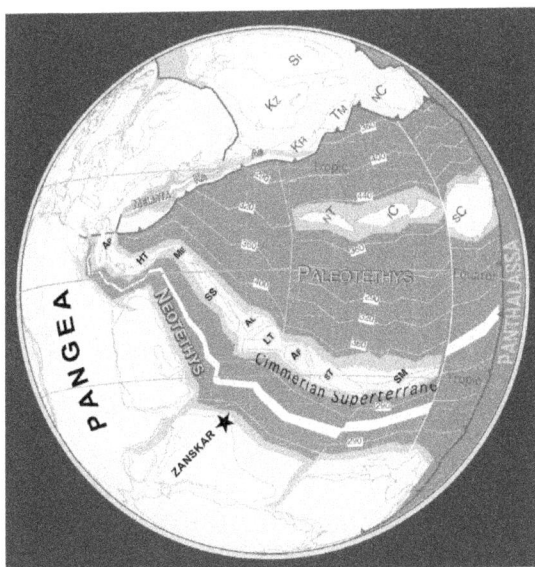

Figure 46: *Plate tectonics- 249[68] million years ago*

Mesozoic Era

The **Mesozoic** extended roughly from 252 to 66[70] million years ago.

After the vigorous convergent plate mountain-building of the late Paleozoic, Mesozoic tectonic deformation was comparatively mild. Nevertheless, the era featured the dramatic rifting of the supercontinent Pangaea. Pangaea gradually split into a northern continent, Laurasia, and a southern continent, Gondwana. This created the passive continental margin that characterizes most of the Atlantic coastline (such as along the U.S. East Coast) today.

Triassic Period

The **Triassic** Period extends from about 252.17 ± 0.06 to 201.3 ± 0.2 Ma. During the Triassic, almost all the Earth's land mass was concentrated into a single supercontinent centered more or less on the equator, called Pangaea ("all the land"). This took the form of a giant "Pac-Man" with an east-facing "mouth" constituting the Tethys sea, a vast gulf that opened farther westward in the mid-Triassic, at the expense of the shrinking Paleo-Tethys Ocean, an ocean that existed during the Paleozoic.

The remainder was the world-ocean known as Panthalassa ("all the sea"). All the deep-ocean sediments laid down during the Triassic have disappeared

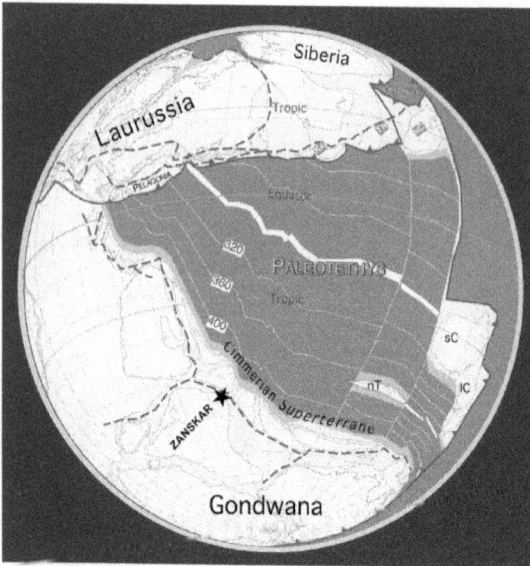

Figure 47: *Plate tectonics- 290[69] million years ago*

through subduction of oceanic plates; thus, very little is known of the Triassic open ocean. The supercontinent Pangaea was rifting during the Triassic—especially late in the period—but had not yet separated. The first nonmarine sediments in the rift that marks the initial break-up of Pangea—which separated New Jersey from Morocco—are of Late Triassic age; in the U.S., these thick sediments comprise the Newark Supergroup. Because of the limited shoreline of one super-continental mass, Triassic marine deposits are globally relatively rare; despite their prominence in Western Europe, where the Triassic was first studied. In North America, for example, marine deposits are limited to a few exposures in the west. Thus Triassic stratigraphy is mostly based on organisms living in lagoons and hypersaline environments, such as *Estheria* crustaceans and terrestrial vertebrates.

Jurassic Period

The **Jurassic** Period extends from about 201.3 ± 0.2 to 145.0 Ma. During the early Jurassic, the supercontinent Pangaea broke up into the northern supercontinent Laurasia and the southern supercontinent Gondwana; the Gulf of Mexico opened in the new rift between North America and what is now Mexico's Yucatan Peninsula. The Jurassic North Atlantic Ocean was relatively narrow, while the South Atlantic did not open until the following Cretaceous Period, when Gondwana itself rifted apart. The Tethys Sea closed, and the

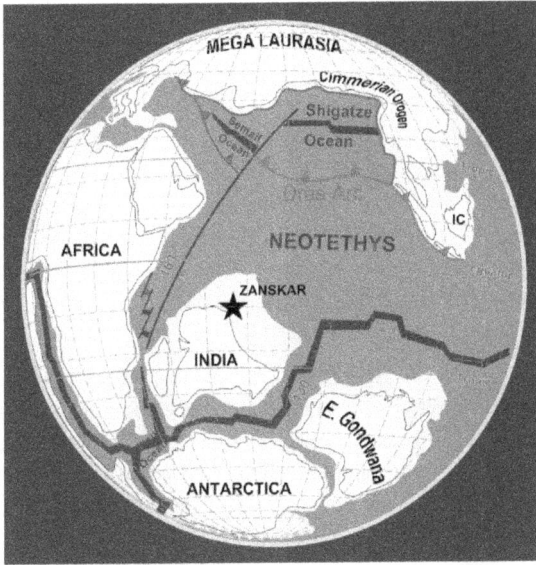

Figure 48: *Plate tectonics- 100 Ma, Cretaceous period*

Neotethys basin appeared. Climates were warm, with no evidence of glaciation. As in the Triassic, there was apparently no land near either pole, and no extensive ice caps existed. The Jurassic geological record is good in western Europe, where extensive marine sequences indicate a time when much of the continent was submerged under shallow tropical seas; famous locales include the Jurassic Coast World Heritage Site and the renowned late Jurassic *lagerstätten* of Holzmaden and Solnhofen. In contrast, the North American Jurassic record is the poorest of the Mesozoic, with few outcrops at the surface. Though the epicontinental Sundance Sea left marine deposits in parts of the northern plains of the United States and Canada during the late Jurassic, most exposed sediments from this period are continental, such as the alluvial deposits of the Morrison Formation. The first of several massive batholiths were emplaced in the northern Cordillera beginning in the mid-Jurassic, marking the Nevadan orogeny.[71] Important Jurassic exposures are also found in Russia, India, South America, Japan, Australasia and the United Kingdom.

Cretaceous Period

The **Cretaceous** Period extends from circa 145[72] million years ago to 66[73] million years ago.

During the Cretaceous, the late Paleozoic-early Mesozoic supercontinent of Pangaea completed its breakup into present day continents, although their positions were substantially different at the time. As the Atlantic Ocean widened, the convergent-margin orogenies that had begun during the Jurassic continued in the North American Cordillera, as the Nevadan orogeny was followed by the Sevier and Laramide orogenies. Though Gondwana was still intact in the beginning of the Cretaceous, Gondwana itself broke up as South America, Antarctica and Australia rifted away from Africa (though India and Madagascar remained attached to each other); thus, the South Atlantic and Indian Oceans were newly formed. Such active rifting lifted great undersea mountain chains along the welts, raising eustatic sea levels worldwide.

To the north of Africa the Tethys Sea continued to narrow. Broad shallow seas advanced across central North America (the Western Interior Seaway) and Europe, then receded late in the period, leaving thick marine deposits sandwiched between coal beds. At the peak of the Cretaceous transgression, one-third of Earth's present land area was submerged.[74] The Cretaceous is justly famous for its chalk; indeed, more chalk formed in the Cretaceous than in any other period in the Phanerozoic. Mid-ocean ridge activity—or rather, the circulation of seawater through the enlarged ridges—enriched the oceans in calcium; this made the oceans more saturated, as well as increased the bioavailability of the element for calcareous nanoplankton. These widespread carbonates and other sedimentary deposits make the Cretaceous rock record especially fine. Famous formations from North America include the rich marine fossils of Kansas's Smoky Hill Chalk Member and the terrestrial fauna of the late Cretaceous Hell Creek Formation. Other important Cretaceous exposures occur in Europe and China. In the area that is now India, massive lava beds called the Deccan Traps were laid down in the very late Cretaceous and early Paleocene.

Cenozoic Era

The **Cenozoic** Era covers the 66 million years since the Cretaceous–Paleogene extinction event up to and including the present day. By the end of the Mesozoic era, the continents had rifted into nearly their present form. Laurasia became North America and Eurasia, while Gondwana split into South America, Africa, Australia, Antarctica and the Indian subcontinent, which collided with the Asian plate. This impact gave rise to the Himalayas. The Tethys Sea, which had separated the northern continents from Africa and India, began to close up, forming the Mediterranean sea.

Paleogene Period

The **Paleogene** (alternatively **Palaeogene**) Period is a unit of geologic time that began 66 and ended 23.03 Ma and comprises the first part of the Cenozoic Era. This period consists of the Paleocene, Eocene and Oligocene Epochs.

Paleocene Epoch

The **Paleocene**, lasted from 66[73] million years ago to 56[75] million years ago.

In many ways, the Paleocene continued processes that had begun during the late Cretaceous Period. During the Paleocene, the continents continued to drift toward their present positions. Supercontinent Laurasia had not yet separated into three continents. Europe and Greenland were still connected. North America and Asia were still intermittently joined by a land bridge, while Greenland and North America were beginning to separate.[76] The Laramide orogeny of the late Cretaceous continued to uplift the Rocky Mountains in the American west, which ended in the succeeding epoch. South and North America remained separated by equatorial seas (they joined during the Neogene); the components of the former southern supercontinent Gondwana continued to split apart, with Africa, South America, Antarctica and Australia pulling away from each other. Africa was heading north toward Europe, slowly closing the Tethys Ocean, and India began its migration to Asia that would lead to a tectonic collision and the formation of the Himalayas.

Eocene Epoch

During the **Eocene** (56[75] million years ago - 33.9[77] million years ago), the continents continued to drift toward their present positions. At the beginning of the period, Australia and Antarctica remained connected, and warm equatorial currents mixed with colder Antarctic waters, distributing the heat around the world and keeping global temperatures high. But when Australia split from the southern continent around 45 Ma, the warm equatorial currents were deflected away from Antarctica, and an isolated cold water channel developed between the two continents. The Antarctic region cooled down, and the ocean surrounding Antarctica began to freeze, sending cold water and ice floes north, reinforcing the cooling. The present pattern of ice ages began about 40[59] million years ago.Wikipedia:Citation needed

The northern supercontinent of Laurasia began to break up, as Europe, Greenland and North America drifted apart. In western North America, mountain building started in the Eocene, and huge lakes formed in the high flat basins among uplifts. In Europe, the Tethys Sea finally vanished, while the uplift of the Alps isolated its final remnant, the Mediterranean, and created another shallow sea with island archipelagos to the north. Though the North

Atlantic was opening, a land connection appears to have remained between North America and Europe since the faunas of the two regions are very similar. India continued its journey away from Africa and began its collision with Asia, creating the Himalayan orogeny.

Oligocene Epoch

The **Oligocene** Epoch extends from about 34^{78} million years ago to 23^{79} million years ago. During the Oligocene the continents continued to drift toward their present positions.

Antarctica continued to become more isolated and finally developed a permanent ice cap. Mountain building in western North America continued, and the Alps started to rise in Europe as the African plate continued to push north into the Eurasian plate, isolating the remnants of Tethys Sea. A brief marine incursion marks the early Oligocene in Europe. There appears to have been a land bridge in the early Oligocene between North America and Europe since the faunas of the two regions are very similar. During the Oligocene, South America was finally detached from Antarctica and drifted north toward North America. It also allowed the Antarctic Circumpolar Current to flow, rapidly cooling the continent.

Neogene Period

The **Neogene** Period is a unit of geologic time starting 23.03 Ma. and ends at 2.588 Mya. The Neogene Period follows the Paleogene Period. The Neogene consists of the Miocene and Pliocene and is followed by the Quaternary Period.

Miocene Epoch

The **Miocene** extends from about 23.03 to 5.333 Ma.

During the Miocene continents continued to drift toward their present positions. Of the modern geologic features, only the land bridge between South America and North America was absent, the subduction zone along the Pacific Ocean margin of South America caused the rise of the Andes and the southward extension of the Meso-American peninsula. India continued to collide with Asia. The Tethys Seaway continued to shrink and then disappeared as Africa collided with Eurasia in the Turkish-Arabian region between 19 and 12 Ma (ICS 2004). Subsequent uplift of mountains in the western Mediterranean region and a global fall in sea levels combined to cause a temporary drying up of the Mediterranean Sea resulting in the Messinian salinity crisis near the end of the Miocene.

Pliocene Epoch

The **Pliocene** extends from 5.333[80] million years ago to 2.588[81] million years ago. During the Pliocene continents continued to drift toward their present positions, moving from positions possibly as far as 250 kilometres (155 mi) from their present locations to positions only 70 km from their current locations.

South America became linked to North America through the Isthmus of Panama during the Pliocene, bringing a nearly complete end to South America's distinctive marsupial faunas. The formation of the Isthmus had major consequences on global temperatures, since warm equatorial ocean currents were cut off and an Atlantic cooling cycle began, with cold Arctic and Antarctic waters dropping temperatures in the now-isolated Atlantic Ocean. Africa's collision with Europe formed the Mediterranean Sea, cutting off the remnants of the Tethys Ocean. Sea level changes exposed the land-bridge between Alaska and Asia. Near the end of the Pliocene, about 2.58[82] million years ago (the start of the Quaternary Period), the current ice age began. The polar regions have since undergone repeated cycles of glaciation and thaw, repeating every 40,000–100,000 years.

Quaternary Period

Pleistocene Epoch

The **Pleistocene** extends from 2.588[81] million years ago to 11,700 years before present. The modern continents were essentially at their present positions during the Pleistocene, the plates upon which they sit probably having moved no more than 100 kilometres (62 mi) relative to each other since the beginning of the period.

Holocene Epoch

The **Holocene** Epoch began approximately 11,700 calendar years before present and continues to the present. During the Holocene, continental motions have been less than a kilometer.

The last glacial period of the current ice age ended about 10,000 years ago. Ice melt caused world sea levels to rise about 35 metres (115 ft) in the early part of the Holocene. In addition, many areas above about 40 degrees north latitude had been depressed by the weight of the Pleistocene glaciers and rose as much as 180 metres (591 ft) over the late Pleistocene and Holocene, and are still rising today. The sea level rise and temporary land depression allowed temporary marine incursions into areas that are now far from the sea. Holocene marine fossils are known from Vermont, Quebec, Ontario and Michigan. Other than higher latitude temporary marine incursions associated with glacial depression, Holocene fossils are found primarily in lakebed, floodplain and cave

Figure 49: *Current Earth - without water (click/enlarge to "spin" 3D-globe).*

deposits. Holocene marine deposits along low-latitude coastlines are rare because the rise in sea levels during the period exceeds any likely upthrusting of
non-glacial origin. Post-glacial rebound in Scandinavia resulted in the emergence of coastal areas around the Baltic Sea, including much of Finland. The
region continues to rise, still causing weak earthquakes across Northern Europe. The equivalent event in North America was the rebound of Hudson
Bay, as it shrank from its larger, immediate post-glacial Tyrrell Sea phase, to
near its present boundaries.

Further reading

<templatestyles src="Template:Refbegin/styles.css" />

- Stanley, Steven M. (1999). *Earth system history* (New ed.). New York:
  W. H. Freeman. ISBN 978-0-7167-3377-5.

External links

- Cosmic Evolution[83] — a detailed look at events from the origin of the
  universe to the present

- Valley, John W. " A Cool Early Earth?[84]" *Scientific American*. 2005 Oct:58–65. – discusses the timing of the formation of the oceans and other major events in Eh's early history.
- Davies, Paul. " Quantum leap of life[85]". *The Guardian*. 2005 Dec 20. – discusses speculation into the role of quantum systems in the origin of life
- Evolution timeline[86] (uses Shockwave). Animated story of life since about 13,700,000,000 shows everything from the big bang to the formation of the earth and the development of bacteria and other organisms to the ascent of man.
- Theory of the Earth & Abstract of the Theory of the Earth[87]
- Paleomaps Since 600 Ma (Mollweide Projection, Longitude 0)[88]
- Paleomaps Since 600 Ma (Mollweide Projection, Longitude 180)[89]
- Ageing the Earth[90] on *In Our Time* at the BBC

Origin of life and evolution

Evolutionary history of life

<indicator name="good-star"> ⊕ </indicator>

Part of a series on
Evolutionary biology

- ⚘ Evolutionary biology portal
- ◯ Category
- ↪ *Book*
- Related topics

- v
- t
- e[91]

The **evolutionary history of life** on Earth traces the processes by which both living organisms and fossil organisms evolved since life emerged on the planet, until the present. Earth formed about 4.5 billion years (Ga) ago and evidence suggests life emerged prior to 3.7 Ga. Although there is some evidence to suggest that life appeared as early as 4.1 to 4.28 Ga this evidence remains controversial due to the non-biological mechanisms that may have formed these potential signatures of past life. The similarities among all present-day organisms indicate the presence of a common ancestor from which all known species have diverged through the process of evolution. More than 99 percent of all species, amounting to over five billion species, that ever lived on Earth are estimated to be extinct. Estimates on the number of Earth's current species range

from 10 million to 14 million, of which about 1.9 million are estimated to have
been named and 1.6 million documented in a central database to date. More
recently, in May 2016, scientists reported that 1 trillion species are estimated
to be on Earth currently with only one-thousandth of one percent described.

Life timeline

Axis scale: million years

Also see: *Human timeline* and *Nature timeline*

The earliest evidences of life on Earth are biogenic carbon signatures and stromatolite fossils discovered in 3.7 billion-year-old metasedimentary rocks discovered in western Greenland. In 2015, "remains of biotic life" were potentially found in 4.1 billion-year-old rocks in Western Australia.[92] In March 2017, researchers reported evidence of possibly the oldest forms of life on Earth. Putative fossilized microorganisms were discovered in hydrothermal vent precipitates in the Nuvvuagittuq Belt of Quebec, Canada, that may have lived as early as 4.280 billion years ago, not long after the oceans formed 4.4 billion years ago, and not long after the formation of the Earth 4.54 billion years ago. According to biologist Stephen Blair Hedges, "If life arose relatively quickly on Earth ... then it could be common in the universe."

Microbial mats of coexisting bacteria and archaea were the dominant form of life in the early Archean and many of the major steps in early evolution are thought to have taken place within them. The evolution of photosynthesis, around 3.5 Ga, eventually led to a buildup of its waste product, oxygen, in the atmosphere, leading to the great oxygenation event, beginning around 2.4 Ga. The earliest evidence of eukaryotes (complex cells with organelles) dates from 1.85 Ga, and while they may have been present earlier, their diversification accelerated when they started using oxygen in their metabolism. Later, around 1.7 Ga, multicellular organisms began to appear, with differentiated cells performing specialised functions. Sexual reproduction, which involves the fusion of male and female reproductive cells (gametes) to create a zygote in a process called fertilization is, in contrast to asexual reproduction, the primary method of reproduction for the vast majority of macroscopic organisms, including almost all eukaryotes (which includes animals and plants). However the origin and evolution of sexual reproduction remain a puzzle for biologists though it did evolve from a common ancestor that was a single celled eukaryotic species. Bilateria, animals with a front and a back, appeared by 555 Ma (million years ago).

The earliest complex land plants date back to around 850 Ma, from carbon isotopes in Precambrian rocks, while algae-like multicellular land plants are dated back even to about 1 billion years ago, although evidence suggests that microorganisms formed the earliest terrestrial ecosystems, at least 2.7 Ga. Microorganisms are thought to have paved the way for the inception of land plants in the Ordovician. Land plants were so successful that they are thought to have contributed to the Late Devonian extinction event. (The long causal chain implied seems to involve the success of early tree archaeopteris (1) drew down CO_2 levels, leading to global cooling and lowered sea levels, (2) roots of archeopteris fostered soil development which *increased* rock weathering, and the subsequent nutrient run-off may have triggered algal blooms resulting in anoxic events which caused marine-life die-offs. Marine species were the primary victims of the Late Devonian extinction.)

Ediacara biota appear during the Ediacaran period, while vertebrates, along with most other modern phyla originated about 525[93] Ma during the Cambrian explosion. During the Permian period, synapsids, including the ancestors of mammals, dominated the land, but most of this group became extinct in the Permian–Triassic extinction event 252[94] Ma. During the recovery from this catastrophe, archosaurs became the most abundant land vertebrates; one archosaur group, the dinosaurs, dominated the Jurassic and Cretaceous periods. After the Cretaceous–Paleogene extinction event 66[95] Ma killed off the non-avian dinosaurs, mammals increased rapidly in size and diversity. Such mass extinctions may have accelerated evolution by providing opportunities for new groups of organisms to diversify.

Earliest history of Earth

History of Earth and its life

Scale: Ma (Millions of years)

The oldest meteorite fragments found on Earth are about 4.54 billion years old; this, coupled primarily with the dating of ancient lead deposits, has put the estimated age of Earth at around that time. The Moon has the same composition as Earth's crust but does not contain an iron-rich core like the Earth's. Many scientists think that about 40 million years after the formation of Earth, it collided with a body the size of Mars, throwing into orbit crust material that formed the Moon. Another hypothesis is that the Earth and Moon started to coalesce at the same time but the Earth, having much stronger gravity than the early Moon, attracted almost all the iron particles in the area.

Until 2001, the oldest rocks found on Earth were about 3.8 billion years old, leading scientists to estimate that the Earth's surface had been molten until then. Accordingly, they named this part of Earth's history the Hadean. However, analysis of zircons formed 4.4 Ga indicates that Earth's crust solidified about 100 million years after the planet's formation and that the planet quickly acquired oceans and an atmosphere, which may have been capable of supporting life.

Evidence from the Moon indicates that from 4 to 3.8 Ga it suffered a Late Heavy Bombardment by debris that was left over from the formation of the Solar System, and the Earth should have experienced an even heavier bombardment due to its stronger gravity. While there is no direct evidence of conditions on Earth 4 to 3.8 Ga, there is no reason to think that the Earth was not also affected by this late heavy bombardment. This event may well have stripped away any previous atmosphere and oceans; in this case gases and water from comet impacts may have contributed to their replacement, although outgassing from volcanoes on Earth would have supplied at least half. However, if subsurface microbial life had evolved by this point, it would have survived the bombardment.

Earliest evidence for life on Earth

The earliest identified organisms were minute and relatively featureless, and their fossils look like small rods, which are very difficult to tell apart from structures that arise through abiotic physical processes. The oldest undisputed evidence of life on Earth, interpreted as fossilized bacteria, dates to 3 Ga. Other finds in rocks dated to about 3.5 Ga have been interpreted as bacteria, with geochemical evidence also seeming to show the presence of life 3.8 Ga. However, these analyses were closely scrutinized, and non-biological processes were found which could produce all of the "signatures of life" that had been reported. While this does not prove that the structures found had a non-biological origin, they cannot be taken as clear evidence for the presence of life.

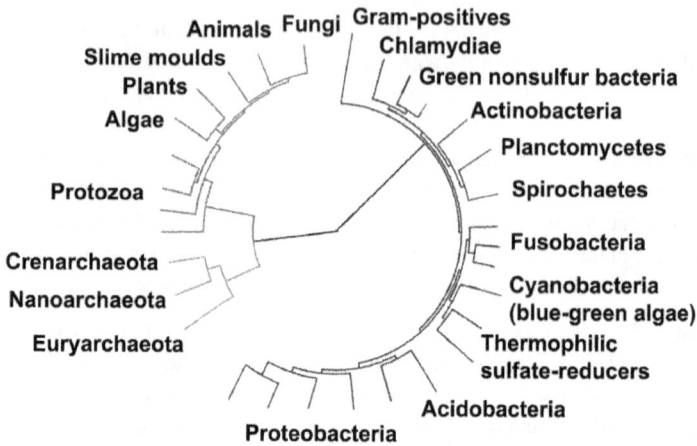

Figure 50: *citeseerx=10.1.1.381.9514 }}</ref> The three domains are colored, with bacteria blue, archaea green, and eukaryotes red.*

Geochemical signatures from rocks deposited 3.4 Ga have been interpreted as evidence for life, although these statements have not been thoroughly examined by critics.

Evidence for fossilized microorganisms considered to be 3,770 million to 4,280 million years was found in the Nuvvuagittuq Greenstone Belt in Quebec, Canada, although the evidence is disputed as inconclusive.

Origins of life on Earth

Biologists reason that all living organisms on Earth must share a single last universal ancestor, because it would be virtually impossible that two or more separate lineages could have independently developed the many complex biochemical mechanisms common to all living organisms.

Independent emergence on Earth

Life on Earth is based on carbon and water. Carbon provides stable frameworks for complex chemicals and can be easily extracted from the environment, especially from carbon dioxide. There is no other chemical element whose properties are similar enough to carbon's to be called an analogue; silicon, the element directly below carbon on the periodic table, does not form very many complex stable molecules, and because most of its compounds are water-insoluble, it would be more difficult for organisms to extract. The elements boron and phosphorus have more complex chemistries, but suffer from other limitations relative to carbon. Water is an excellent solvent and has two other useful properties: the fact that ice floats enables aquatic organisms to survive beneath it in winter; and its molecules have electrically negative and positive ends, which enables it to form a wider range of compounds than other solvents can. Other good solvents, such as ammonia, are liquid only at such low temperatures that chemical reactions may be too slow to sustain life, and lack water's other advantages. Organisms based on alternative biochemistry may, however, be possible on other planets.

Research on how life might have emerged from non-living chemicals focuses on three possible starting points: self-replication, an organism's ability to produce offspring that are very similar to itself; metabolism, its ability to feed and repair itself; and external cell membranes, which allow food to enter and waste products to leave, but exclude unwanted substances. Research on abiogenesis still has a long way to go, since theoretical and empirical approaches are only beginning to make contact with each other.

Replication first: RNA world

Even the simplest members of the three modern domains of life use DNA to record their "recipes" and a complex array of RNA and protein molecules to "read" these instructions and use them for growth, maintenance and self-replication. The discovery that some RNA molecules can catalyze both their own replication and the construction of proteins led to the hypothesis of earlier life-forms based entirely on RNA. These ribozymes could have formed an RNA world in which there were individuals but no species, as mutations and horizontal gene transfers would have meant that the offspring in each generation were quite likely to have different genomes from those that their parents started with. RNA would later have been replaced by DNA, which is more stable and therefore can build longer genomes, expanding the range of capabilities a single organism can have. Ribozymes remain as the main components of ribosomes, modern cells' "protein factories." Evidence suggests the first RNA molecules formed on Earth prior to 4.17 Ga.

Although short self-replicating RNA molecules have been artificially produced in laboratories, doubts have been raised about where natural non-biological synthesis of RNA is possible. The earliest "ribozymes" may have been formed of simpler nucleic acids such as PNA, TNA or GNA, which would have been replaced later by RNA.

In 2003, it was proposed that porous metal sulfide precipitates would assist RNA synthesis at about 100 °C (212 °F) and ocean-bottom pressures near hydrothermal vents. Under this hypothesis, lipid membranes would be the last major cell components to appear and, until then, the protocells would be confined to the pores.

Metabolism first: Iron–sulfur world

A series of experiments starting in 1997 showed that early stages in the formation of proteins from inorganic materials including carbon monoxide and hydrogen sulfide could be achieved by using iron sulfide and nickel sulfide as catalysts. Most of the steps required temperatures of about 100 °C (212 °F) and moderate pressures, although one stage required 250 °C (482 °F) and a pressure equivalent to that found under 7 kilometres (4.3 mi) of rock. Hence it was suggested that self-sustaining synthesis of proteins could have occurred near hydrothermal vents.

Membranes first: Lipid world

File:Liposome cross section.png

Cross-section through a liposome

It has been suggested that double-walled "bubbles" of lipids like those that form the external membranes of cells may have been an essential first step. Experiments that simulated the conditions of the early Earth have reported the formation of lipids, and these can spontaneously form liposomes, double-walled "bubbles," and then reproduce themselves. Although they are not intrinsically information-carriers as nucleic acids are, they would be subject to natural selection for longevity and reproduction. Nucleic acids such as RNA might then have formed more easily within the liposomes than they would have outside.

The clay hypothesis

RNA is complex and there are doubts about whether it can be produced non-biologically in the wild. Some clays, notably montmorillonite, have properties that make them plausible accelerators for the emergence of an RNA world: they grow by self-replication of their crystalline pattern; they are subject to an analog of natural selection, as the clay "species" that grows fastest in a particular environment rapidly becomes dominant; and they can catalyze the formation of RNA molecules. Although this idea has not become the scientific consensus, it still has active supporters.[96]

Research in 2003 reported that montmorillonite could also accelerate the conversion of fatty acids into "bubbles," and that the "bubbles" could encapsulate RNA attached to the clay. These "bubbles" can then grow by absorbing additional lipids and then divide. The formation of the earliest cells may have been aided by similar processes.

A similar hypothesis presents self-replicating iron-rich clays as the progenitors of nucleotides, lipids and amino acids.

Life "seeded" from elsewhere

The Panspermia hypothesis does not explain how life arose in the first place, but simply examines the possibility of it coming from somewhere other than the Earth. The idea that life on Earth was "seeded" from elsewhere in the Universe dates back at least to the Greek philosopher Anaximander in the sixth century BCE. In the twentieth century it was proposed by the physical chemist Svante Arrhenius, by the astronomers Fred Hoyle and Chandra Wickramasinghe, and by molecular biologist Francis Crick and chemist Leslie Orgel.

There are three main versions of the "seeded from elsewhere" hypothesis: from elsewhere in our Solar System via fragments knocked into space by a large meteor impact, in which case the most credible sources are Mars and Venus; by alien visitors, possibly as a result of accidental contamination by microorganisms that they brought with them; and from outside the Solar System but by natural means.

Experiments in low Earth orbit, such as EXOSTACK, demonstrated that some microorganism spores can survive the shock of being catapulted into space and some can survive exposure to outer space radiation for at least 5.7 years. Scientists are divided over the likelihood of life arising independently on Mars, or on other planets in our galaxy.

Environmental and evolutionary impact of microbial mats

Microbial mats are multi-layered, multi-species colonies of bacteria and other organisms that are generally only a few millimeters thick, but still contain a wide range of chemical environments, each of which favors a different set of microorganisms. To some extent each mat forms its own food chain, as the by-products of each group of microorganisms generally serve as "food" for adjacent groups.

Stromatolites are stubby pillars built as microorganisms in mats slowly migrate upwards to avoid being smothered by sediment deposited on them by water. There has been vigorous debate about the validity of alleged fossils from before 3 Ga, with critics arguing that so-called stromatolites could have been formed by non-biological processes. In 2006, another find of stromatolites was reported from the same part of Australia as previous ones, in rocks dated to 3.5 Ga.

In modern underwater mats the top layer often consists of photosynthesizing cyanobacteria which create an oxygen-rich environment, while the bottom layer is oxygen-free and often dominated by hydrogen sulfide emitted by the

Figure 51: *Modern stromatolites in Shark Bay, Western Australia*

organisms living there. It is estimated that the appearance of oxygenic photosynthesis by bacteria in mats increased biological productivity by a factor of between 100 and 1,000. The reducing agent used by oxygenic photosynthesis is water, which is much more plentiful than the geologically produced reducing agents required by the earlier non-oxygenic photosynthesis. From this point onwards life itself produced significantly more of the resources it needed than did geochemical processes. Oxygen is toxic to organisms that are not adapted to it, but greatly increases the metabolic efficiency of oxygen-adapted organisms.[97] Oxygen became a significant component of Earth's atmosphere about 2.4 Ga. Although eukaryotes may have been present much earlier, the oxygenation of the atmosphere was a prerequisite for the evolution of the most complex eukaryotic cells, from which all multicellular organisms are built. The boundary between oxygen-rich and oxygen-free layers in microbial mats would have moved upwards when photosynthesis shut down overnight, and then downwards as it resumed on the next day. This would have created selection pressure for organisms in this intermediate zone to acquire the ability to tolerate and then to use oxygen, possibly via endosymbiosis, where one organism lives inside another and both of them benefit from their association.

Cyanobacteria have the most complete biochemical "toolkits" of all the mat-forming organisms. Hence they are the most self-sufficient of the mat organisms and were well-adapted to strike out on their own both as floating mats and as the first of the phytoplankton, providing the basis of most marine food chains.

Diversification of eukaryotes

<templatestyles src="Template:Clade/styles.css" />

Eu-
karya   <templatestyles src="Template:Clade/styles.css" />

    Diaphoret-   <templatestyles src="Template:Clade/styles.css" />
    ickes
        Archaeplastida (Land plants, green algae, red algae, and glaucophytes)

        Hacrobia

        SAR (Stramenopiles, Alveolata, and Rhizaria)

        Excavata

    Amorphea   <templatestyles src="Template:Clade/styles.css" />
        Amoebozoa

        Sulcozoa

        Opisthokonta   <templatestyles src="Template:Clade/styles.css" />
        Metazoa (Animals)

        Choanozoa

        Fungi

One possible family tree of eukaryotes

Chromatin, nucleus, endomembrane system, and mitochondria

Eukaryotes may have been present long before the oxygenation of the atmosphere, but most modern eukaryotes require oxygen, which their mitochondria use to fuel the production of ATP, the internal energy supply of all known cells. In the 1970s it was proposed and, after much debate, widely accepted that eukaryotes emerged as a result of a sequence of endosymbiosis between "prokaryotes." For example: a predatory microorganism invaded a large prokaryote, probably an archaean, but the attack was neutralized, and the attacker took up residence and evolved into the first of the mitochondria; one of these chimeras later tried to swallow a photosynthesizing cyanobacterium, but the victim survived inside the attacker and the new combination became the ancestor of plants; and so on. After each endosymbiosis began, the partners would have eliminated unproductive duplication of genetic functions by re-arranging their genomes, a process which sometimes involved transfer of genes between them. Another hypothesis proposes that mitochondria were originally sulfur- or hydrogen-metabolising endosymbionts, and became oxygen-consumers later. On the other hand, mitochondria might have been part of eukaryotes' original equipment.

There is a debate about when eukaryotes first appeared: the presence of steranes in Australian shales may indicate that eukaryotes were present 2.7 Ga; however, an analysis in 2008 concluded that these chemicals infiltrated the rocks less than 2.2 Ga and prove nothing about the origins of eukaryotes. Fossils of the algae *Grypania* have been reported in 1.85 billion-year-old rocks (originally dated to 2.1 Ga but later revised), and indicates that eukaryotes with organelles had already evolved. A diverse collection of fossil algae were found in rocks dated between 1.5 and 1.4 Ga. The earliest known fossils of fungi date from 1.43 Ga.

Plastids

Plastids, the superclass of organelles of which chloroplasts are the best-known exemplar, are thought to have originated from endosymbiotic cyanobacteria. The symbiosis evolved around 1.5 Ga and enabled eukaryotes to carry out oxygenic photosynthesis. Three evolutionary lineages have since emerged in which the plastids are named differently: chloroplasts in green algae and plants, rhodoplasts in red algae and cyanelles in the glaucophytes.

Sexual reproduction and multicellular organisms

Evolution of sexual reproduction

The defining characteristics of sexual reproduction in eukaryotes are meiosis and fertilization. There is much genetic recombination in this kind of reproduction, in which offspring receive 50% of their genes from each parent, in contrast with asexual reproduction, in which there is no recombination. Bacteria also exchange DNA by bacterial conjugation, the benefits of which include resistance to antibiotics and other toxins, and the ability to utilize new metabolites. However, conjugation is not a means of reproduction, and is not limited to members of the same species – there are cases where bacteria transfer DNA to plants and animals.

On the other hand, bacterial transformation is clearly an adaptation for transfer of DNA between bacteria of the same species. Bacterial transformation is a complex process involving the products of numerous bacterial genes and can

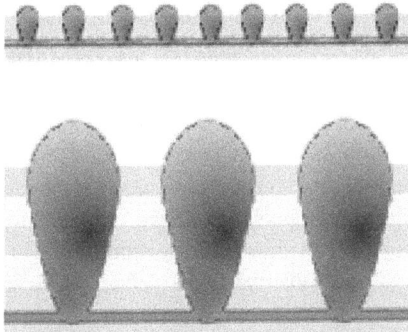

Figure 52: *Horodyskia may have been an early metazoan, or a colonial foraminiferan. It apparently re-arranged itself into fewer but larger main masses as the sediment grew deeper round its base.*

be regarded as a bacterial form of sex. This process occurs naturally in at least 67 prokaryotic species (in seven different phyla). Sexual reproduction in eukaryotes may have evolved from bacterial transformation. (Also see Evolution of sexual reproduction#Origin of sexual reproduction.)

The disadvantages of sexual reproduction are well-known: the genetic reshuffle of recombination may break up favorable combinations of genes; and since males do not directly increase the number of offspring in the next generation, an asexual population can out-breed and displace in as little as 50 generations a sexual population that is equal in every other respect. Nevertheless, the great majority of animals, plants, fungi and protists reproduce sexually. There is strong evidence that sexual reproduction arose early in the history of eukaryotes and that the genes controlling it have changed very little since then. How sexual reproduction evolved and survived is an unsolved puzzle.

The Red Queen hypothesis suggests that sexual reproduction provides protection against parasites, because it is easier for parasites to evolve means of overcoming the defenses of genetically identical clones than those of sexual species that present moving targets, and there is some experimental evidence for this. However, there is still doubt about whether it would explain the survival of sexual species if multiple similar clone species were present, as one of the clones may survive the attacks of parasites for long enough to out-breed the sexual species. Furthermore, contrary to the expectations of the Red Queen hypothesis, Kathryn A. Hanley et al. found that the prevalence, abundance and mean intensity of mites was significantly higher in sexual geckos than in asexuals sharing the same habitat. In addition, biologist Matthew Parker, after reviewing numerous genetic studies on plant disease resistance, failed to find a single example consistent with the concept that pathogens are the primary selective agent responsible for sexual reproduction in the host.

Alexey Kondrashov's *deterministic mutation hypothesis* (DMH) assumes that each organism has more than one harmful mutation and the combined effects of these mutations are more harmful than the sum of the harm done by each individual mutation. If so, sexual recombination of genes will reduce the harm that bad mutations do to offspring and at the same time eliminate some bad mutations from the gene pool by isolating them in individuals that perish quickly because they have an above-average number of bad mutations. However, the evidence suggests that the DMH's assumptions are shaky, because many species have on average less than one harmful mutation per individual and no species that has been investigated shows evidence of synergy between harmful mutations. (Further criticisms of this hypothesis are discussed in the article Evolution of sexual reproduction#Removal of deleterious genes)

The random nature of recombination causes the relative abundance of alternative traits to vary from one generation to another. This genetic drift is insufficient on its own to make sexual reproduction advantageous, but a combination of genetic drift and natural selection may be sufficient. When chance produces combinations of good traits, natural selection gives a large advantage to lineages in which these traits become genetically linked. On the other hand, the benefits of good traits are neutralized if they appear along with bad traits. Sexual recombination gives good traits the opportunities to become linked with other good traits, and mathematical models suggest this may be more than enough to offset the disadvantages of sexual reproduction. Other combinations of hypotheses that are inadequate on their own are also being examined.

The adaptive function of sex today remains a major unresolved issue in biology. The competing models to explain the adaptive function of sex were reviewed by John A. Birdsell and Christopher Wills. The hypotheses discussed above all depend on possible beneficial effects of random genetic variation produced by genetic recombination. An alternative view is that sex arose, and is maintained, as a process for repairing DNA damage, and that the genetic variation produced is an occasionally beneficial byproduct.

Multicellularity

The simplest definitions of "multicellular," for example "having multiple cells," could include colonial cyanobacteria like *Nostoc*. Even a technical definition such as "having the same genome but different types of cell" would still include some genera of the green algae Volvox, which have cells that specialize in reproduction. Multicellularity evolved independently in organisms as

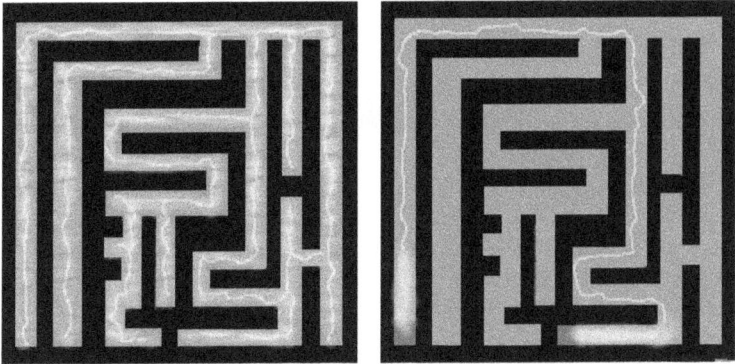

Figure 53: *A slime mold solves a maze. The mold (yellow) explored and filled the maze (left). When the researchers placed sugar (red) at two separate points, the mold concentrated most of its mass there and left only the most efficient connection between the two points (right).*

diverse as sponges and other animals, fungi, plants, brown algae, cyanobacteria, slime molds and myxobacteria. For the sake of brevity, this article focuses on the organisms that show the greatest specialization of cells and variety of cell types, although this approach to the evolution of biological complexity could be regarded as "rather anthropocentric."

The initial advantages of multicellularity may have included: more efficient sharing of nutrients that are digested outside the cell, increased resistance to predators, many of which attacked by engulfing; the ability to resist currents by attaching to a firm surface; the ability to reach upwards to filter-feed or to obtain sunlight for photosynthesis; the ability to create an internal environment that gives protection against the external one; and even the opportunity for a group of cells to behave "intelligently" by sharing information. These features would also have provided opportunities for other organisms to diversify, by creating more varied environments than flat microbial mats could.

Multicellularity with differentiated cells is beneficial to the organism as a whole but disadvantageous from the point of view of individual cells, most of which lose the opportunity to reproduce themselves. In an asexual multicellular organism, rogue cells which retain the ability to reproduce may take over and reduce the organism to a mass of undifferentiated cells. Sexual reproduction eliminates such rogue cells from the next generation and therefore appears to be a prerequisite for complex multicellularity.

The available evidence indicates that eukaryotes evolved much earlier but remained inconspicuous until a rapid diversification around 1 Ga. The only respect in which eukaryotes clearly surpass bacteria and archaea is their capacity for variety of forms, and sexual reproduction enabled eukaryotes to exploit that advantage by producing organisms with multiple cells that differed in form and function.

By comparing the composition of transcription factor families and regulatory network motifs between unicellular organisms and multicellular organisms, scientists found there are many novel transcription factor families and three novel types of regulatory network motifs in multicellular organisms, and novel family transcription factors are preferentially wired into these novel network motifs which are essential for multicullular development. These results propose a plausible mechanism for the contribution of novel-family transcription factors and novel network motifs to the origin of multicellular organisms at transcriptional regulatory level.

Fossil evidence

The Francevillian biota fossils, dated to 2.1 Ga, are the earliest known fossil organisms that are clearly multicellular. They may have had differentiated cells. Another early multicellular fossil, *Qingshania*, dated to 1.7 Ga, appears to consist of virtually identical cells. The red algae called *Bangiomorpha*, dated at 1.2 Ga, is the earliest known organism that certainly has differentiated, specialized cells, and is also the oldest known sexually reproducing organism. The 1.43 billion-year-old fossils interpreted as fungi appear to have been multicellular with differentiated cells. The "string of beads" organism *Horodyskia*, found in rocks dated from 1.5 Ga to 900 Ma, may have been an early metazoan; however, it has also been interpreted as a colonial foraminiferan.

Emergence of animals

<templatestyles src="Template:Clade/styles.css" />

<templatestyles src="Template:Clade/styles.css" />

<templatestyles src="Template:Clade/styles.css" />
　　<templatestyles src="Template:Clade/styles.css" />
　　　<templatestyles src="Template:Clade/styles.css" />
　　　Bilate-　　<templatestyles src="Template:Clade/styles.css" />
　　　rians　　　　<templatestyles src="Template:Clade/styles.css" />
　　　　　　　　　　Deuterostomes (chordates, hemichordates, echinoderms)

　　　　　Protostomes　<templatestyles src="Template:Clade/styles.css" />

　　　　　　　　　　Ecdysozoa (arthropods, nematodes, tardigrades, etc.)

　　　　　　　　　Lophotrochozoa (molluscs, annelids, brachiopods, etc.)

　　　　Acoelomorpha

　　　　Cnidaria (jellyfish, sea anemones, corals)

　　　　Ctenophora (comb jellies)

　　　Placozoa

　　　Porifera (sponges): Calcarea

　　Porifera: Hexactinellida & Demospongiae

　Choanoflagellata

　Mesomycetozoea

A family tree of the animals

Animals are multicellular eukaryotes,[98] and are distinguished from plants, algae, and fungi by lacking cell walls. All animals are motile, if only at certain life stages. All animals except sponges have bodies differentiated into separate tissues, including muscles, which move parts of the animal by contracting, and nerve tissue, which transmits and processes signals.

The earliest widely accepted animal fossils are the rather modern-looking cnidarians (the group that includes jellyfish, sea anemones and *Hydra*), possibly from around 580[99] Ma, although fossils from the Doushantuo Formation can only be dated approximately. Their presence implies that the cnidarian and bilaterian lineages had already diverged.

The Ediacara biota, which flourished for the last 40 million years before the start of the Cambrian, were the first animals more than a very few centimetres long. Many were flat and had a "quilted" appearance, and seemed so strange that there was a proposal to classify them as a separate kingdom, Vendozoa. Others, however, have been interpreted as early molluscs (*Kimberella*), echinoderms (*Arkarua*), and arthropods (*Spriggina*,[100] *Parvancorina*). There is

Figure 54: *Opabinia made the largest single contribution to modern interest in the Cambrian explosion.*

still debate about the classification of these specimens, mainly because the diagnostic features which allow taxonomists to classify more recent organisms, such as similarities to living organisms, are generally absent in the Ediacarans. However, there seems little doubt that *Kimberella* was at least a triploblastic bilaterian animal, in other words, an animal significantly more complex than the cnidarians.

The small shelly fauna are a very mixed collection of fossils found between the Late Ediacaran and Middle Cambrian periods. The earliest, *Cloudina*, shows signs of successful defense against predation and may indicate the start of an evolutionary arms race. Some tiny Early Cambrian shells almost certainly belonged to molluscs, while the owners of some "armor plates," *Halkieria* and *Microdictyon*, were eventually identified when more complete specimens were found in Cambrian lagerstätten that preserved soft-bodied animals.

In the 1970s there was already a debate about whether the emergence of the modern phyla was "explosive" or gradual but hidden by the shortage of Precambrian animal fossils. A re-analysis of fossils from the Burgess Shale lagerstätte increased interest in the issue when it revealed animals, such as *Opabinia*, which did not fit into any known phylum. At the time these were interpreted as evidence that the modern phyla had evolved very rapidly in the

Figure 55: *Acanthodians were among the earliest vertebrates with jaws.*

Cambrian explosion and that the Burgess Shale's "weird wonders" showed that the Early Cambrian was a uniquely experimental period of animal evolution. Later discoveries of similar animals and the development of new theoretical approaches led to the conclusion that many of the "weird wonders" were evolutionary "aunts" or "cousins" of modern groups—for example that *Opabinia* was a member of the lobopods, a group which includes the ancestors of the arthropods, and that it may have been closely related to the modern tardigrades. Nevertheless, there is still much debate about whether the Cambrian explosion was really explosive and, if so, how and why it happened and why it appears unique in the history of animals.

Deuterostomes and the first vertebrates

Most of the animals at the heart of the Cambrian explosion debate are protostomes, one of the two main groups of complex animals. The other major group, the deuterostomes, contains invertebrates such as starfish and sea urchins (echinoderms), as well as chordates (see below). Many echinoderms have hard calcite "shells," which are fairly common from the Early Cambrian small shelly fauna onwards. Other deuterostome groups are soft-bodied, and most of the significant Cambrian deuterostome fossils come from the Chengjiang fauna, a lagerstätte in China. The chordates are another major deuterostome group: animals with a distinct dorsal nerve cord. Chordates include soft-bodied invertebrates such as tunicates as well as vertebrates—animals with a backbone. While tunicate fossils predate the Cambrian explosion, the Chengjiang fossils *Haikouichthys* and *Myllokunmingia* appear to be true vertebrates, and *Haikouichthys* had distinct vertebrae, which may have been slightly mineralized. Vertebrates with jaws, such as the acanthodians, first appeared in the Late Ordovician.

Colonization of land

Adaptation to life on land is a major challenge: all land organisms need to avoid drying-out and all those above microscopic size must create special structures to withstand gravity; respiration and gas exchange systems have to change; reproductive systems cannot depend on water to carry eggs and sperm towards each other. Although the earliest good evidence of land plants and animals dates back to the Ordovician period (488 to 444[101] Ma), and a number of microorganism lineages made it onto land much earlier, modern land ecosystems only appeared in the Late Devonian, about 385 to 359[102] Ma. In May 2017, evidence of the earliest known life on land may have been found in 3.48-billion-year-old geyserite and other related mineral deposits (often found around hot springs and geysers) uncovered in the Pilbara Craton of Western Australia.

Evolution of terrestrial antioxidants

Oxygen is a potent oxidant whose accumulation in terrestrial atmosphere resulted from the development of photosynthesis over 3 Ga, in cyanobacteria (blue-green algae), which were the most primitive oxygenic photosynthetic organisms. Brown algae accumulate inorganic mineral antioxidants such as rubidium, vanadium, zinc, iron, copper, molybdenum, selenium and iodine which is concentrated more than 30,000 times the concentration of this element in seawater. Protective endogenous antioxidant enzymes and exogenous dietary antioxidants helped to prevent oxidative damage. Most marine mineral antioxidants act in the cells as essential trace elements in redox and antioxidant metalloenzymes.

When plants and animals began to enter rivers and land about 500 Ma, environmental deficiency of these marine mineral antioxidants was a challenge to the evolution of terrestrial life. Terrestrial plants slowly optimized the production of "new" endogenous antioxidants such as ascorbic acid, polyphenols, flavonoids, tocopherols, etc. A few of these appeared more recently, in last 200–50 Ma, in fruits and flowers of angiosperm plants.

In fact, angiosperms (the dominant type of plant today) and most of their antioxidant pigments evolved during the Late Jurassic period. Plants employ antioxidants to defend their structures against reactive oxygen species produced during photosynthesis. Animals are exposed to the same oxidants, and they have evolved endogenous enzymatic antioxidant systems. Iodine is the most primitive and abundant electron-rich essential element in the diet of marine and terrestrial organisms, and as iodide acts as an electron donor and has this ancestral antioxidant function in all iodide-concentrating cells from primitive marine algae to more recent terrestrial vertebrates.

Figure 56: *Lichens growing on concrete*

Evolution of soil

Before the colonization of land, soil, a combination of mineral particles and decomposed organic matter, did not exist. Land surfaces would have been either bare rock or unstable sand produced by weathering. Water and any nutrients in it would have drained away very quickly. In the Sub-Cambrian peneplain in Sweden for example maximum depth of kaolinitization by Neoproterozoic weathering is about 5 m, in contrast nearby kaolin deposits developed in the Mesozoic are much thicker. It has been argued that in the late Neoproterozoic sheet wash was a dominant process of erosion of surface material due to the lack of plants on land.

Films of cyanobacteria, which are not plants but use the same photosynthesis mechanisms, have been found in modern deserts, and only in areas that are unsuitable for vascular plants. This suggests that microbial mats may have been the first organisms to colonize dry land, possibly in the Precambrian. Mat-forming cyanobacteria could have gradually evolved resistance to desiccation as they spread from the seas to intertidal zones and then to land.[103] Lichens, which are symbiotic combinations of a fungus (almost always an ascomycete) and one or more photosynthesizers (green algae or cyanobacteria), are also important colonizers of lifeless environments, and their ability to break down rocks contributes to soil formation in situations where plants cannot survive. The earliest known ascomycete fossils date from 423 to 419[104] Ma in the Silurian.

Figure 57: *Reconstruction of Cooksonia, a vascular plant from the Silurian*

Soil formation would have been very slow until the appearance of burrowing animals, which mix the mineral and organic components of soil and whose feces are a major source of the organic components. Burrows have been found in Ordovician sediments, and are attributed to annelids ("worms") or arthropods.

Plants and the Late Devonian wood crisis

In aquatic algae, almost all cells are capable of photosynthesis and are nearly independent. Life on land required plants to become internally more complex and specialized: photosynthesis was most efficient at the top; roots were required in order to extract water from the ground; the parts in between became supports and transport systems for water and nutrients.

Spores of land plants, possibly rather like liverworts, have been found in Middle Ordovician rocks dated to about 476[105] Ma. In Middle Silurian rocks 430[106] Ma, there are fossils of actual plants including clubmosses such as *Baragwanathia*; most were under 10 centimetres (3.9 in) high, and some appear closely related to vascular plants, the group that includes trees.

By the Late Devonian 370[107] Ma, trees such as *Archaeopteris* were so abundant that they changed river systems from mostly braided to mostly meandering, because their roots bound the soil firmly.[108] In fact, they caused the "Late Devonian wood crisis"[109] because:

Figure 58: *Fossilized trees from the Middle Devonian Gilboa Fossil Forest*

- They removed more carbon dioxide from the atmosphere, reducing the greenhouse effect and thus causing an ice age in the Carboniferous period. In later ecosystems the carbon dioxide "locked up" in wood is returned to the atmosphere by decomposition of dead wood. However, the earliest fossil evidence of fungi that can decompose wood also comes from the Late Devonian.
- The increasing depth of plants' roots led to more washing of nutrients into rivers and seas by rain. This caused algal blooms whose high consumption of oxygen caused anoxic events in deeper waters, increasing the extinction rate among deep-water animals.

Land invertebrates

Animals had to change their feeding and excretory systems, and most land animals developed internal fertilization of their eggs. The difference in refractive index between water and air required changes in their eyes. On the other hand, in some ways movement and breathing became easier, and the better transmission of high-frequency sounds in air encouraged the development of hearing.

The oldest known air-breathing animal is *Pneumodesmus*, an archipolypodan millipede from the Middle Silurian, about 428[110] Ma. Its air-breathing, terrestrial nature is evidenced by the presence of spiracles, the openings to tracheal systems. However, some earlier trace fossils from the Cambrian-Ordovician boundary about 490[111] Ma are interpreted as the tracks of large amphibious

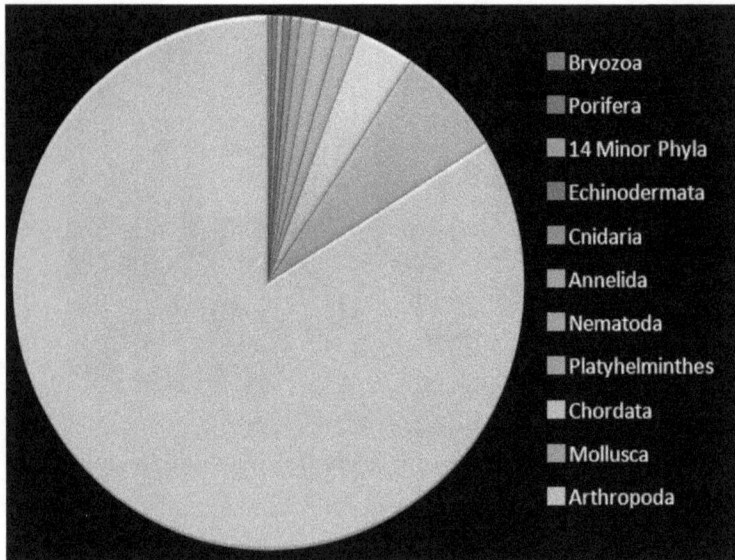

Figure 59: *The relative number of species contributed to the total by each phylum of animals. Nematoda is the phylum with the most individual organisms while arthropod has the most species.*

arthropods on coastal sand dunes, and may have been made by euthycarcinoids, which are thought to be evolutionary "aunts" of myriapods. Other trace fossils from the Late Ordovician a little over 445[112] Ma probably represent land invertebrates, and there is clear evidence of numerous arthropods on coasts and alluvial plains shortly before the Silurian-Devonian boundary, about 415[113] Ma, including signs that some arthropods ate plants. Arthropods were well pre-adapted to colonise land, because their existing jointed exoskeletons provided protection against desiccation, support against gravity and a means of locomotion that was not dependent on water.

The fossil record of other major invertebrate groups on land is poor: none at all for non-parasitic flatworms, nematodes or nemerteans; some parasitic nematodes have been fossilized in amber; annelid worm fossils are known from the Carboniferous, but they may still have been aquatic animals; the earliest fossils of gastropods on land date from the Late Carboniferous, and this group may have had to wait until leaf litter became abundant enough to provide the moist conditions they need.[114]

The earliest confirmed fossils of flying insects date from the Late Carboniferous, but it is thought that insects developed the ability to fly in the Early

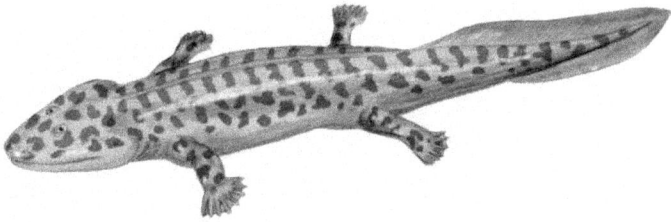

Figure 60: *Acanthostega changed views about the early evolution of tetrapods.*

Carboniferous or even Late Devonian. This gave them a wider range of eco-logical niches for feeding and breeding, and a means of escape from predators and from unfavorable changes in the environment. About 99% of modern insect species fly or are descendants of flying species.

Early land vertebrates

<templatestyles src="Template:Clade/styles.css" />

"Fish" <templatestyles src="Template:Clade/styles.css" />
        Osteolepiformes

            <templatestyles src="Template:Clade/styles.css" />
            Panderichthyidae

        <templatestyles src="Template:Clade/styles.css" />
        Obruchevichthidae

            <templatestyles src="Template:Clade/styles.css" />
            Acanthostega

            <templatestyles src="Template:Clade/styles.css" />
            Ichthyostega

                <templatestyles src="Template:Clade/styles.css" />
                Tulerpeton

                <templatestyles src="Template:Clade/styles.css" />
                Early labyrinthodonts

                    <templatestyles src="Template:Clade/styles.css" />
                    Anthracosauria

                Amniotes

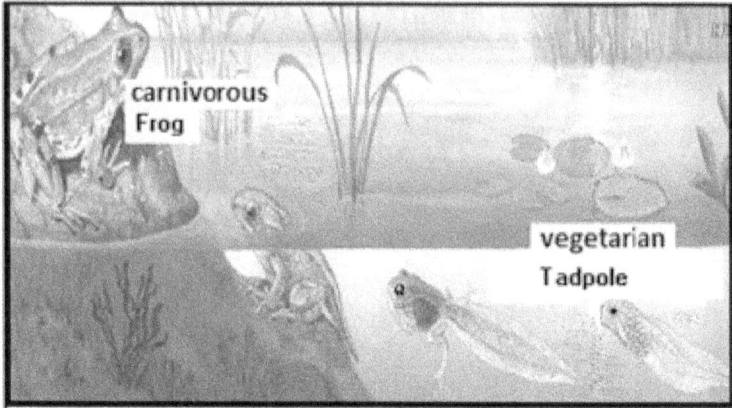

Figure 61: *Amphibian Metamorphosis*

Family tree of tetrapods

Tetrapods, vertebrates with four limbs, evolved from other rhipidistian fish over a relatively short timespan during the Late Devonian (370 to 360[115] Ma). The early groups are grouped together as Labyrinthodontia. They retained aquatic, fry-like tadpoles, a system still seen in modern amphibians.

Iodine and T4/T3 stimulate the amphibian metamorphosis and the evolution of nervous systems transforming the aquatic, vegetarian tadpole into a "more evoluted" terrestrial, carnivorous frog with better neurological, visuospatial, olfactory and cognitive abilities for hunting. The new hormonal action of T3 was made possible by the formation of T3-receptors in the cells of vertebrates. Firstly, about 600-500 million years ago, in primitive Chordata appeared the alpha T3-receptors with a metamorphosing action and then, about 250-150 million years ago, in the Birds and Mammalia appeared the beta T3-receptors with metabolic and thermogenetic actions.

From the 1950s to the early 1980s it was thought that tetrapods evolved from fish that had already acquired the ability to crawl on land, possibly in order to go from a pool that was drying out to one that was deeper. However, in 1987, nearly complete fossils of *Acanthostega* from about 363[116] Ma showed that this Late Devonian transitional animal had legs and both lungs and gills, but could never have survived on land: its limbs and its wrist and ankle joints were too weak to bear its weight; its ribs were too short to prevent its lungs from being squeezed flat by its weight; its fish-like tail fin would have been damaged by dragging on the ground. The current hypothesis is that *Acanthostega*, which was about 1 metre (3.3 ft) long, was a wholly aquatic predator that hunted in shallow water. Its skeleton differed from that of most fish, in ways that enabled it to raise its head to breathe air while its body remained submerged, including: its jaws show modifications that would have enabled it to gulp air; the bones at the back of its skull are locked together, providing strong attachment points for muscles that raised its head; the head is not joined to the shoulder girdle and it has a distinct neck.

The Devonian proliferation of land plants may help to explain why air breathing would have been an advantage: leaves falling into streams and rivers would have encouraged the growth of aquatic vegetation; this would have attracted grazing invertebrates and small fish that preyed on them; they would have been attractive prey but the environment was unsuitable for the big marine predatory fish; air-breathing would have been necessary because these waters would have been short of oxygen, since warm water holds less dissolved oxygen than cooler marine water and since the decomposition of vegetation would have used some of the oxygen.

Later discoveries revealed earlier transitional forms between *Acanthostega* and completely fish-like animals. Unfortunately, there is then a gap (Romer's gap) of about 30 Ma between the fossils of ancestral tetrapods and Middle Carboniferous fossils of vertebrates that look well-adapted for life on land. Some of these look like early relatives of modern amphibians, most of which need to keep their skins moist and to lay their eggs in water, while others are accepted as early relatives of the amniotes, whose waterproof skin and egg membranes enable them to live and breed far from water.

Dinosaurs, birds and mammals

<templatestyles src="Template:Clade/styles.css" />

Am-   <templatestyles src="Template:Clade/styles.css" />
niotes

Synap-   <templatestyles src="Template:Clade/styles.css" />
sids   Early synapsids (extinct)

       Pely-   <templatestyles src="Template:Clade/styles.css" />
       cosaurs   Extinct pelycosaurs

              Ther-   <templatestyles src="Template:Clade/styles.css" />
              apsids   Extinct therapsids

                     Mammali-   <templatestyles src="Template:Clade/styles.css" />
                     aformes
                                       Extinct mammaliaforms

                                 Mammals

Saurop- <templatestyles src="Template:Clade/styles.css" />
sids   <templatestyles src="Template:Clade/styles.css" />
             Anapsids; whether turtles belong here is debated

              <templatestyles src="Template:Clade/styles.css" />
              Captorhinidae and Protorothyrididae (extinct)

       Di-   <templatestyles src="Template:Clade/styles.css" />
       ap-   Araeoscelidia (extinct)
       sids

              <templatestyles src="Template:Clade/styles.css" />
              Squamata (lizards and snakes)

          Archosaurs   <templatestyles src="Template:Clade/styles.css" />
                       Extinct archosaurs

              Crocodilians

              <templatestyles src="Template:Clade/styles.css" />
              Pterosaurs (extinct)

          Dinosaurs   <templatestyles src="Template:Clade/styles.css" />
                      <templatestyles src="Template:Clade/styles.css"
                      />
                      Thero-   <templatestyles
                      pods   src="Template:Clade/styles.css"
                                />

                                Extinct
                                theropods

                                Birds

                      Sauropods
                      (extinct)

              Ornithischians (extinct)

Possible family tree of dinosaurs, birds and mammals

Amniotes, whose eggs can survive in dry environments, probably evolved in the Late Carboniferous period (330 to 298.9[117] Ma). The earliest fossils of the two surviving amniote groups, synapsids and sauropsids, date from around 313[118] Ma. The synapsid pelycosaurs and their descendants the therapsids are the most common land vertebrates in the best-known Permian (298.9 to 251.902[119] Ma) fossil beds. However, at the time these were all in temperate zones at middle latitudes, and there is evidence that hotter, drier environments nearer the Equator were dominated by sauropsids and amphibians.

The Permian–Triassic extinction event wiped out almost all land vertebrates, as well as the great majority of other life. During the slow recovery from this catastrophe, estimated to have taken 30 million years, a previously obscure sauropsid group became the most abundant and diverse terrestrial vertebrates: a few fossils of archosauriformes ("ruling lizard forms") have been found in Late Permian rocks, but, by the Middle Triassic, archosaurs were the dominant land vertebrates. Dinosaurs distinguished themselves from other archosaurs in the Late Triassic, and became the dominant land vertebrates of the Jurassic and Cretaceous periods (201.3 to 66[120] Ma).

During the Late Jurassic, birds evolved from small, predatory theropod dinosaurs. The first birds inherited teeth and long, bony tails from their dinosaur ancestors, but some had developed horny, toothless beaks by the very Late Jurassic and short pygostyle tails by the Early Cretaceous.

While the archosaurs and dinosaurs were becoming more dominant in the Triassic, the mammaliaform successors of the therapsids evolved into small, mainly nocturnal insectivores. This ecological role may have promoted the evolution of mammals, for example nocturnal life may have accelerated the development of endothermy ("warm-bloodedness") and hair or fur. By 195[121] Ma in the Early Jurassic there were animals that were very like today's mammals in a number of respects. Unfortunately, there is a gap in the fossil record throughout the Middle Jurassic. However, fossil teeth discovered in Madagascar indicate that the split between the lineage leading to monotremes and the one leading to other living mammals had occurred by 167[122] Ma. After dominating land vertebrate niches for about 150 Ma, the non-avian dinosaurs perished in the Cretaceous–Paleogene extinction event (66[95] Ma) along with many other groups of organisms. Mammals throughout the time of the dinosaurs had been restricted to a narrow range of taxa, sizes and shapes, but increased rapidly in size and diversity after the extinction, with bats taking to the air within 13 million years, and cetaceans to the sea within 15 million years.

Flowering plants

<templatestyles src="Template:Clade/-styles.css" />

Gym-nosperms   <templatestyles src="Template:Clade/styles.css" />

<templatestyles src="Template:Clade/-styles.css" />

<templatestyles src="Template:Clade/-styles.css" />

<templatestyles src="Template:Clade/styles.css" />

Gnetales
(gymnosperm)

Welwitschia
(gymnosperm)

Ephedra
(gymnosperm)

Bennettitales

Angiosperms
(flowering plants)

One possible family tree of flowering plants

<templatestyles src="Template:Clade/-styles.css" />

Gym-nosperms   <templatestyles src="Template:Clade/styles.css" />

Angiosperms
(flowering plants)

<templatestyles src="Template:Clade/-styles.css" />

<templatestyles src="Template:Clade/-styles.css" />

<templatestyles src="Template:Clade/styles.css" />

Cycads
(gymnosperm)

Bennettitales

Ginkgo

<templatestyles src="Template:Clade/-styles.css" />

Gnetales
(gymnosperm)

Conifers
(gymnosperm)

Another possible family tree

The first flowering plants appeared around 130 Ma. The 250,000 to 400,000 species of flowering plants outnumber all other ground plants combined, and are the dominant vegetation in most terrestrial ecosystems. There is fossil evidence that flowering plants diversified rapidly in the Early Cretaceous, from 130 to 90[123] Ma,[124] and that their rise was associated with that of pollinating insects. Among modern flowering plants *Magnolia* are thought to be close to the common ancestor of the group. However, paleontologists have not succeeded in identifying the earliest stages in the evolution of flowering plants.

Social insects

The social insects are remarkable because the great majority of individuals in each colony are sterile. This appears contrary to basic concepts of evolution such as natural selection and the selfish gene. In fact, there are very few eusocial insect species: only 15 out of approximately 2,600 living families of insects contain eusocial species, and it seems that eusociality has evolved independently only 12 times among arthropods, although some eusocial lineages have diversified into several families. Nevertheless, social insects have been spectacularly successful; for example although ants and termites account for

Figure 62: *These termite mounds have survived a bush fire.*

only about 2% of known insect species, they form over 50% of the total mass of insects. Their ability to control a territory appears to be the foundation of their success.

The sacrifice of breeding opportunities by most individuals has long been explained as a consequence of these species' unusual haplodiploid method of sex determination, which has the paradoxical consequence that two sterile worker daughters of the same queen share more genes with each other than they would with their offspring if they could breed. However, E. O. Wilson and Bert Hölldobler argue that this explanation is faulty: for example, it is based on kin selection, but there is no evidence of nepotism in colonies that have multiple queens. Instead, they write, eusociality evolves only in species that are under strong pressure from predators and competitors, but in environments where it is possible to build "fortresses"; after colonies have established this security, they gain other advantages through co-operative foraging. In support of this explanation they cite the appearance of eusociality in bathyergid mole rats, which are not haplodiploid.

The earliest fossils of insects have been found in Early Devonian rocks from about 400[125] Ma, which preserve only a few varieties of flightless insect. The Mazon Creek lagerstätten from the Late Carboniferous, about 300[126] Ma, include about 200 species, some gigantic by modern standards, and indicate that insects had occupied their main modern ecological niches as herbivores, detritivores and insectivores. Social termites and ants first appear in the Early Cretaceous, and advanced social bees have been found in Late Cretaceous rocks but did not become abundant until the Middle Cenozoic.

Humans

Hominin timeline

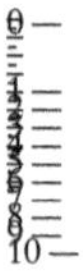

Axis scale: million years

Also see: *Life timeline* and *Nature timeline*

The idea that, along with other life forms, modern-day humans evolved from an ancient, common ancestor was proposed by Robert Chambers in 1844 and taken up by Charles Darwin in 1871. Modern humans evolved from a lineage of upright-walking apes that has been traced back over 6[127] Ma to *Sahelanthropus*. The first known stone tools were made about 2.5[128] Ma, apparently by *Australopithecus garhi*, and were found near animal bones that bear scratches made by these tools. The earliest hominines had chimpanzee-sized brains, but there has been a fourfold increase in the last 3 Ma; a statistical analysis suggests that hominine brain sizes depend almost completely on the date of the fossils, while the species to which they are assigned has only slight influence. There is a long-running debate about whether modern humans evolved all over the world simultaneously from existing advanced hominines or are descendants of a single small population in Africa, which then migrated all over the world less than 200,000 years ago and replaced previous hominine species. There is also debate about whether anatomically modern humans had an intellectual, cultural and technological "Great Leap Forward" under 100,000 years ago and, if so, whether this was due to neurological changes that are not visible in fossils.

Mass extinctions

Apparent extinction intensity, i.e. the fraction of genera going extinct at any given time, as reconstructed from the fossil record. (Graph not meant to include the recent, ongoing Holocene extinction event).

Life on Earth has suffered occasional mass extinctions at least since 542[129] Ma. Although they were disasters at the time, mass extinctions have sometimes accelerated the evolution of life on Earth. When dominance of particular ecological niches passes from one group of organisms to another, it is rarely because the new dominant group is "superior" to the old and usually because an extinction event eliminates the old dominant group and makes way for the new one.

File:Phanerozoic biodiversity blank 01.png

Phanerozoic biodiversity as shown by the fossil record

The fossil record appears to show that the gaps between mass extinctions are becoming longer and the average and background rates of extinction are decreasing. Both of these phenomena could be explained in one or more ways:

- The oceans may have become more hospitable to life over the last 500 Ma and less vulnerable to mass extinctions: dissolved oxygen became more widespread and penetrated to greater depths; the development of life on land reduced the run-off of nutrients and hence the risk of eutrophication and anoxic events; and marine ecosystems became more diversified so that food chains were less likely to be disrupted.
- Reasonably complete fossils are very rare, most extinct organisms are represented only by partial fossils, and complete fossils are rarest in the oldest rocks. So paleontologists have mistakenly assigned parts of the same organism to different genera, which were often defined solely to accommodate these finds—the story of *Anomalocaris* is an example of this. The risk of this mistake is higher for older fossils because these are often both unlike parts of any living organism and poorly conserved. Many of the "superfluous" genera are represented by fragments which are not found again and the "superfluous" genera appear to become extinct very quickly.

Biodiversity in the fossil record, which is "...the number of distinct genera alive at any given time; that is, those whose first occurrence predates and whose last occurrence postdates that time" shows a different trend: a fairly swift rise from 542 to 400[130] Ma; a slight decline from 400 to 200[131] Ma, in which the devastating Permian–Triassic extinction event is an important factor; and a swift rise from 200[132] Ma to the present.

Bibliography

<templatestyles src="Template:Refbegin/styles.css" />

- Arrhenius, Svante (1980) [Arrhenius paper originally published 1903]. "The Propagation of Life in Space". In Goldsmith, Donald. *The Quest for Extraterrestrial Life: A Book of Readings*. Foreword by Sir Fred Hoyle. Mill Valley, CA: University Science Books. Bibcode: 1980qel..book...32A[133]. ISBN 978-0-935702-02-6. LCCN 79057423[134]. OCLC 7121102[135].
- Bengtson, Stefan (2004). "Early Skeletal Fossils"[136] (PDF). In Lipps, Jere H.; Waggoner, Benjamin M. *Neoproterozoic-Cambrian Biological Revolutions: Presented as a Paleontological Society Short Course at the Annual Meeting of the Geological Society of America, Denver, Colorado, November 6, 2004*. Paleontological Society Papers. **10**. New Haven, CT: Yale University Reprographics & Imaging Service; Paleontological Society. OCLC 57481790[137]. Archived from the original[138] (PDF) on February 11, 2017. Retrieved 2015-02-06.

- Bennett, Jeffrey O. (2008). *Beyond UFOs: The Search for Extraterrestrial Life and Its Astonishing Implications for Our Future*. Princeton, NJ: Princeton University Press. ISBN 978-0-691-13549-6. LCCN 2007037872[139]. OCLC 172521761[140].
- Benton, Michael J. (1997). *Vertebrate Palaeontology* (2nd ed.). London: Chapman & Hall. ISBN 978-0-412-73800-5. OCLC 37378512[141].
- Benton, Michael J. (2005) [Originally published 2003]. *When Life Nearly Died: The Greatest Mass Extinction of All Time* (1st paperback ed.). London: Thames & Hudson. ISBN 978-0-500-28573-2. LCCN 2002109744[142]. OCLC 62145244[143].
- Benton, Michael J. (2005a). *Vertebrate Palaeontology* (3rd ed.). Malden, MA: Blackwell Science. ISBN 978-0-632-05637-8. LCCN 2003028152[144]. OCLC 53970617[145].
- Bernstein, Harris; Bernstein, Carol; Michod, Richard E. (2012). "DNA Repair as the Primary Adaptive Function of Sex in Bacteria and Eukaryotes"[146]. In Kimura, Sakura; Shimizu, Sora. *DNA Repair: New Research*. Hauppauge, NY: Nova Science Publishers. ISBN 978-1-62100-808-8. LCCN 2011038504[147]. OCLC 828424701[148].
- Bernstein, Harris; Hopf, Frederic A.; Michod, Richard E. (1987). "The Molecular Basis of the Evolution of Sex". In Scandalios, John G. *Molecular Genetics of Development*. Advances in Genetics. San Diego, CA: Academic Press. ISBN 978-0-12-017624-3. OCLC 646754753[149]. PMID 3324702[150].
- Birdsell, John A.; Wills, Christopher (2003). "The Evolutionary Origin and Maintenance of Sexual Recombination: A Review of Contemporary Models". In MacIntyre, Ross J.; Clegg, Michael T. *Evolutionary Biology*. Evolutionary Biology. 33. New York: Springer Science+Business Media. ISBN 978-1-4419-3385-0. OCLC 751583918[151].
- Briggs, Derek E. G.; Crowther, Peter R., eds. (2001). *Palaeobiology II*. Foreword by E. N. K. Clarkson. Malden, MA: Blackwell Science. ISBN 978-0-632-05149-6. LCCN 0632051477[152]. OCLC 43945263[153].
- Cairns-Smith, A. G. (1968). "An Approach to a Blueprint for a Primitive Organism". In Waddington, C. H. *Towards a Theoretical Biology*. 1. Edinburgh, Scotland: Edinburgh University Press. ISBN 978-0-85224-018-2. LCCN 71419832[154]. OCLC 230043266[155].
- Clancy, Paul; Brack, André; Horneck, Gerda (2005). *Looking for Life, Searching the Solar System*. Cambridge; New York: Cambridge University Press. ISBN 978-0-521-82450-7. LCCN 2006271630[156]. OCLC 57574490[157].
- Cowen, Richard (2000). *History of Life* (3rd ed.). Malden, MA: Blackwell Science. ISBN 978-0-632-04444-3. LCCN 99016542[158]. OCLC 47011068[159].

- Dalrymple, G. Brent (1991). *The Age of the Earth*. Stanford, CA: Stanford University Press. ISBN 978-0-8047-1569-0. LCCN 90047051[160]. OCLC 22347190[161].
- Futuyma, Douglas J. (2005). *Evolution*. Sunderland, MA: Sinauer Associates. ISBN 978-0-87893-187-3. LCCN 2004029808[162]. OCLC 57311264[163].
- Gauthier, Jacques; Cannatella, David C.; de Queiroz, Kevin; et al. (1989). "Tetrapod phylogeny". In Fernholm, Bo; Bremer, Kåre; Jörnvall, Hans. *The Hierarchy of Life: Molecules and Morphology in Phylogenetic Analysis*. International Congress Series. **824**. Amsterdam, the Netherlands; New York: Excerpta Medica/Elsevier Science Publishers B.V. (Biomedical Division). ISBN 978-0-444-81073-1. LCCN 89001132[164]. OCLC 19129518[165]. "Proceedings from Nobel Symposium 70 held at Alfred Nobel's Björkborn, Karlskoga, Sweden, August 29 – September 2, 1988"
- Gee, Henry, ed. (2000). *Shaking the Tree: Readings from* Nature *in the History of Life*. Chicago, IL: University of Chicago Press. ISBN 978-0-226-28497-2. LCCN 99049796[166]. OCLC 42476104[167].
- Gould, Stephen Jay (1989). *Wonderful Life: The Burgess Shale and the Nature of History* (1st ed.). New York: W. W. Norton & Company. ISBN 978-0-393-02705-1. LCCN 88037469[168]. OCLC 18983518[169].
- Grimaldi, David; Engel, Michael S. (2005). *Evolution of the Insects*. Cambridge; New York: Cambridge University Press. ISBN 978-0-521-82149-0. LCCN 2004054605[170]. OCLC 56057971[171].
- Hinde, Rosalind T. (2001). "The Cnidaria and Ctenophora". In Anderson, D. T. *Invertebrate Zoology* (2nd ed.). Melbourne; New York: Oxford University Press. ISBN 978-0-19-551368-4. LCCN 2002276846[172]. OCLC 49663129[173].
- Holmes, Randall K.; Jobling, Michael G. (1996). "Genetics"[174]. In Baron, Samuel. *Medical Microbiology* (4th ed.). Galveston, TX: University of Texas Medical Branch. Exchange of Genetic Information. ISBN 978-0-9631172-1-2. LCCN 95050499[175]. OCLC 33838234[176]. PMID 21413277[177]. Retrieved 2015-01-24.
- Krumbein, Wolfgang E.; Brehm, Ulrike; Gerdes, Gisela; et al. (2003). "Biofilm, Biodictyon, Biomat Microbialites, Oolites, Stromatolites Geophysiology, Global Mechanism, Parahistology"[178] (PDF). In Krumbein, Wolfgang E.; Paterson, David M.; Zavarzin, Georgii A. *Fossil and Recent Biofilms: A Natural History of Life on Earth*. Dordrecht, the Netherlands: Kluwer Academic Publishers. ISBN 978-1-4020-1597-7. LCCN 2003061870[179]. OCLC 52901566[180]. Archived from the original[181] (PDF) on 2007-01-06. Retrieved 2008-07-09.
- Labandeira, Conrad C.; Eble, Gunther J. (1999). "The Fossil Record of Insect Diversity and Disparity"[182] (PDF). In Anderson, John M.;

Thackeray, John Francis; et al. *Towards Gondwana Alive: Promoting biodiversity & stemming the Sixth Extinction*. Pretoria: Gondwana Alive Society. ISBN 978-1-919795-43-0. LCCN 2001385090[183]. OCLC 44822625[184]. "Preview booklet for 'Gondwana alive : biodiversity and the evolving terrestrial biosphere', book planned for September 2000, and associated projects."

- Leakey, Richard (1994). *The Origin of Humankind*. Science Masters Series. New York: Basic Books. ISBN 978-0-465-03135-1. LCCN 94003617[185]. OCLC 30739453[186].

- Margulis, Lynn (1981). *Symbiosis in Cell Evolution: Life and its Environment on the Early Earth*. San Francisco, CA: W. H. Freeman and Company. ISBN 978-0-7167-1256-5. LCCN 80026695[187]. OCLC 6982472[188].

- McKinney, Michael L. (1997). "How do rare species avoid extinction? A paleontological view". In Kunin, William E.; Gaston, Kevin J. *The Biology of Rarity: Causes and consequences of rare—common differences* (1st ed.). London; New York: Chapman & Hall. ISBN 978-0-412-63380-5. LCCN 96071014[189]. OCLC 36442106[190].

- Miller, G. Tyler; Spoolman, Scott E. (2012). *Environmental Science* (14th ed.). Belmont, CA: Brooks/Cole. ISBN 978-1-111-98893-7. LCCN 2011934330[191]. OCLC 741539226[192].

- Newman, William L. (July 9, 2007). "Age of the Earth"[193]. *Geologic Time*. Reston, VA: Publications Services, USGS. OCLC 18792528[194]. Retrieved 2008-08-29.

- O'Leary, Margaret R. (2008). *Anaxagoras and the Origin of Panspermia Theory*. Bloomington, IN: iUniverse. ISBN 978-0-595-49596-2. OCLC 757322661[195].

- Padian, Kevin (2004). "Basal Avialae". In Weishampel, David B.; Dodson, Peter; Osmólska, Halszka. *The Dinosauria* (2nd ed.). Berkeley: University of California Press. ISBN 978-0-520-24209-8. LCCN 2004049804[196]. OCLC 55000644[197].

- Sansom, Ivan J.; Smith, Moya M.; Smith, M. Paul (2001). "The Ordovician radiation of vertebrates". In Ahlberg, Per Erik. *Major Events in Early Vertebrate Evolution: Palaeontology, phylogeny, genetics and development*. Systematics Association special volume series. **61**. London; New York: Taylor & Francis. ISBN 978-0-415-23370-5. LCCN 00062919[198]. OCLC 51667292[199].

- Thewissen, J. G. M.; Madar, S. I.; Hussain, S. T. (1996). *Ambulocetus natans, an Eocene cetacean (Mammalia) from Pakistan*. Courier Forschungsinstitut Senckenberg. **191**. Frankfurt: Senckenbergische Naturforschende Gesellschaft. ISBN 978-3-929907-32-2. LCCN 97151576[200]. OCLC 36463214[201].

Further reading

- Dawkins, Richard (1989). *The Selfish Gene* (New ed.). Oxford; New York: Oxford University Press. ISBN 978-0-19-286092-7. LCCN 89016077[202]. OCLC 20012195[203].
- Dawkins, Richard (2004). *The Ancestor's Tale: A Pilgrimage to the Dawn of Life*. Boston: Houghton Mifflin Company. ISBN 978-0-618-00583-3. LCCN 2004059864[204]. OCLC 56617123[205].
- Ruse, Michael; Travis, Joseph, eds. (2009). *Evolution: The First Four Billion Years*. Foreword by Edward O. Wilson. Cambridge, MA: Belknap Press of Harvard University Press. ISBN 978-0-674-03175-3. LCCN 2008030270[206]. OCLC 225874308[207].
- Smith, John Maynard; Szathmáry, Eörs (1997) [Originally published 1995; Oxford: W. H. Freeman/Spektrum]. *The Major Transitions in Evolution*. Oxford; New York: Oxford University Press. ISBN 978-0-19-850294-4. LCCN 94026965[208]. OCLC 715217397[209].

External links

General information

- "Evolution"[210]. *The Virtual Fossil Museum*. Retrieved 2015-02-22. General information on evolution compiled by Roger Perkins
- "Understanding Evolution: your one-stop resource for information on evolution"[211]. University of California, Berkeley. Retrieved 2015-02-22.
- "Evolution Resources"[212]. Washington, D.C.: National Academies. Retrieved 2015-02-23.
- "Tree of Life"[213]. Archived from the original[214] on 2015-02-10. Retrieved 2015-02-23. Tree of life diagram by Neal Olander
- "Evolution"[215]. *New Scientist*. Retrieved 2015-02-23.
- Brain, Marshall. How Evolution Works[216] at HowStuffWorks
- "Modern Theories of Evolution: An Introduction to the Concepts and Theories That Led to Our Current Understanding of Evolution"[217]. Palomar College. Retrieved 2015-02-23. Tutorial created by Dennis O'Neil

History of evolutionary thought

- van Wyhe, John (ed.). "The Complete Work of Charles Darwin Online"[218]. Retrieved 2015-02-23.
- Price, R. G. "Understanding Evolution: History, Theory, Evidence, and Implications"[219]. *rationalrevolution.net*. Retrieved 2015-02-23.

Future

Future of Earth

<indicator name="good-star"> ⊕ </indicator>

The biological and geological **future of Earth** can be extrapolated based upon the estimated effects of several long-term influences. These include the chemistry at Earth's surface, the rate of cooling of the planet's interior, the gravitational interactions with other objects in the Solar System, and a steady increase in the Sun's luminosity. An uncertain factor in this extrapolation is the ongoing influence of technology introduced by humans, such as climate engineering, which could cause significant changes to the planet. The current Holocene extinction is being caused by technology[220] and the effects may last for up to five million years.[221] In turn, technology may result in the extinction of humanity, leaving the planet to gradually return to a slower evolutionary pace resulting solely from long-term natural processes.

Over time intervals of hundreds of millions of years, random celestial events pose a global risk to the biosphere, which can result in mass extinctions. These include impacts by comets or asteroids, and the possibility of a massive stellar explosion, called a supernova, within a 100-light-year radius of the Sun; known as a near-Earth supernova. Other large-scale geological events are more predictable. If the long-term effects of global warming are disregarded, Milankovitch theory predicts that the planet will continue to undergo glacial periods at least until the Quaternary glaciation comes to an end. These periods are caused by variations in eccentricity, axial tilt, and precession of the Earth's orbit. As part of the ongoing supercontinent cycle, plate tectonics will probably result in a supercontinent in 250–350 million years. Some time in the next 1.5–4.5 billion years, the axial tilt of the Earth may begin to undergo chaotic variations, with changes in the axial tilt of up to 90°.

The luminosity of the Sun will steadily increase, resulting in a rise in the solar radiation reaching the Earth. This will result in a higher rate of weathering of

Figure 63: *Conjectured illustration of the scorched Earth after the Sun has entered the red giant phase, about 7 billion years from now.*

silicate minerals, which will cause a decrease in the level of carbon dioxide in the atmosphere. In about 600 million years from now, the level of CO_2 will fall below the level needed to sustain C_3 carbon fixation photosynthesis used by trees. Some plants use the C_4 carbon fixation method, allowing them to persist at CO

$_2$ concentrations as low as 10 parts per million. However, the long-term trend is for plant life to die off altogether. The extinction of plants will be the demise of almost all animal life, since plants are the base of the food chain on Earth.

In about one billion years, the solar luminosity will be 10% higher than at present. This will cause the atmosphere to become a "moist greenhouse", resulting in a runaway evaporation of the oceans. As a likely consequence, plate tectonics will come to an end, and with them the entire carbon cycle. Following this event, in about 2–3 billion years, the planet's magnetic dynamo may cease, causing the magnetosphere to decay and leading to an accelerated loss of volatiles from the outer atmosphere. Four billion years from now, the increase in the Earth's surface temperature will cause a runaway greenhouse effect, heating the surface enough to melt it. By that point, all life on the Earth will be extinct.[222,223] The most probable fate of the planet is absorption by the Sun in about 7.5 billion years, after the star has entered the red giant phase and expanded beyond the planet's current orbit.

Human influence

Humans play a key role in the biosphere, with the large human population dominating many of Earth's ecosystems. This has resulted in a widespread, ongoing mass extinction of other species during the present geological epoch, now known as the Holocene extinction. The large-scale loss of species caused by human influence since the 1950s has been called a biotic crisis, with an estimated 10% of the total species lost as of 2007.[220] At current rates, about 30% of species are at risk of extinction in the next hundred years. The Holocene extinction event is the result of habitat destruction, the widespread distribution of invasive species, hunting, and climate change.[224] In the present day, human activity has had a significant impact on the surface of the planet. More than a third of the land surface has been modified by human actions, and humans use about 20% of global primary production. The concentration of carbon dioxide in the atmosphere has increased by close to 30% since the start of the Industrial Revolution.

The consequences of a persistent biotic crisis have been predicted to last for at least five million years.[221] It could result in a decline in biodiversity and homogenization of biotas, accompanied by a proliferation of species that are opportunistic, such as pests and weeds. Novel species may also emerge; in particular taxa that prosper in human-dominated ecosystems may rapidly diversify into many new species. Microbes are likely to benefit from the increase in nutrient-enriched environmental niches. No new species of existing large vertebrates are likely to arise and food chains will probably be shortened.

There are multiple scenarios for known risks that can have a global impact on the planet. From the perspective of humanity, these can be subdivided into survivable risks and terminal risks. Risks that humanity pose to itself include climate change, the misuse of nanotechnology, a nuclear holocaust, warfare with a programmed superintelligence, a genetically engineered disease, or a disaster caused by a physics experiment. Similarly, several natural events may pose a doomsday threat, including a highly virulent disease, the impact of an asteroid or comet, runaway greenhouse effect, and resource depletion. There may also be the possibility of an infestation by an extraterrestrial lifeform. The actual odds of these scenarios are difficult if not impossible to deduce.

Should the human race become extinct, then the various features assembled by humanity will begin to decay. The largest structures have an estimated decay half-life of about 1,000 years. The last surviving structures would most likely be open pit mines, large landfills, major highways, wide canal cuts, and earthfill flank dams. A few massive stone monuments like the pyramids at the Giza Necropolis or the sculptures at Mount Rushmore may still survive in some form after a million years.[225]

Figure 64: *The Barringer Meteorite Crater in Flagstaff, Arizona, showing evidence of the impact of celestial objects upon the Earth*

Random events

As the Sun orbits the Milky Way, wandering stars may approach close enough to have a disruptive influence on the Solar System. A close stellar encounter may cause a significant reduction in the perihelion distances of comets in the Oort cloud—a spherical region of icy bodies orbiting within half a light year of the Sun. Such an encounter can trigger a 40-fold increase in the number of comets reaching the inner Solar System. Impacts from these comets can trigger a mass extinction of life on Earth. These disruptive encounters occur at an average of once every 45 million years. The mean time for the Sun to collide with another star in the solar neighborhood is approximately 3×10^{13} years, which is much longer than the estimated age of the Milky Way galaxy, at $\sim 1.3 \times 10^{10}$ years. This can be taken as an indication of the low likelihood of such an event occurring during the lifetime of the Earth.[226]

The energy release from the impact of an asteroid or comet with a diameter of 5–10 km (3.1–6.2 mi) or larger is sufficient to create a global environmental disaster and cause a statistically significant increase in the number of species extinctions. Among the deleterious effects resulting from a major impact event is a cloud of fine dust ejecta blanketing the planet, which lowers land temperatures by about 15 °C (27 °F) within a week and halts photosynthesis for several months. The mean time between major impacts is estimated to be at least 100 million years. During the last 540 million years, simulations demonstrated that such an impact rate is sufficient to cause 5–6 mass extinctions and 20–30 lower severity events. This matches the geologic record of significant extinctions during the Phanerozoic Eon. Such events can be expected to continue into the future.

A supernova is a cataclysmic explosion of a star. Within the Milky Way galaxy, supernova explosions occur on average once every 40 years. During the history of the Earth, multiple such events have likely occurred within a distance of 100 light years. Explosions inside this distance can contaminate the planet with radioisotopes and possibly impact the biosphere. Gamma rays emitted by a supernova react with nitrogen in the atmosphere, producing nitrous oxides. These molecules cause a depletion of the ozone layer that protects the surface from ultraviolet radiation from the Sun. An increase in UV-B radiation of only 10–30% is sufficient to cause a significant impact to life; particularly to the phytoplankton that form the base of the oceanic food chain. A supernova explosion at a distance of 26 light years will reduce the ozone column density by half. On average, a supernova explosion occurs within 32 light years once every few hundred million years, resulting in a depletion of the ozone layer lasting several centuries.[227] Over the next two billion years, there will be about 20 supernova explosions and one gamma ray burst that will have a significant impact on the planet's biosphere.

The incremental effect of gravitational perturbations between the planets causes the inner Solar System as a whole to behave chaotically over long time periods. This does not significantly affect the stability of the Solar System over intervals of a few million years or less, but over billions of years the orbits of the planets become unpredictable. Computer simulations of the Solar System's evolution over the next five billion years suggest that there is a small (less than 1%) chance that a collision could occur between Earth and either Mercury, Venus, or Mars. During the same interval, the odds that the Earth will be scattered out of the Solar System by a passing star are on the order of one part in 10^5. In such a scenario, the oceans would freeze solid within several million years, leaving only a few pockets of liquid water about 14 km (8.7 mi) underground. There is a remote chance that the Earth will instead be captured by a passing binary star system, allowing the planet's biosphere to remain intact. The odds of this happening are about one chance in three million.[228]

Orbit and rotation

The gravitational perturbations of the other planets in the Solar System combine to modify the orbit of the Earth and the orientation of its spin axis. These changes can influence the planetary climate.[229,230] Despite such interactions, highly accurate simulations show that overall, Earth's orbit is likely to remain dynamically stable for billions of years into the future. In all 1,600 simulations, the planet's semimajor axis, eccentricity, and inclination remained nearly constant.

Glaciation

Historically, there have been cyclical ice ages in which glacial sheets periodically covered the higher latitudes of the continents. Ice ages may occur because of changes in ocean circulation and continentality induced by plate tectonics.[231] The Milankovitch theory predicts that glacial periods occur during ice ages because of astronomical factors in combination with climate feedback mechanisms. The primary astronomical drivers are a higher than normal orbital eccentricity, a low axial tilt (or obliquity), and the alignment of summer solstice with the aphelion. Each of these effects occur cyclically. For example, the eccentricity changes over time cycles of about 100,000 and 400,000 years, with the value ranging from less than 0.01 up to 0.05. This is equivalent to a change of the semiminor axis of the planet's orbit from 99.95% of the semimajor axis to 99.88%, respectively.

The Earth is passing through an ice age known as the quaternary glaciation, and is presently in the Holocene interglacial period. This period would normally be expected to end in about 25,000 years.[230] However, the increased rate of carbon dioxide release into the atmosphere by humans may delay the onset of the next glacial period until at least 50,000–130,000 years from now. On the other hand, a global warming period of finite duration (based on the assumption that fossil fuel use will cease by the year 2200) will probably only impact the glacial period for about 5,000 years. Thus, a brief period of global warming induced through a few centuries worth of greenhouse gas emission would only have a limited impact in the long term.

Obliquity

The tidal acceleration of the Moon slows the rotation rate of the Earth and increases the Earth-Moon distance. Friction effects—between the core and mantle and between the atmosphere and surface—can dissipate the Earth's rotational energy. These combined effects are expected to increase the length of the day by more than 1.5 hours over the next 250 million years, and to increase the obliquity by about a half degree. The distance to the Moon will increase by about 1.5 Earth radii during the same period.

Based on computer models, the presence of the Moon appears to stabilize the obliquity of the Earth, which may help the planet to avoid dramatic climate changes. This stability is achieved because the Moon increases the precession

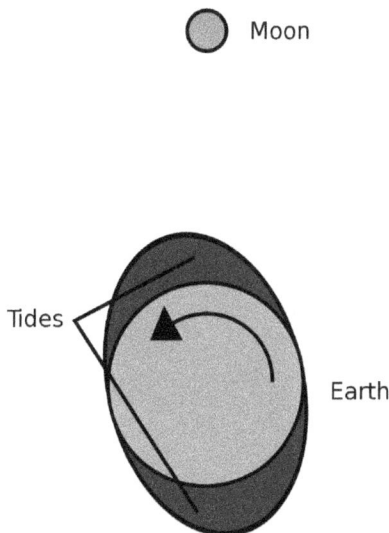

Figure 65: *The rotational offset of the tidal bulge exerts a net torque on the Moon, boosting it while slowing the Earth's rotation. This image is not to scale.*

rate of the Earth's spin axis, thereby avoiding resonances between the precession of the spin and precession of the planet's orbital plane (that is, the precession motion of the ecliptic). However, as the semimajor axis of the Moon's orbit continues to increase, this stabilizing effect will diminish. At some point, perturbation effects will probably cause chaotic variations in the obliquity of the Earth, and the axial tilt may change by angles as high as 90° from the plane of the orbit. This is expected to occur between 1.5 and 4.5 billion years from now.

A high obliquity would probably result in dramatic changes in the climate and may destroy the planet's habitability.[229] When the axial tilt of the Earth exceeds 54°, the yearly insolation at the equator is less than that at the poles. The planet could remain at an obliquity of 60° to 90° for periods as long as 10 million years.

Geodynamics

Tectonics-based events will continue to occur well into the future and the surface will be steadily reshaped by tectonic uplift, extrusions, and erosion. Mount Vesuvius can be expected to erupt about 40 times over the next 1,000 years. During the same period, about five to seven earthquakes of magnitude

Figure 66: *Pangaea was the last supercontinent to form before the present.*

8 or greater should occur along the San Andreas Fault, while about 50 magnitude 9 events may be expected worldwide. Mauna Loa should experience about 200 eruptions over the next 1,000 years, and the Old Faithful Geyser will likely cease to operate. The Niagara Falls will continue to retreat upstream, reaching Buffalo in about 30,000–50,000 years.

In 10,000 years, the post-glacial rebound of the Baltic Sea will have reduced the depth by about 90 m (300 ft). The Hudson Bay will decrease in depth by 100 m over the same period. After 100,000 years, the island of Hawaii will have shifted about 9 km (5.6 mi) to the northwest. The planet may be entering another glacial period by this time.

Continental drift

The theory of plate tectonics demonstrates that the continents of the Earth are moving across the surface at the rate of a few centimeters per year. This is expected to continue, causing the plates to relocate and collide. Continental drift is facilitated by two factors: the energy generation within the planet and the presence of a hydrosphere. With the loss of either of these, continental drift will come to a halt. The production of heat through radiogenic processes is sufficient to maintain mantle convection and plate subduction for at least the next 1.1 billion years.

At present, the continents of North and South America are moving westward from Africa and Europe. Researchers have produced several scenarios about how this will continue in the future.[232] These geodynamic models can be distinguished by the subduction flux, whereby the oceanic crust moves under a continent. In the introversion model, the younger, interior, Atlantic ocean becomes preferentially subducted and the current migration of North and South America is reversed. In the extroversion model, the older, exterior, Pacific ocean remains preferentially subducted and North and South America migrate toward eastern Asia.

As the understanding of geodynamics improves, these models will be subject to revision. In 2008, for example, a computer simulation was used to predict that a reorganization of the mantle convection will occur over the next 100 million years, causing a supercontinent composed of Africa, Eurasia, Australia, Antarctica and South America to form around Antarctica.

Regardless of the outcome of the continental migration, the continued subduction process causes water to be transported to the mantle. After a billion years from the present, a geophysical model gives an estimate that 27% of the current ocean mass will have been subducted. If this process were to continue unmodified into the future, the subduction and release would reach an equilibrium after 65% of the current ocean mass has been subducted.

Introversion

Christopher Scotese and his colleagues have mapped out the predicted motions several hundred million years into the future as part of the Paleomap Project.[232] In their scenario, 50 million years from now the Mediterranean sea may vanish and the collision between Europe and Africa will create a long mountain range extending to the current location of the Persian Gulf. Australia will merge with Indonesia, and Baja California will slide northward along the coast. New subduction zones may appear off the eastern coast of North and South America, and mountain chains will form along those coastlines. To the south, the migration of Antarctica to the north will cause all of its ice sheets to melt. This, along with the melting of the Greenland ice sheets, will raise the average ocean level by 90 m (300 ft). The inland flooding of the continents will result in climate changes.[232]

As this scenario continues, by 100 million years from the present the continental spreading will have reached its maximum extent and the continents will then begin to coalesce. In 250 million years, North America will collide with Africa while South America will wrap around the southern tip of Africa. The result will be the formation of a new supercontinent (sometimes called Pangaea Ultima), with the Pacific Ocean stretching across half the planet. The continent of Antarctica will reverse direction and return to the South Pole, building up a new ice cap.[233]

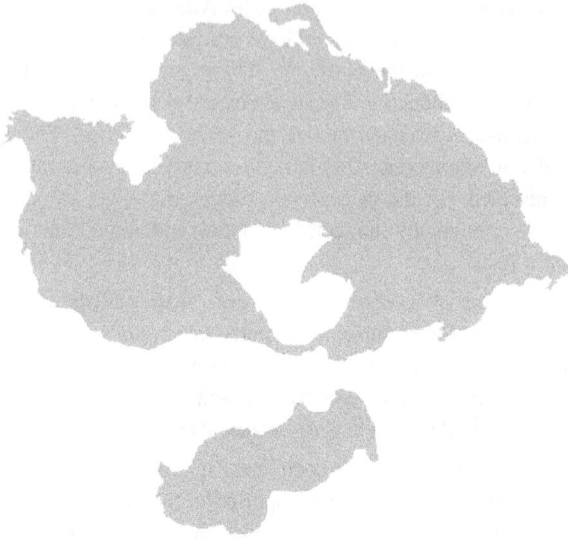

Figure 67: *A rough approximation of Pangaea Ultima, one of the three models for a future supercontinent.*

Extroversion

The first scientist to extrapolate the current motions of the continents was
Canadian geologist Paul F. Hoffman of Harvard University. In 1992, Hoffman predicted that the continents of North and South America would continue
to advance across the Pacific Ocean, pivoting about Siberia until they begin
to merge with Asia. He dubbed the resulting supercontinent, Amasia.[234,235]
Later, in the 1990s, Roy Livermore calculated a similar scenario. He predicted that Antarctica would start to migrate northward, and east Africa and
Madagascar would move across the Indian Ocean to collide with Asia.

In an extroversion model, the closure of the Pacific Ocean would be complete
in about 350 million years. This marks the completion of the current supercontinent cycle, wherein the continents split apart and then rejoin each other about
every 400–500 million years. Once the supercontinent is built, plate tectonics
may enter a period of inactivity as the rate of subduction drops by an order of
magnitude. This period of stability could cause an increase in the mantle temperature at the rate of 30–100 °C (54–180 °F) every 100 million years, which
is the minimum lifetime of past supercontinents. As a consequence, volcanic
activity may increase.

Supercontinent

The formation of a supercontinent can dramatically affect the environment. The collision of plates will result in mountain building, thereby shifting weather patterns. Sea levels may fall because of increased glaciation.[236] The rate of surface weathering can rise, resulting in an increase in the rate that organic material is buried. Supercontinents can cause a drop in global temperatures and an increase in atmospheric oxygen. This, in turn, can affect the climate, further lowering temperatures. All of these changes can result in more rapid biological evolution as new niches emerge.[237]

The formation of a supercontinent insulates the mantle. The flow of heat will be concentrated, resulting in volcanism and the flooding of large areas with basalt. Rifts will form and the supercontinent will split up once more.[238] The planet may then experience a warming period, as occurred during the Cretaceous period.[237]

Solidification of the outer core

The iron-rich core region of the Earth is divided into a 1,220 km (760 mi) radius solid inner core and a 3,480 km (2,160 mi) radius liquid outer core. The rotation of the Earth creates convective eddies in the outer core region that cause it to function as a dynamo.[239] This generates a magnetosphere about the Earth that deflects particles from the solar wind, which prevents significant erosion of the atmosphere from sputtering. As heat from the core is transferred outward toward the mantle, the net trend is for the inner boundary of the liquid outer core region to freeze, thereby releasing thermal energy and causing the solid inner core to grow. This iron crystallization process has been ongoing for about a billion years. In the modern era, the radius of the inner core is expanding at an average rate of roughly 0.5 mm (0.02 in) per year, at the expense of the outer core. Nearly all of the energy needed to power the dynamo is being supplied by this process of inner core formation.

The growth of the inner core may be expected to consume most of the outer core by some 3–4 billion years from now, resulting in a nearly solid core composed of iron and other heavy elements. The surviving liquid envelope will mainly consist of lighter elements that will undergo less mixing.[240] Alternatively, if at some point plate tectonics comes to an end, the interior will cool less efficiently, which may end the growth of the inner core. In either case, this can result in the loss of the magnetic dynamo. Without a functioning dynamo, the magnetic field of the Earth will decay in a geologically short time period of roughly 10,000 years.[241] The loss of the magnetosphere will cause an increase in erosion of light elements, particularly hydrogen, from the Earth's outer atmosphere into space, resulting in less favorable conditions for life.

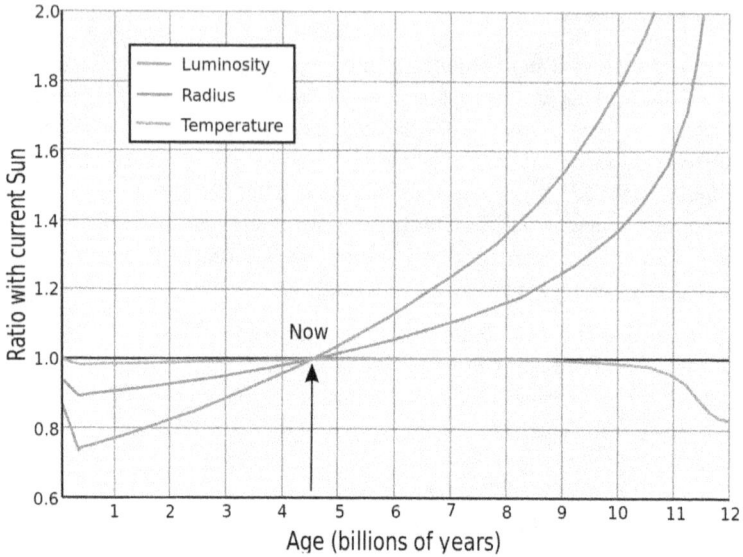

Figure 68: *Evolution of the Sun's luminosity, radius and effective temperature compared to the present Sun. After Ribas (2010).*

Solar evolution

The energy generation of the Sun is based upon thermonuclear fusion of hydrogen into helium. This occurs in the core region of the star using the proton–proton chain reaction process. Because there is no convection in the solar core, the helium concentration builds up in that region without being distributed throughout the star. The temperature at the core of the Sun is too low for nuclear fusion of helium atoms through the triple-alpha process, so these atoms do not contribute to the net energy generation that is needed to maintain hydrostatic equilibrium of the Sun.

At present, nearly half the hydrogen at the core has been consumed, with the remainder of the atoms consisting primarily of helium. As the number of hydrogen atoms per unit mass decreases, so too does their energy output provided through nuclear fusion. This results in a decrease in pressure support, which causes the core to contract until the increased density and temperature bring the core pressure into equilibrium with the layers above. The higher temperature causes the remaining hydrogen to undergo fusion at a more rapid rate, thereby generating the energy needed to maintain the equilibrium.

The result of this process has been a steady increase in the energy output of the Sun. When the Sun first became a main sequence star, it radiated only

70% of the current luminosity. The luminosity has increased in a nearly linear fashion to the present, rising by 1% every 110 million years. Likewise, in three billion years the Sun is expected to be 33% more luminous. The hydrogen fuel at the core will finally be exhausted in five billion years, when the Sun will be 67% more luminous than at present. Thereafter the Sun will continue to burn hydrogen in a shell surrounding its core, until the luminosity reaches 121% above the present value. This marks the end of the Sun's main sequence lifetime, and thereafter it will pass through the subgiant stage and evolve into a red giant.

By this time, the collision of the Milky Way and Andromeda galaxies should be underway. Although this could result in the Solar System being ejected from the newly combined galaxy, it is considered unlikely to have any adverse effect on the Sun or its planets.

Climate impact

The rate of weathering of silicate minerals will increase as rising temperatures speed up chemical processes. This in turn will decrease the level of carbon dioxide in the atmosphere, as these weathering processes convert carbon dioxide gas into solid carbonates. Within the next 600 million years from the present, the concentration of CO_2 will fall below the critical threshold needed to sustain C_3 photosynthesis: about 50 parts per million. At this point, trees and forests in their current forms will no longer be able to survive, the last living trees being evergreen conifers. However, C_4 carbon fixation can continue at much lower concentrations, down to above 10 parts per million. Thus plants using C_4 photosynthesis may be able to survive for at least 0.8 billion years and possibly as long as 1.2 billion years from now, after which rising temperatures will make the biosphere unsustainable. Currently, C_4 plants represent about 5% of Earth's plant biomass and 1% of its known plant species. For example, about 50% of all grass species (Poaceae) use the C_4 photosynthetic pathway,[242] as do many species in the herbaceous family Amaranthaceae.

When the levels of carbon dioxide fall to the limit where photosynthesis is barely sustainable, the proportion of carbon dioxide in the atmosphere is expected to oscillate up and down. This will allow land vegetation to flourish each time the level of carbon dioxide rises due to tectonic activity and animal life. However, the long term trend is for the plant life on land to die off altogether as most of the remaining carbon in the atmosphere becomes sequestered in the Earth.[243] Some microbes are capable of photosynthesis at concentrations of CO_2 of a few parts per million, so these life forms would probably disappear only because of rising temperatures and the loss of the biosphere.

Plants—and, by extension, animals—could survive longer by evolving other strategies such as requiring less CO
$_2$ for photosynthetic processes, becoming carnivorous, adapting to desiccation, or associating with fungi. These adaptations are likely to appear near the beginning of the moist greenhouse (see further).

The loss of plant life will also result in the eventual loss of oxygen as well as ozone due to the respiration of animals, chemical reactions in the atmosphere, and volcanic eruptions. This will result in less attenuation of DNA-damaging ultraviolet radiation, as well as the death of animals; the first animals to disappear would be large mammals, followed by small mammals, birds, amphibians and large fish, reptiles and small fish, and finally invertebrates. Before this happened it's expected that life would concentrate at refugia of lower temperature such as high elevations where less land surface area is available, thus restricting population sizes. Smaller animals would survive better than larger ones because of lesser oxygen requirements, while birds would fare better than mammals thanks to their ability to travel large distances looking for colder temperatures.

In their work *The Life and Death of Planet Earth*, authors Peter D. Ward and Donald Brownlee have argued that some form of animal life may continue even after most of the Earth's plant life has disappeared. Ward and Brownlee use fossil evidence from the Burgess Shale in British Columbia, Canada, to determine the climate of the Cambrian Explosion, and use it to predict the climate of the future when rising global temperatures caused by a warming Sun and declining oxygen levels result in the final extinction of animal life. Initially, they expect that some insects, lizards, birds and small mammals may persist, along with sea life. However, without oxygen replenishment by plant life, they believe that animals would probably die off from asphyxiation within a few million years. Even if sufficient oxygen were to remain in the atmosphere through the persistence of some form of photosynthesis, the steady rise in global temperature would result in a gradual loss of biodiversity.[243]

As temperatures continue to rise, the last animal life will be driven back toward the poles, and possibly underground. They would become primarily active during the polar night, aestivating during the polar day due to the intense heat. Much of the surface would become a barren desert and life would primarily be found in the oceans.[243] However, due to a decrease of the amount or organic matter coming to the oceans from the land as well as oxygen in the water, life would disappear there too following a similar path to that on Earth's surface. This process would start with the loss of freshwater species and conclude with invertebrates, particularly those that do not depend on living plants such as termites or those near hydrothermal vents such as worms of the genus *Riftia*. As a result of these processes, multi-cellular lifeforms may be extinct

Figure 69: *The atmosphere of Venus is in a "supergreenhouse" state.*

in about 800 million years, and eukaryotes in 1.3 billion years, leaving only the prokaryotes.

Loss of oceans

One billion years from now, about 27% of the modern ocean will have been subducted into the mantle. If this process were allowed to continue uninterrupted, it would reach an equilibrium state where 65% of the current surface reservoir would remain at the surface. Once the solar luminosity is 10% higher than its current value, the average global surface temperature will rise to 320 K (47 °C; 116 °F). The atmosphere will become a "moist greenhouse" leading to a runaway evaporation of the oceans.[244] At this point, models of the Earth's future environment demonstrate that the stratosphere would contain increasing levels of water. These water molecules will be broken down through photodissociation by solar ultraviolet radiation, allowing hydrogen to escape the atmosphere. The net result would be a loss of the world's sea water by about 1.1 billion years from the present. This will be a simple dramatic step in annihilating all life on Earth.

There will be two variations of this future warming feedback: the "moist greenhouse" where water vapor dominates the troposphere while water vapor starts to accumulate in the stratosphere (if the oceans evaporate very quickly),

and the "runaway greenhouse" where water vapor becomes a dominant component of the atmosphere (if the oceans evaporate too slowly). The Earth
will undergo rapid warming that could send its surface temperature to over
900 °C (1,650 °F) as the atmosphere will be totally overwhelmed by water
vapor, causing its entire surface to melt and killing all life, perhaps in about
three billion years. In this ocean-free era, there will continue to be surface
reservoirs as water is steadily released from the deep crust and mantle, where
it is estimated there is an amount of water equivalent to several times that currently present in the Earth's oceans.[245] Some water may be retained at the
poles and there may be occasional rainstorms, but for the most part the planet
would be a dry desert with large dunefields covering its equator, and a few
salt flats on what was once the ocean floor, similar to the ones in the Atacama
Desert in Chile.

With no water to serve as a lubricant, plate tectonics would very likely stop and
the most visible signs of geological activity would be shield volcanoes located
above mantle hotspots.[244] In these arid conditions the planet may retain some
microbial and possibly even multi-cellular life.[244] Most of these microbes will
be halophiles and life could find refuge in the atmosphere as has been proposed
that could have happened on Venus. However, the increasingly extreme conditions will likely lead to the extinction of the prokaryotes between 1.6 billion
years and 2.8 billion years from now, with the last of them living in residual
ponds of water at high latitudes and heights or in caverns with trapped ice.
However, underground life could last longer. What happens next depends on
the level of tectonic activity. A steady release of carbon dioxide by volcanic
eruption could cause the atmosphere to enter a "supergreenhouse" state like
that of the planet Venus. But, as stated above, without surface water, plate tectonics would probably come to a halt and most of the carbonates would remain
securely buried until the Sun became a red giant and its increased luminosity
heated the rock to the point of releasing the carbon dioxide.[245]

The loss of the oceans could be delayed until two billion years in the future if
the total atmospheric pressure were to decline. A lower atmospheric pressure
would reduce the greenhouse effect, thereby lowering the surface temperature. This could occur if natural processes were to remove the nitrogen from
the atmosphere. Studies of organic sediments has shown that at least 100 kilo-
pascals (0.99 atm) of nitrogen has been removed from the atmosphere over the
past four billion years; enough to effectively double the current atmospheric
pressure if it were to be released. This rate of removal would be sufficient to
counter the effects of increasing solar luminosity for the next two billion years.

By 2.8 billion years from now, the surface temperature of the Earth will have
reached 422 K (149 °C; 300 °F), even at the poles. At this point, any remaining life will be extinguished due to the extreme conditions. If the Earth loses

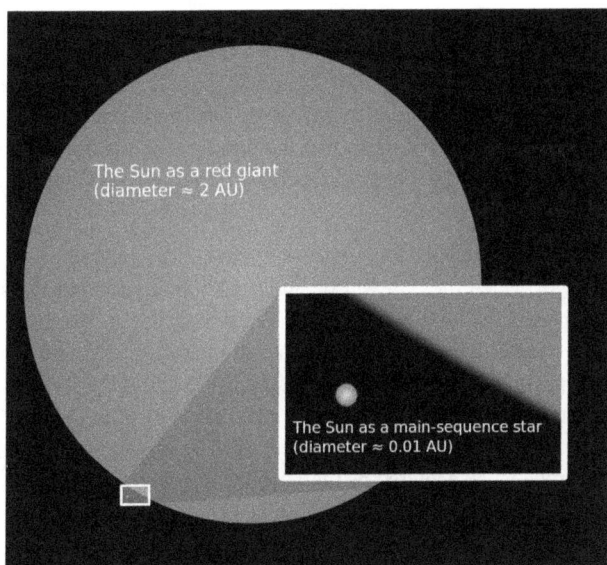

Figure 70: *The size of the current Sun (now in the main sequence) compared to its estimated size during its red giant phase*

its surface water by this point, the planet will stay in the same conditions until the Sun becomes a red giant.[244] If this scenario doesn't happen, then in about 3–4 billion years the amount of water vapour in the lower atmosphere will rise to 40% and a moist greenhouse effect will commence once the luminosity from the Sun reaches 35–40% more than its present-day value. A "runaway greenhouse" effect will ensue, causing the atmosphere to heat up and raising the surface temperature to around 1,600 K (1,330 °C; 2,420 °F). This is sufficient to melt the surface of the planet.[244] However, most of the atmosphere will be retained until the Sun has entered the red giant stage.

With the extinction of life, 2.8 billion years from now it is also expected that Earth biosignatures will disappear, to be replaced by signatures caused by non-biological processes.

Red giant stage

Once the Sun changes from burning hydrogen at its core to burning hydrogen around its shell, the core will start to contract and the outer envelope will expand. The total luminosity will steadily increase over the following billion years until it reaches 2,730 times the Sun's current luminosity at the age of 12.167 billion years. Most of Earth's atmosphere will be lost to space and its surface will consist of a lava ocean with floating continents of metals and metal oxides as well as icebergs of refractory materials, with its surface temperature reaching more than 2,400 K (2,130 °C; 3,860 °F). The Sun will experience more rapid mass loss, with about 33% of its total mass shed with the solar wind. The loss of mass will mean that the orbits of the planets will expand. The orbital distance of the Earth will increase to at most 150% of its current value.

The most rapid part of the Sun's expansion into a red giant will occur during the final stages, when the Sun will be about 12 billion years old. It is likely to expand to swallow both Mercury and Venus, reaching a maximum radius of 1.2 AU (180,000,000 km). The Earth will interact tidally with the Sun's outer atmosphere, which would serve to decrease Earth's orbital radius. Drag from the chromosphere of the Sun would also reduce the Earth's orbit. These effects will act to counterbalance the effect of mass loss by the Sun, and the Earth will probably be engulfed by the Sun.

The drag from the solar atmosphere may cause the orbit of the Moon to decay. Once the orbit of the Moon closes to a distance of 18,470 km (11,480 mi), it will cross the Earth's Roche limit. This means that tidal interaction with the Earth would break apart the Moon, turning it into a ring system. Most of the orbiting ring will then begin to decay, and the debris will impact the Earth. Hence, even if the Earth is not swallowed up by the Sun, the planet may be left moonless. The ablation and vaporization caused by its fall on a decaying trajectory towards the Sun may remove Earth's mantle, leaving just its core, which will finally be destroyed after at most 200 years. Following this event, Earth's sole legacy will be a very slight increase (0.01%) of the solar metallicity.§IIC

Post red-giant stage

After fusing helium in its core to carbon, the Sun will begin to collapse again, evolving into a compact white dwarf star after ejecting its outer atmosphere as a planetary nebula. In 50 billion years, if the Earth and Moon are not engulfed by the Sun, they will become tidelocked, with each showing only one face to the other. Thereafter, the tidal action of the Sun will extract angular momentum from the system, causing the lunar orbit to decay and the Earth's spin to accelerate.

In about 65 billion years, it is estimated that the Moon may end up colliding with the Earth, assuming they are not destroyed by the red giant Sun, due to the remaining energy of the Earth–Moon system being sapped by the remnant Sun, causing the Moon to slowly move inwards toward the Earth.

Figure 71: *The Helix nebula, a planetary nebula similar to what the Sun will produce in 8 billion years.*

Over time intervals of around 30 trillion years, the Sun will undergo a close encounter with another star. As a consequence, the orbits of their planets can become disrupted, potentially ejecting them from the system entirely. If Earth is not destroyed by the expanding red giant Sun in 7.6 billion years and not ejected from its orbit by a stellar encounter, its ultimate fate will be that it collides with the black dwarf Sun due to the decay of its orbit via gravitational radiation, in 10^{20} (100 quintillion) years.

References

Bibliography

<templatestyles src="Template:Refbegin/styles.css" />

- Adams, Fred C. (2008), "Long term astrophysical processes", in Bostrom, Nick; Ćirković, Milan M., *Global catastrophic risks*[246], Oxford University Press, ISBN 0-19-857050-3.
- Brownlee, Donald E. (2010), "Planetary habitability on astronomical time scales"[247], in Schrijver, Carolus J.; Siscoe, George L., *Heliophysics: Evolving Solar Activity and the Climates of Space and Earth*, Cambridge University Press, ISBN 0-521-11294-X.

- Calkin, P. E.; Young, G. M. (1996), "Global glaciation chronologies and causes of glaciation", in Menzies, John, *Past glacial environments: sediments, forms, and techniques*, Glacial environments, **2**, Butterworth-Heinemann, ISBN 0-7506-2352-7.
- Cowie, Jonathan (2007), *Climate change: biological and human aspects*, Cambridge University Press, ISBN 0-521-69619-4.
- Fishbaugh, Kathryn E.; Des Marais, David J.; Korablev, Oleg; Raulin, François; Lognonné, Phillipe (2007), *Geology and habitability of terrestrial planets*, Space Sciences Series of Issi, **24**, Springer, ISBN 0-387-74287-5.
- Gonzalez, Guillermo; Richards, Jay Wesley (2004), *The privileged planet: how our place in the cosmos is designed for discovery*[248], Regnery Publishing, ISBN 0-89526-065-4.
- Hanslmeier, Arnold (2009), "Habitability and cosmic catastrophes"[249], *Advances in Astrobiology and Biogeophysics*, Springer, ISBN 3-540-76944-7.
- Hoffman, Paul F. (1992), "Supercontinents"[250] (PDF), *Encyclopedia of Earth System Sciences*, Academic press, Inc.
- Lunine, Jonathan Irving; Lunine, Cynthia J. (1999), *Earth: evolution of a habitable world*[251], Cambridge University Press, ISBN 0-521-64423-2.
- Meadows, Arthur Jack (2007), *The future of the universe*[252], Springer, ISBN 1-85233-946-2.
- Nield, Ted (2007), *Supercontinent: ten billion dates in the life of our planet*, Harvard University Press, ISBN 0-674-02659-4.
- Myers, Norman (2000), "Biodiversity Loss", in Peter H. Raven and Tania Williams, *Nature and human society: the quest for a sustainable world : proceedings of the 1997 Forum on Biodiversity*, National Academies, pp. 63–70, ISBN 0-309-06555-0.
- Palmer, Douglas (2003), *Prehistoric past revealed: the four billion date history of life on Earth*, University of California Press, ISBN 0-520-24105-3.
- Reaka-Kudla, Marjorie L.; Wilson, Don E.; Wilson, Edward O. (1997), *Biodiversity 2* (2nd ed.), Joseph Henry Press, ISBN 0-309-05584-9.
- Roberts, Neil (1998), *The Holocene: an environmental history* (2nd ed.), Wiley-Blackwell, ISBN 0-631-18638-7.
- Stevenson, D. J. (2002), "Introduction to planetary interiors", in Hemley, Russell Julian; Chiarotti, G.; Bernasconi, M.; Ulivi, L., *Fenomeni ad alte pressioni*[253], IOS Press, ISBN 1-58603-269-0.
- Tayler, Roger John (1993), *Galaxies, structure and evolution*[254] (2nd ed.), Cambridge University Press, ISBN 0-521-36710-7.
- Thompson, Russell D.; Perry, Allen Howard (1997), *Applied Climatology: Principles and Practice*, Routledge, pp. 127–28, ISBN 0-415-14100-

1.

- van der Maarel, E. (2005), *Vegetation ecology*, Wiley-Blackwell, ISBN 0-632-05761-0.
- Ward, Peter Douglas (2006), *Out of thin air: dinosaurs, birds, and Earth's ancient atmosphere*, National Academies Press, ISBN 0-309-10061-5.
- Ward, Peter Douglas; Brownlee, Donald (2003), *The life and death of planet Earth: how the new science of astrobiology charts the ultimate fate of our world*, Macmillan, ISBN 0-8050-7512-7.

Further reading

- Scotese, Christopher R., *PALEOMAP Project*[255], retrieved 2009-08-28.
- Tonn, B. E. (March 2002), "Distant futures and the environment", *Futures*, **34** (2): 117–132, doi: 10.1016/S0016-3287(01)00050-7[256].

Shape

Figure of the Earth

Geodesy

Fundamentals

- Geodesy
- Geodynamics
- Geomatics
- Cartography
- History

Concepts

- Geographical distance
- Geoid
- Figure of the Earth
- Geodetic datum
- Geodesic
- Geographic coordinate system
- Horizontal position representation
- Latitude / Longitude
- Map projection
- Reference ellipsoid
- Satellite geodesy
- Spatial reference system

Technologies

- Global Navigation Satellite System (GNSS)
- Global Positioning System (GPS)
- GLONASS (Russian)
- BeiDou (BDS) (Chinese)

- Galileo (European)
- Indian Regional Navigation
  Satellite System (IRNSS) (India)
- Quasi-Zenith Satellite System (QZSS) (Japan)
- Legenda (satellite system)

Standards (History)	
NGVD 29	Sea Level Datum 1929
OSGB36	Ordnance Survey Great Britain 1936
SK-42	Systema Koordinat 1942 goda
ED50	European Datum 1950
SAD69	South American Datum 1969
GRS 80	Geodetic Reference System 1980
NAD 83	North American Datum 1983
WGS 84	World Geodetic System 1984
NAVD 88	N. American Vertical Datum 1988
ETRS89	European Terrestrial Ref. Sys. 1989
GCJ-02	Chinese obfuscated datum 2002

- International Terrestrial Reference System
- Spatial Reference System Identifier (SRID)
- Universal Transverse Mercator (UTM)

- $\underline{v}$
- $\underline{t}$
- $\underline{e}^{257}$

The **figure of the Earth** is the size and shape of the Earth in geodesy. Its specific meaning depends on the way it is used and the precision with which the Earth's size and shape is to be defined. While the sphere is a close approximation of the true figure of the Earth and satisfactory for many purposes, geodesists have developed several models that more closely approximate the shape of the Earth so that coordinate systems can serve the precise needs of navigation, surveying, cadastre, land use, and various other concerns.

Need for models of the figure of the Earth

The actual topographic surface is most apparent with its variety of land forms and water areas. This is, in fact, the surface on which actual Earth measurements are made. However, it is not feasible for exact mathematical analysis, because the formulas which would be required to take the irregularities into account would necessitate a prohibitive amount of computation. The topographic surface is generally the concern of topographers and hydrographers.

The Pythagorean concept of a spherical Earth offers a simple surface that is mathematically easy to deal with. Many astronomical and navigational computations use it as a surface representing the Earth. While the sphere is a close approximation of the true figure of the Earth and satisfactory for many purposes, to the geodesists interested in the measurement of long distances on the scale of continents and oceans, a more exact figure is necessary. Closer approximations range from modelling the shape of the surface of the entire Earth as an oblate spheroid or an oblate ellipsoid, to the use of spherical harmonics to approximate the geoid, or local approximations in terms of local reference ellipsoids.

The idea of a planar or flat surface for Earth, however, is still sufficient for surveys of small areas, as the local topography is far more significant than the curvature. Plane-table surveys are made for relatively small areas, and no account is taken of the curvature of the Earth. A survey of a city would likely be computed as though the Earth were a plane surface the size of the city. For such small areas, nearly exact positions can be determined relative to each other without considering the size and shape of the entire Earth.

In the mid- to late 20th century, research across the geosciences contributed to drastic improvements in the accuracy of the figure of the Earth. The primary utility (and the motivation for funding, mainly from the military) of this improved accuracy was to provide geographical and gravitational data for the inertial guidance systems of ballistic missiles. This funding also drove the expansion of geoscientific disciplines, fostering the creation and growth of various geoscience departments at many universities.

Models of the figure of the Earth

The models for the figure of the Earth vary in the way they are used, in their complexity, and in the accuracy with which they represent the size and shape of the Earth.

Sphere

The simplest model for the shape of the entire Earth is a sphere. The Earth's radius is the distance from Earth's center to its surface, about 6,371 kilometers (3,959 mi). While "radius" normally is a characteristic of perfect spheres, the Earth deviates from spherical by only a third of a percent, sufficiently close to treat it as a sphere in many contexts and justifying the term "the radius of the Earth".

The concept of a spherical Earth dates back to around the 6th century BC, but remained a matter of philosophical speculation until the 3rd century BC. The first scientific estimation of the radius of the earth was given by Eratosthenes about 240 BC, with estimates of the accuracy of Eratosthenes's measurement ranging from 2% to 15%.

Figure 72: *A view across a 20-km-wide bay in the coast of Spain. Note the curvature of the Earth hiding the base of the buildings on the far shore.*

The Earth is only approximately spherical, so no single value serves as its natural radius. Distances from points on the surface to the center range from 6,353 km to 6,384 km (3,947 – 3,968 mi). Several different ways of modeling the Earth as a sphere each yield a mean radius of 6,371 kilometers (3,959 mi). Regardless of the model, any radius falls between the polar minimum of about 6,357 km and the equatorial maximum of about 6,378 km (3,950 – 3,963 mi). The difference 21 kilometers (13 mi) correspond to the polar radius being approximately 0.3% shorter than the equator radius.

Ellipsoid of revolution

Since the Earth is flattened at the poles and bulges at the Equator, geodesy represents the figure of the Earth as an oblate spheroid. The oblate spheroid, or oblate ellipsoid, is an ellipsoid of revolution obtained by rotating an ellipse about its shorter axis. It is the regular geometric shape that most nearly approximates the shape of the Earth. A spheroid describing the figure of the Earth or other celestial body is called a reference ellipsoid. The reference ellipsoid for Earth is called an Earth ellipsoid.

An ellipsoid of revolution is uniquely defined by two quantities. Several conventions for expressing the two quantities are used in geodesy, but they are all equivalent to and convertible with each other:

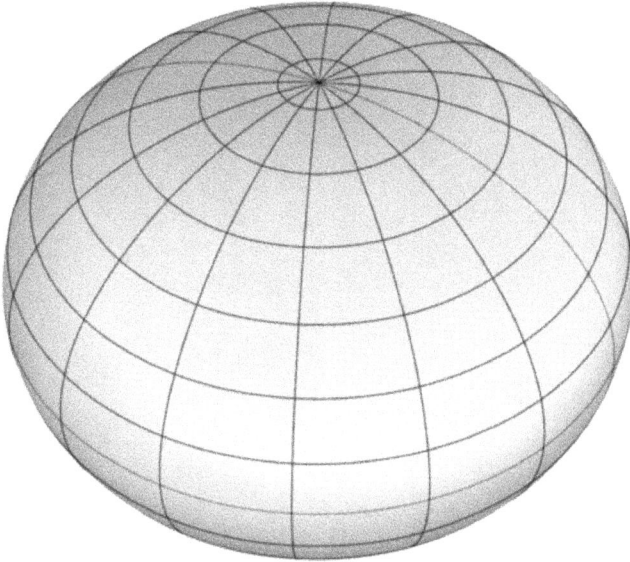

Figure 73: *An oblate spheroid, highly exaggerated relative to the actual Earth*

- Equatorial radius a (called *semimajor axis*), and polar radius b (called *semiminor axis*);
- a and eccentricity e ;
- a and flattening f .

Eccentricity and flattening are different ways of expressing how squashed the ellipsoid is. When flattening appears as one of the defining quantities in geodesy, generally it is expressed by its reciprocal. For example, in the WGS 84 spheroid used by today's GPS systems, the reciprocal of the flattening $1/f$ is set to be exactly 298.257223563.

The difference between a sphere and a reference ellipsoid for Earth is small, only about one part in 300. Historically, flattening was computed from grade measurements. Nowadays, geodetic networks and satellite geodesy are used. In practice, many reference ellipsoids have been developed over the centuries from different surveys. The flattening value varies slightly from one reference ellipsoid to another, reflecting local conditions and whether the reference ellipsoid is intended to model the entire Earth or only some portion of it.

A sphere has a single radius of curvature, which is simply the radius of the sphere. More complex surfaces have radii of curvature that vary over the surface. The radius of curvature describes the radius of the sphere that best approximates the surface at that point. Oblate ellipsoids have constant radius of

Figure 74: *A scale diagram of the oblateness of the 2003 IERS reference ellipsoid, with north at the top. The outer edge of the dark blue line is an ellipse with the same eccentricity as that of Earth. For comparison, the light blue circle within has a diameter equal to the ellipse's minor axis. The red curve represents the Karman line 100 km (62 mi) above sea level, while the yellow band denotes the altitude range of the ISS in low Earth orbit.*

curvature east to west along parallels, if a graticule is drawn on the surface, but varying curvature in any other direction. For an oblate ellipsoid, the polar radius of curvature r_p is larger than the equatorial

$$r_p = \frac{a^2}{b},$$

because the pole is flattened: the flatter the surface, the larger the sphere must be to approximate it. Conversely, the ellipsoid's north-south radius of curvature at the equator r_e is smaller than the polar

$$r_e = \frac{b^2}{a}$$

where a is the distance from the center of the ellipsoid to the equator (semi-major axis), and b is the distance from the center to the pole. (semi-minor axis)

More complicated shapes

The possibility that the Earth's equator is an ellipse rather than a circle and therefore that the ellipsoid is triaxial has been a matter of scientific controversy for many years. Modern technological developments have furnished new and rapid methods for data collection and, since the launch of *Sputnik 1*, orbital data have been used to investigate the theory of ellipticity.

A second theory, more complicated than triaxiality, proposed that observed long periodic orbital variations of the first Earth satellites indicate an additional depression at the south pole accompanied by a bulge of the same degree at the north pole. It is also contended that the northern middle latitudes were slightly flattened and the southern middle latitudes bulged in a similar amount. This concept suggested a slightly pear-shaped Earth and was the subject of much public discussion.Wikipedia:Citation needed Modern geodesy tends to retain the ellipsoid of revolution as a reference ellipsoid and treat triaxiality and pear shape as a part of the geoid figure: they are represented by the spherical harmonic coefficients C_{22}, S_{22} and C_{30} , respectively, corresponding to degree and order numbers 2.2 for the triaxiality and 3.0 for the pear shape.

Geoid

It was stated earlier that measurements are made on the apparent or topographic surface of the Earth and it has just been explained that computations are performed on an ellipsoid. One other surface is involved in geodetic measurement: the geoid. In geodetic surveying, the computation of the geodetic coordinates of points is commonly performed on a reference ellipsoid closely approximating the size and shape of the Earth in the area of the survey. The actual measurements made on the surface of the Earth with certain instruments are however referred to the geoid. The ellipsoid is a mathematically defined regular surface with specific dimensions. The geoid, on the other hand, co-incides with that surface to which the oceans would conform over the entire Earth if free to adjust to the combined effect of the Earth's mass attraction (gravitation) and the centrifugal force of the Earth's rotation. As a result of the uneven distribution of the Earth's mass, the geoidal surface is irregular and, since the ellipsoid is a regular surface, the separations between the two, referred to as geoid undulations, geoid heights, or geoid separations, will be irregular as well.

The geoid is a surface along which the gravity potential is everywhere equal and to which the direction of gravity is always perpendicular (see equipotential surface). The latter is particularly important because optical instruments containing gravity-reference leveling devices are commonly used to make geodetic measurements. When properly adjusted, the vertical axis of the instrument coincides with the direction of gravity and is, therefore, perpendicular to the geoid. The angle between the plumb line which is perpendicular to the geoid (sometimes called "the vertical") and the perpendicular to the ellipsoid (sometimes called "the ellipsoidal normal") is defined as the deflection of the vertical. It has two components: an east-west and a north-south component.[258]

Earth rotation and Earth's interior

Determining the exact figure of the Earth is not only a geodetic operation or a task of geometry, but is also related to geophysics. Without any idea of the Earth's interior, we can state a "constant density" of 5.515 g/cm^3 and, according to theoretical arguments (see Leonhard Euler, Albert Wangerin, etc.), such a body rotating like the Earth would have a flattening of 1:230.

In fact, the measured flattening is 1:298.25, which is closer to a sphere and a strong argument that the Earth's core is *very compact*. Therefore, the density must be a function of the depth, ranging from 2.6 g/cm^3 at the surface (rock density of granite, etc.), up to 13 g/cm^3 within the inner core, see Structure of the Earth.

Global and regional gravity field

Also with implications for the physical exploration of the Earth's interior is the gravitational field, which can be measured very accurately at the surface and remotely by satellites. True vertical generally does not correspond to theoretical vertical (deflection ranges up to 50") because topography and all *geological masses* disturb the gravitational field. Therefore, the gross structure of the earth's crust and mantle can be determined by geodetic-geophysical models of the subsurface.

Volume

Earth is an ellipsoid, so its volume is V = 4/3πabc, where a, b and c are its three orthogonal principal semi-axes, or radii. Using the parameters of the WGS84 ellipsoid of revolution, an oblate spheroid where a = b > c, a = b = 6,378.137 km and c = 6,357.7523142km, V = $1.0832073198 \times 10^{12}$ km^3 $\approx$ 1.08321×10^{12} km^3 (2.5988×10^{11} cu mi).

Notes and references

- Guy Bomford, *Geodesy*, Oxford 1962 and 1880.
- Guy Bomford, *Determination of the European geoid by means of vertical deflections*. Rpt of Comm. 14, IUGG 10th Gen. Ass., Rome 1954.
- Karl Ledersteger and Gottfried Gerstbach, *Die horizontale Isostasie / Das isostatische Geoid 31. Ordnung*. Geowissenschaftliche Mitteilungen Band 5, TU Wien 1975.
- Helmut Moritz and Bernhard Hofmann, *Physical Geodesy*. Springer, Wien & New York 2005.
- *Geodesy for the Layman*, Defense Mapping Agency, St. Louis, 1983.

External links

- Reference Ellipsoids (PCI Geomatics)[259]
- Reference Ellipsoids (ScanEx)[260]
- Changes in earth shape due to climate changes[261]
- Jos Leys "The shape of Planet Earth"[262]

Chemical composition

Abundance of the chemical elements

The **abundance of the chemical elements** is a measure of the occurrence of the chemical elements relative to all other elements in a given environment. Abundance is measured in one of three ways: by the mass-fraction (the same as weight fraction); by the mole-fraction (fraction of atoms by numerical count, or sometimes fraction of molecules in gases); or by the volume-fraction. Volume-fraction is a common abundance measure in mixed gases such as planetary atmospheres, and is similar in value to molecular mole-fraction for gas mixtures at relatively low densities and pressures, and ideal gas mixtures. Most abundance values in this article are given as mass-fractions.

For example, the abundance of oxygen in pure water can be measured in two ways: the *mass fraction* is about 89%, because that is the fraction of water's mass which is oxygen. However, the *mole-fraction* is 33.3333...% because only 1 atom of 3 in water, H_2O, is oxygen. As another example, looking at the *mass-fraction* abundance of hydrogen and helium in both the Universe as a whole and in the atmospheres of gas-giant planets such as Jupiter, it is 74% for hydrogen and 23–25% for helium; while the *(atomic) mole-fraction* for hydrogen is 92%, and for helium is 8%, in these environments. Changing the given environment to Jupiter's outer atmosphere, where hydrogen is diatomic while helium is not, changes the *molecular* mole-fraction (fraction of total gas molecules), as well as the fraction of atmosphere by volume, of hydrogen to about 86%, and of helium to 13%.[263]

The abundance of chemical elements in the universe is dominated by the large amounts of hydrogen and helium which were produced in the Big Bang. Remaining elements, making up only about 2% of the universe, were largely produced by supernovae and certain red giant stars. Lithium, beryllium and boron are rare because although they are produced by nuclear fusion, they are then destroyed by other reactions in the stars. The elements from carbon to iron are relatively more common in the universe because of the ease of making them in supernova nucleosynthesis. Elements of higher atomic number than iron (element 26) become progressively more rare in the universe, because they increasingly absorb stellar energy in being produced. Elements with even atomic numbers are generally more common than their neighbors in the periodic table, also due to favorable energetics of formation.

The abundance of elements in the Sun and outer planets is similar to that in the universe. Due to solar heating, the elements of Earth and the inner rocky planets of the Solar System have undergone an additional depletion of volatile hydrogen, helium, neon, nitrogen, and carbon (which volatilizes as methane). The crust, mantle, and core of the Earth show evidence of chemical segregation plus some sequestration by density. Lighter silicates of aluminum are found in the crust, with more magnesium silicate in the mantle, while metallic iron and nickel compose the core. The abundance of elements in specialized environments, such as atmospheres, or oceans, or the human body, are primarily a product of chemical interactions with the medium in which they reside.

Universe

**Ten most common elements in the Milky
Way Galaxy estimated spectroscopically**

Z	Element	Mass fraction (ppm)
1	Hydrogen	739,000
2	Helium	240,000
8	Oxygen	10,400
6	Carbon	4,600
10	Neon	1,340
26	Iron	1,090
7	Nitrogen	960
14	Silicon	650
12	Magnesium	580
16	Sulfur	440

The elements – that is, ordinary (baryonic) matter made of protons, neutrons, and electrons, are only a small part of the content of the Universe. Cosmological observations suggest that only 4.6% of the universe's energy (including the mass contributed by energy, $E = mc^2 \leftrightarrow m = E / c^2$) comprises the visible baryonic matter that constitutes stars, planets, and living beings. The rest is thought to be made up of dark energy (68%) and dark matter (27%).[264] These are forms of matter and energy believed to exist on the basis of scientific theory and observational deductions, but they have not been directly observed and their nature is not well understood.

Most standard (baryonic) matter is found in intergalactic gas, stars, and interstellar clouds, in the form of atoms or ions (plasma), although it can be found in degenerate forms in extreme astrophysical settings, such as the high densities inside white dwarfs and neutron stars.

Hydrogen is the most abundant element in the Universe; helium is second. However, after this, the rank of abundance does not continue to correspond to the atomic number; oxygen has abundance rank 3, but atomic number 8. All others are substantially less common.

The abundance of the lightest elements is well predicted by the standard cosmological model, since they were mostly produced shortly (i.e., within a few hundred seconds) after the Big Bang, in a process known as Big Bang nucleosynthesis. Heavier elements were mostly produced much later, inside of stars.

Hydrogen and helium are estimated to make up roughly 74% and 24% of all baryonic matter in the universe respectively. Despite comprising only a very small fraction of the universe, the remaining "heavy elements" can greatly influence astronomical phenomena. Only about 2% (by mass) of the Milky Way galaxy's disk is composed of heavy elements.

These other elements are generated by stellar processes. In astronomy, a "metal" is any element other than hydrogen or helium. This distinction is significant because hydrogen and helium are the only elements that were produced in significant quantities in the Big Bang. Thus, the metallicity of a galaxy or other object is an indication of stellar activity, after the Big Bang.

In general, elements up to iron are made in large stars in the process of becoming supernovae. Iron-56 is particularly common, since it is the most stable element that can easily be made from alpha particles (being a product of decay of radioactive nickel-56, ultimately made from 14 helium nuclei). Elements heavier than iron are made in energy-absorbing processes in large stars, and their abundance in the universe (and on Earth) generally decreases with increasing atomic number.

File:Nucleosynthesis periodic table.svg

Periodic table showing the cosmogenic origin of each element.

Solar system

Most abundant nuclides

in the Solar System

Nuclide	A	Mass fraction in parts per million	Atom fraction in parts per million
Hydrogen-1	1	705,700	909,964
Helium-4	4	275,200	88,714
Oxygen-16	16	9,592	477
Carbon-12	12	3,032	326
Nitrogen-14	14	1,105	102
Neon-20	20	1,548	100
Other nuclides:		3,879	149
Silicon-28	28	653	30
Magnesium-24	24	513	28
Iron-56	56	1,169	27
Sulfur-32	32	396	16
Helium-3	3	35	15
Hydrogen-2	2	23	15
Neon-22	22	208	12
Magnesium-26	26	79	4
Carbon-13	13	37	4
Magnesium-25	25	69	4

Aluminium-27	27	58	3
Argon-36	36	77	3
Calcium-40	40	60	2
Sodium-23	23	33	2
Iron-54	54	72	2
Silicon-29	29	34	2
Nickel-58	58	49	1
Silicon-30	30	23	1
Iron-57	57	28	1

The following graph (note log scale) shows abundance of elements in the Solar System. The table shows the twelve most common elements in our galaxy (estimated spectroscopically), as measured in parts per million, by mass. Nearby galaxies that have evolved along similar lines have a corresponding enrichment of elements heavier than hydrogen and helium. The more distant galaxies are being viewed as they appeared in the past, so their abundances of elements appear closer to the primordial mixture. Since physical laws and processes are uniform throughout the universe, however, it is expected that these galaxies will likewise have evolved similar abundances of elements.

The abundance of elements is in keeping with their origin from the Big Bang and nucleosynthesis in a number of progenitor supernova stars. Very abundant hydrogen and helium are products of the Big Bang, while the next three elements are rare since they had little time to form in the Big Bang and are not made in stars (they are, however, produced in small quantities by breakup of heavier elements in interstellar dust, as a result of impact by cosmic rays).

Beginning with carbon, elements have been produced in stars by buildup from alpha particles (helium nuclei), resulting in an alternatingly larger abundance of elements with even atomic numbers (these are also more stable). The effect of odd-numbered chemical elements generally being more rare in the universe was empirically noticed in 1914, and is known as the Oddo-Harkins rule.

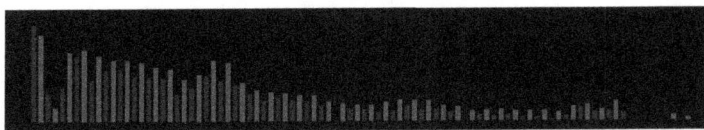

File:Elements_abundance-bars.svg

Estimated abundances of the chemical elements in the Solar System (logarithmic scale).

Relation to nuclear binding energy

Loose correlations have been observed between estimated elemental abundances in the universe and the nuclear binding energy curve. Roughly speaking, the relative stability of various atomic nuclides has exerted a strong influence on the relative abundance of elements formed in the Big Bang, and during the development of the universe thereafter. See the article about nucleosynthesis for the explanation on how certain nuclear fusion processes in stars (such as carbon burning, etc.) create the elements heavier than hydrogen and helium.

A further observed peculiarity is the jagged alternation between relative abundance and scarcity of adjacent atomic numbers in the elemental abundance curve, and a similar pattern of energy levels in the nuclear binding energy curve. This alternation is caused by the higher relative binding energy (corresponding to relative stability) of even atomic numbers compared with odd atomic numbers and is explained by the Pauli Exclusion Principle. The semi-empirical mass formula (SEMF), also called *Weizsäcker's formula* or the *Bethe-Weizsäcker mass formula*, gives a theoretical explanation of the overall shape of the curve of nuclear binding energy.

Earth

The Earth formed from the same cloud of matter that formed the Sun, but the planets acquired different compositions during the formation and evolution of the solar system. In turn, the natural history of the Earth caused parts of this planet to have differing concentrations of the elements.

The mass of the Earth is approximately 5.98×10^{24} kg. In bulk, by mass, it is composed mostly of iron (32.1%), oxygen (30.1%), silicon (15.1%), magnesium (13.9%), sulfur (2.9%), nickel (1.8%), calcium (1.5%), and aluminium (1.4%); with the remaining 1.2% consisting of trace amounts of other elements.

The bulk composition of the Earth by elemental-mass is roughly similar to the gross composition of the solar system, with the major differences being that Earth is missing a great deal of the volatile elements hydrogen, helium, neon, and nitrogen, as well as carbon which has been lost as volatile hydrocarbons. The remaining elemental composition is roughly typical of the "rocky" inner planets, which formed in the thermal zone where solar heat drove volatile compounds into space. The Earth retains oxygen as the second-largest component of its mass (and largest atomic-fraction), mainly from this element being retained in silicate minerals which have a very high melting point and low vapor pressure.

Estimated abundances of chemical elements in the Earth.[265] The right two columns give the fraction of the mass in parts per million (ppm) and the fraction by number of atoms in parts per billion (ppb).

Atomic Number	Name	Symbol	Mass fraction (ppm)	Atomic fraction (ppb)
8	oxygen	O	297000	482,000,000
12	magnesium	Mg	154000	164,000,000
14	silicon	Si	161000	150,000,000
26	iron	Fe	319000	148,000,000
13	aluminum	Al	15900	15,300,000
20	calcium	Ca	17100	11,100,000
28	nickel	Ni	18220	8,010,000
1	hydrogen	H	260	6,700,000
16	sulfur	S	6350	5,150,000
24	chromium	Cr	4700	2,300,000
11	sodium	Na	1800	2,000,000
6	carbon	C	730	1,600,000
15	phosphorus	P	1210	1,020,000
25	manganese	Mn	1700	800,000
22	titanium	Ti	810	440,000
27	cobalt	Co	880	390,000
19	potassium	K	160	110,000
17	chlorine	Cl	76	56,000
23	vanadium	V	105	53,600
7	nitrogen	N	25	46,000
29	copper	Cu	60	25,000
30	zinc	Zn	40	16,000
9	fluorine	F	10	14,000
21	scandium	Sc	11	6,300
3	lithium	Li	1.10	4,100
38	strontium	Sr	13	3,900
32	germanium	Ge	7.00	2,500
40	zirconium	Zr	7.10	2,000
31	gallium	Ga	3.00	1,000
34	selenium	Se	2.70	890
56	barium	Ba	4.50	850

39	yttrium	Y	2.90	850
33	arsenic	As	1.70	590
5	boron	B	0.20	480
42	molybdenum	Mo	1.70	460
44	ruthenium	Ru	1.30	330
78	platinum	Pt	1.90	250
46	palladium	Pd	1.00	240
58	cerium	Ce	1.13	210
60	neodymium	Nd	0.84	150
4	beryllium	Be	0.05	140
41	niobium	Nb	0.44	120
76	osmium	Os	0.90	120
77	iridium	Ir	0.90	120
37	rubidium	Rb	0.40	120
35	bromine	Br	0.30	97
57	lanthanum	La	0.44	82
66	dysprosium	Dy	0.46	74
64	gadolinium	Gd	0.37	61
52	tellurium	Te	0.30	61
45	rhodium	Rh	0.24	61
50	tin	Sn	0.25	55
62	samarium	Sm	0.27	47
68	erbium	Er	0.30	47
70	ytterbium	Yb	0.30	45
59	praseodymium	Pr	0.17	31
82	lead	Pb	0.23	29
72	hafnium	Hf	0.19	28
74	tungsten	W	0.17	24
79	gold	Au	0.16	21
48	cadmium	Cd	0.08	18
63	europium	Eu	0.10	17
67	holmium	Ho	0.10	16
47	silver	Ag	0.05	12
65	terbium	Tb	0.07	11
51	antimony	Sb	0.05	11
75	rhenium	Re	0.08	10
53	iodine	I	0.05	10

69	thulium	Tm	0.05	7
55	cesium	Cs	0.04	7
71	lutetium	Lu	0.05	7
90	thorium	Th	0.06	6
73	tantalum	Ta	0.03	4
80	mercury	Hg	0.02	3
92	uranium	U	0.02	2
49	indium	In	0.01	2
81	thallium	Tl	0.01	2
83	bismuth	Bi	0.01	1

Crust

The mass-abundance of the nine most abundant elements in the Earth's crust is approximately: oxygen 46%, silicon 28%, aluminum 8.2%, iron 5.6%, calcium 4.2%, sodium 2.5%, magnesium 2.4%, potassium 2.0%, and titanium 0.61%. Other elements occur at less than 0.15%. For a complete list, see abundance of elements in Earth's crust.

The graph at right illustrates the relative atomic-abundance of the chemical elements in Earth's upper continental crust— the part that is relatively accessible for measurements and estimation.

Many of the elements shown in the graph are classified into (partially overlapping) categories:

1. rock-forming elements (major elements in green field, and minor elements in light green field);
2. rare earth elements (lanthanides, La-Lu, and Y; labeled in blue);
3. major industrial metals (global production $> \sim 3 \times 10^7$ kg/year; labeled in red);
4. precious metals (labeled in purple);
5. the nine rarest "metals" — the six platinum group elements plus Au, Re, and Te (a metalloid) — in the yellow field. These are rare in the crust from being soluble in iron and thus concentrated in the Earth's core.

Note that there are two breaks where the unstable (radioactive) elements technetium (atomic number: 43) and promethium (atomic number: 61) would be. These elements are surrounded by stable elements, yet both have relatively short half lives ($\sim$ 4 million years and $\sim$ 18 years respectively). These are thus extremely rare, since any primordial initial fractions of these in pre-Solar System materials have long since decayed and disappeared. These two elements are now only produced naturally through the spontaneous fission of

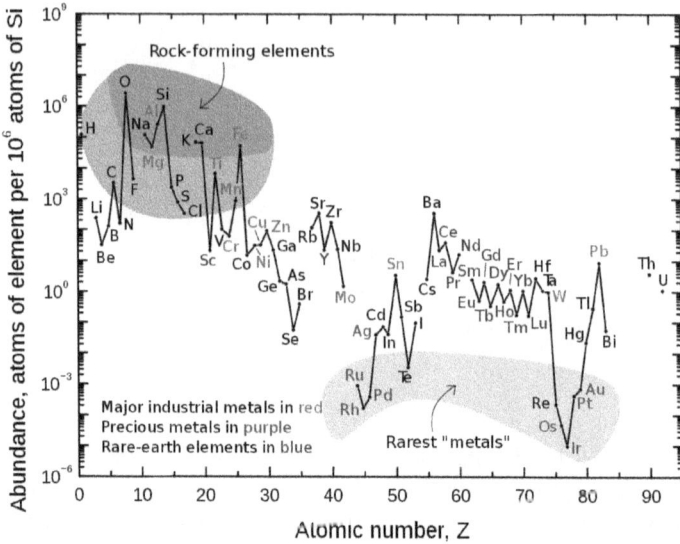

Figure 75: *Abundance (atom fraction) of the chemical elements in Earth's upper continental crust as a function of atomic number. The rarest elements in the crust (shown in yellow) are rare due to a combination of factors: all but 1 of them are the densest siderophiles (iron-loving) elements in the Goldschmidt classification of elements, meaning they have a tendency to mix well with iron, depleting them by being relocated deeper into the Earth's core. Their abundance in meteoroids is higher. Additionally, tellurium has been depleted from the crust due to formation of volatile hydrides.*

very heavy radioactive elements (for example, uranium, thorium, or the trace amounts of plutonium that exist in uranium ores), or by the interaction of certain other elements with cosmic rays. Both technetium and promethium have been identified spectroscopically in the atmospheres of stars, where they are produced by ongoing nucleosynthetic processes.

There are also breaks in the abundance graph where the six noble gases would be, since they are not chemically bound in the Earth's crust, and they are only generated by decay chains from radioactive elements in the crust, and are therefore extremely rare there.

The eight naturally occurring very rare, highly radioactive elements (polonium, astatine, francium, radium, actinium, protactinium, neptunium, and plutonium) are not included, since any of these elements that were present at the formation of the Earth have decayed away eons ago, and their quantity today is negligible and is only produced from the radioactive decay of uranium and thorium.

Oxygen and silicon are notably the most common elements in the crust. On Earth and in rocky planets in general, silicon and oxygen are far more common than their cosmic abundance. The reason is that they combine with each other to form silicate minerals. In this way, they are the lightest of all of the two-percent "astronomical metals" (i.e., non-hydrogen and helium elements) to form a solid that is refractory to the Sun's heat, and thus cannot boil away into space. All elements lighter than oxygen have been removed from the crust in this way.

Rare-earth elements

"Rare" earth elements is a historical misnomer. The persistence of the term reflects unfamiliarity rather than true rarity. The more abundant rare earth elements are similarly concentrated in the crust compared to commonplace industrial metals such as chromium, nickel, copper, zinc, molybdenum, tin, tungsten, or lead. The two least abundant rare earth elements (thulium and lutetium) are nearly 200 times more common than gold. However, in contrast to the ordinary base and precious metals, rare earth elements have very little tendency to become concentrated in exploitable ore deposits. Consequently, most of the world's supply of rare earth elements comes from only a handful of sources. Furthermore, the rare earth metals are all quite chemically similar to each other, and they are thus quite difficult to separate into quantities of the pure elements.

Differences in abundances of individual rare earth elements in the upper continental crust of the Earth represent the superposition of two effects, one nuclear and one geochemical. First, the rare earth elements with even atomic numbers ($_{58}$Ce, $_{60}$Nd, ...) have greater cosmic and terrestrial abundances than the adjacent rare earth elements with odd atomic numbers ($_{57}$La, $_{59}$Pr, ...). Second, the lighter rare earth elements are more incompatible (because they have larger ionic radii) and therefore more strongly concentrated in the continental crust than the heavier rare earth elements. In most rare earth ore deposits, the first four rare earth elements – lanthanum, cerium, praseodymium, and neodymium – constitute 80% to 99% of the total amount of rare earth metal that can be found in the ore.

Mantle

The mass-abundance of the eight most abundant elements in the Earth's mantle (see main article above) is approximately: oxygen 45%, magnesium 23%, silicon 22%, iron 5.8%, calcium 2.3%, aluminum 2.2%, sodium 0.3%, potassium 0.3%.

The mantle differs in elemental composition from the crust in having a great deal more magnesium and significantly more iron, while having much less aluminum and sodium.

Core

Due to mass segregation, the core of the Earth is believed to be primarily composed of iron (88.8%), with smaller amounts of nickel (5.8%), sulfur (4.5%), and less than 1% trace elements.

Ocean

The most abundant elements in the ocean by proportion of mass in percent are oxygen (85.84), hydrogen (10.82), chlorine (1.94), sodium (1.08), magnesium (0.1292), sulfur (0.091), calcium (0.04), potassium (0.04), bromine (0.0067), carbon (0.0028), and boron (0.00043).

Atmosphere

The order of elements by volume-fraction (which is approximately molecular mole-fraction) in the atmosphere is nitrogen (78.1%), oxygen (20.9%), argon (0.96%), followed by (in uncertain order) carbon and hydrogen because water vapor and carbon dioxide, which represent most of these two elements in the air, are variable components. Sulfur, phosphorus, and all other elements are present in significantly lower proportions.

According to the abundance curve graph (above right), argon, a significant if not major component of the atmosphere, does not appear in the crust at all. This is because the atmosphere has a far smaller mass than the crust, so argon remaining in the crust contributes little to mass-fraction there, while at the same time buildup of argon in the atmosphere has become large enough to be significant.

Urban soils

For a complete list of the abundance of elements in urban soils, see Abundances of the elements (data page)#Urban soils.

Human body

Elemental abundance in the human body

Element	Proportion (by mass)
Oxygen	65
Carbon	18
Hydrogen	10
Nitrogen	3
Calcium	1.5
Phosphorus	1.2
Potassium	0.2
Sulfur	0.2
Chlorine	0.2
Sodium	0.1
Magnesium	0.05
Iron	< 0.05
Cobalt	< 0.05
Copper	< 0.05
Zinc	< 0.05
Iodine	< 0.05
Selenium	< 0.01

By mass, human cells consist of 65–90% water (H_2O), and a significant portion of the remainder is composed of carbon-containing organic molecules. Oxygen therefore contributes a majority of a human body's mass, followed by carbon. Almost 99% of the mass of the human body is made up of six elements: oxygen, carbon, hydrogen, nitrogen, calcium, and phosphorus. The next 0.75% is made up of the next five elements: potassium, sulfur, chlorine, sodium, and magnesium. Only 17 elements are known for certain to be necessary to human life, with one additional element (fluorine) thought to be helpful for tooth enamel strength. A few more trace elements may play some role in the health of mammals. Boron and silicon are notably necessary for plants but have uncertain roles in animals. The elements aluminium and silicon, although very common in the earth's crust, are conspicuously rare in the human body.[266]

Below is a periodic table highlighting nutritional elements.

Nutritional elements in the periodic table

H																	He
Li	Be											B	C	N	O	F	Ne
Na	Mg											Al	Si	P	S	Cl	Ar
K	Ca	Sc		Ti	V	Cr	Mn	Fe	Co	Ni	Cu	Zn	Ga	Ge	As	Se	Br	Kr
Rb	Sr	Y		Zr	Nb	Mo	Tc	Ru	Rh	Pd	Ag	Cd	In	Sn	Sb	Te	I	Xe
Cs	Ba	La	*	Hf	Ta	W	Re	Os	Ir	Pt	Au	Hg	Tl	Pb	Bi	Po	At	Rn
Fr	Ra	Ac	**	Rf	Db	Sg	Bh	Hs	Mt	Ds	Rg	Cn	Nh	Fl	Mc	Lv	Ts	Og

	*	Ce	Pr	Nd	Pm	Sm	Eu	Gd	Tb	Dy	Ho	Er	Tm	Yb	Lu
	**	Th	Pa	U	Np	Pu	Am	Cm	Bk	Cf	Es	Fm	Md	No	Lr

The four basic organic elements

Quantity elements

Essential trace elements

Deemed essential trace element by U.S., not by European Union

Suggested function from deprivation effects or active metabolic handling, but no clearly-identified bio-chemical function in humans

Limited circumstantial evidence for trace benefits or biological action in mammals

No evidence for biological action in mammals, but essential in some lower organisms.
(In the case of lanthanum, the definition of an essential nutrient as being indispensable and irreplaceable is not completely applicable due to the extreme similarity of the lanthanides. Thus Ce, Pr, and Nd may be substituted for La without ill effects for organisms using La, and the smaller Sm, Eu, and Gd may also be similarly substituted but cause slower growth.)

References

Notations

- "Rare Earth Elements—Critical Resources for High Technology | USGS Fact Sheet 087-02"[267]. *geopubs.wr.usgs.gov.*
- "Imagine the Universe! Dictionary"[268]. 3 December 2003. Archived from the original[269] on 3 December 2003.

External links

- List of elements in order of abundance in the Earth's crust[270] (only correct for the twenty most common elements)
- Cosmic abundance of the elements and nucleosynthesis[271]
- WebElements.com[272] Lists of elemental abundances for the Universe, Sun, meteorites, Earth, ocean, streamwater, etc.

Internal structure

Structure of the Earth

The internal **structure of the Earth** is layered in spherical shells: an outer silicate solid crust, a highly viscous asthenosphere and mantle, a liquid outer core that is much less viscous than the mantle, and a solid inner core. Scientific understanding of the internal structure of the Earth is based on observations of topography and bathymetry, observations of rock in outcrop, samples brought to the surface from greater depths by volcanoes or volcanic activity, analysis of the seismic waves that pass through the Earth, measurements of the gravitational and magnetic fields of the Earth, and experiments with crystalline solids at pressures and temperatures characteristic of the Earth's deep interior.

Mass

The force exerted by Earth's gravity can be used to calculate its mass. Astronomers can also calculate Earth's mass by observing the motion of orbiting satellites. Earth's average density can be determined through gravimetric experiments, which have historically involved pendulums.

The mass of Earth is about 6×10^{24} kg.[273]

Structure

<templatestyles src="Multiple_image/styles.css" />

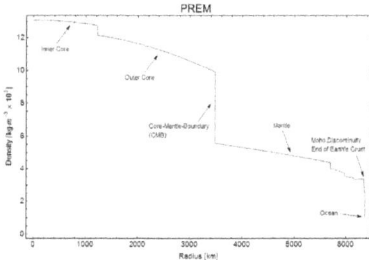

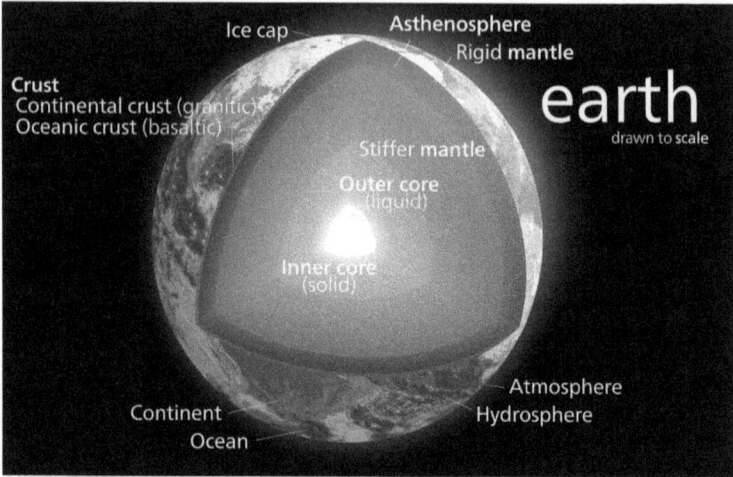

Figure 76: *Structure of the Earth*

Earth's radial density distribution according to the preliminary reference earth model (PREM).

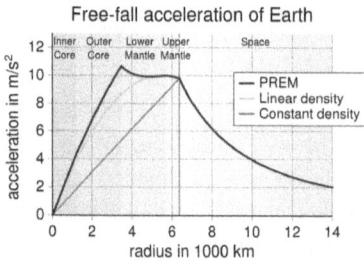

Earth's gravity according to the preliminary reference earth model (PREM). Comparison to approximations using constant and linear density for Earth's interior.

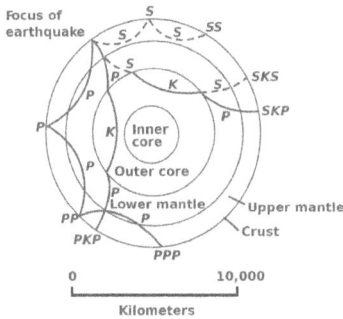

Mapping the interior of Earth with earthquake waves.

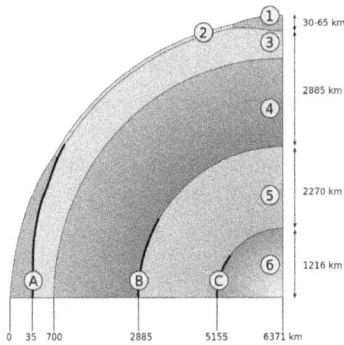

Schematic view of the interior of Earth. 1. continental crust – 2. oceanic crust – 3. upper mantle – 4. lower mantle – 5. outer core – 6. inner core – A: Mohorovičić discontinuity – B: Gutenberg Discontinuity – C: Lehmann–Bullen discontinuity.

The structure of Earth can be defined in two ways: by mechanical properties such as rheology, or chemically. Mechanically, it can be divided into lithosphere, asthenosphere, mesospheric mantle, outer core, and the inner core. Chemically, Earth can be divided into the crust, upper mantle, lower mantle, outer core, and inner core. The geologic component layers of Earth-Wikipedia:Verifiability are at the following depths below the surface:

Depth		Layer
Kilometres	Miles	
0–60	0–37	Lithosphere (locally varies between 5 and 200 km)
0–35	0–22	... Crust (locally varies between 5 and 70 km)
35–60	22–37	... Uppermost part of mantle
35–2,890	22–1,790	Mantle
210-270	130-168	... Upper mesosphere (upper mantle)
660–2,890	410–1,790	... Lower mesosphere (lower mantle)
2,890–5,150	1,790–3,160	Outer core
5,150–6,360	3,160–3,954	Inner core

The layering of Earth has been inferred indirectly using the time of travel of refracted and reflected seismic waves created by earthquakes. The core does not allow shear waves to pass through it, while the speed of travel (seismic velocity) is different in other layers. The changes in seismic velocity between different layers causes refraction owing to Snell's law, like light bending as it passes through a prism. Likewise, reflections are caused by a large increase in seismic velocity and are similar to light reflecting from a mirror.

Crust

The crust ranges from 5–70 kilometres (3.1–43.5 mi) in depth and is the outermost layer. The thin parts are the oceanic crust, which underlie the ocean basins (5–10 km) and are composed of dense (mafic) iron magnesium silicate igneous rocks, like basalt. The thicker crust is continental crust, which is less dense and composed of (felsic) sodium potassium aluminium silicate rocks, like granite. The rocks of the crust fall into two major categories – sial and sima (Suess,1831–1914). It is estimated that sima starts about 11 km below the Conrad discontinuity (a second order discontinuity). The uppermost mantle together with the crust constitutes the lithosphere. The crust-mantle boundary occurs as two physically different events. First, there is a discontinuity in the seismic velocity, which is most commonly known as the Mohorovičić discontinuity or Moho. The cause of the Moho is thought to be a change in rock composition from rocks containing plagioclase feldspar (above) to rocks that contain no feldspars (below). Second, in oceanic crust, there is a chemical discontinuity between ultramafic cumulates and tectonized harzburgites, which has been observed from deep parts of the oceanic crust that have been obducted onto the continental crust and preserved as ophiolite sequences.

Many rocks now making up Earth's crust formed less than 100 million (1×10^8) years ago; however, the oldest known mineral grains are about 4.4

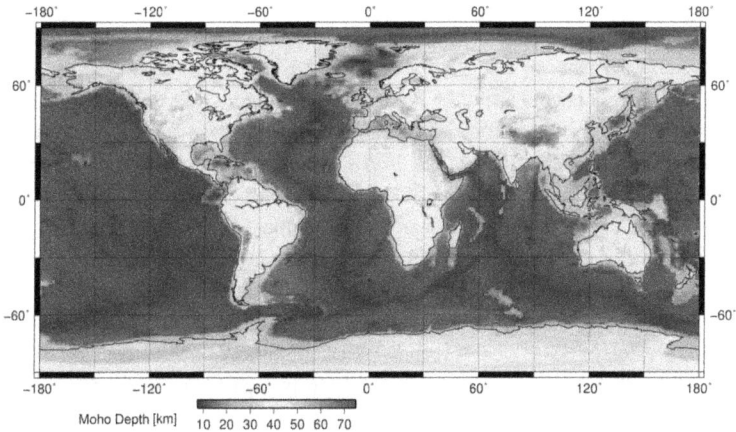

Figure 77: *World map showing the position of the Moho.*

billion (4.4×10^9) years old, indicating that Earth has had a solid crust for at least 4.4 billion years.[274]

Mantle

Earth's mantle extends to a depth of 2,890 km, making it the thickest layer of Earth. The mantle is divided into upper and lower mantle. The upper and lower mantle are separated by the transition zone. The lowest part of the mantle next to the core-mantle boundary is known as the D" (pronounced dee-double-prime) layer. The pressure at the bottom of the mantle is ≈140 GPa (1.4 Matm). The mantle is composed of silicate rocks that are rich in iron and magnesium relative to the overlying crust. Although solid, the high temperatures within the mantle cause the silicate material to be sufficiently ductile that it can flow on very long timescales. Convection of the mantle is expressed at the surface through the motions of tectonic plates. As there is intense and increasing pressure as one travels deeper into the mantle, the lower part of the mantle flows less easily than does the upper mantle (chemical changes within the mantle may also be important). The viscosity of the mantle ranges between 10^{21} and 10^{24} Pa·s, depending on depth.[275] In comparison, the viscosity of water is approximately 10^{-3} Pa·s and that of pitch is 10^7 Pa·s. The source of heat that drives plate tectonics is the primordial heat left over from the planet's formation as well as the radioactive decay of uranium, thorium, and potassium in Earth's crust and mantle.[276]

Core

The average density of Earth is 5.515 g/cm^3. Because the average density of surface material is only around 3.0 g/cm^3, we must conclude that denser materials exist within Earth's core. This result has been known since the Schiehallion experiment, performed in the 1770s. Charles Hutton in his 1778 report concluded that the mean density of the Earth must be about $\frac{9}{5}$ that of surface rock, concluding that the interior of the Earth must be metallic. Hutton estimated this metallic portion to occupy some 65% of the diameter of the Earth. Hutton's estimate on the mean density of the Earth was still about 20% too low, at 4.5 g/cm^3 Henry Cavendish in his torsion balance experiment of 1798 found a value of 5.45 g/cm^3, within 1% of the modern value. Seismic measurements show that the core is divided into two parts, a "solid" inner core with a radius of ≈1,220 km and a liquid outer core extending beyond it to a radius of ≈3,400 km. The densities are between 9,900 and 12,200 kg/m^3 in the outer core and 12,600–13,000 kg/m^3 in the inner core.

The inner core was discovered in 1936 by Inge Lehmann and is generally believed to be composed primarily of iron and some nickel. Since this layer is able to transmit shear waves (transverse seismic waves), it must be solid. Experimental evidence has at times been critical of crystal models of the core. Other experimental studies show a discrepancy under high pressure: diamond anvil (static) studies at core pressures yield melting temperatures that are approximately 2000 K below those from shock laser (dynamic) studies. The laser studies create plasma, and the results are suggestive that constraining inner core conditions will depend on whether the inner core is a solid or is a plasma with the density of a solid. This is an area of active research.

In early stages of Earth's formation about 4.6 billion years ago, melting would have caused denser substances to sink toward the center in a process called planetary differentiation (see also the iron catastrophe), while less-dense materials would have migrated to the crust. The core is thus believed to largely be composed of iron (80%), along with nickel and one or more light elements, whereas other dense elements, such as lead and uranium, either are too rare to be significant or tend to bind to lighter elements and thus remain in the crust (see felsic materials). Some have argued that the inner core may be in the form of a single iron crystal.

Under laboratory conditions a sample of iron–nickel alloy was subjected to the corelike pressures by gripping it in a vise between 2 diamond tips (diamond anvil cell), and then heating to approximately 4000 K. The sample was observed with x-rays, and strongly supported the theory that Earth's inner core was made of giant crystals running north to south.[277]

The liquid outer core surrounds the inner core and is believed to be composed of iron mixed with nickel and trace amounts of lighter elements.

Recent speculation suggests that the innermost part of the core is enriched in gold, platinum and other siderophile elements.

The matter that comprises Earth is connected in fundamental ways to matter of certain chondrite meteorites, and to matter of outer portion of the Sun. There is good reason to believe that Earth is, in the main, like a chondrite meteorite. Beginning as early as 1940, scientists, including Francis Birch, built geophysics upon the premise that Earth is like ordinary chondrites, the most common type of meteorite observed impacting Earth, while totally ignoring another, albeit less abundant type, called enstatite chondrites. The principal difference between the two meteorite types is that enstatite chondrites formed under circumstances of extremely limited available oxygen, leading to certain normally oxyphile elements existing either partially or wholly in the alloy portion that corresponds to the core of Earth.

Dynamo theory suggests that convection in the outer core, combined with the Coriolis effect, gives rise to Earth's magnetic field. The solid inner core is too hot to hold a permanent magnetic field (see Curie temperature) but probably acts to stabilize the magnetic field generated by the liquid outer core. The average magnetic field strength in Earth's outer core is estimated to be 25 Gauss (2.5 mT), 50 times stronger than the magnetic field at the surface.[278]

Recent evidence has suggested that the inner core of Earth may rotate slightly faster than the rest of the planet; however, more recent studies in 2011Wikipedia:Avoid weasel words found this hypothesis to be inconclusive. Options remain for the core which may be oscillatory in nature or a chaotic system.Wikipedia:Citation needed In August 2005 a team of geophysicists announced in the journal *Science* that, according to their estimates, Earth's inner core rotates approximately 0.3 to 0.5 degrees per year faster relative to the rotation of the surface.[279]

The current scientific explanation for Earth's temperature gradient is a combination of heat left over from the planet's initial formation, decay of radioactive elements, and freezing of the inner core.

Further reading

<templatestyles src="Template:Refbegin/styles.css" />

- Drollette, Daniel (October 1996). "A Spinning Crystal Ball". *Scientific American*. **275** (4): 28–33.
- Kruglinski, Susan (June 2007). "Journey to the Center of the Earth"[280]. *Discover*. Retrieved 9 July 2016.

- Lehmann, I (1936). "Inner Earth". *Bur. Cent. Seismol. Int.* **14**: 3–31.
- Wegener, Alfred (1966). *The origin of continents and oceans.* New York: Dover Publications. ISBN 9780486617084.

External links

The Wikibook *Historical Geology* has a page on the topic of: ***Structure of the Earth***

Media related to Structure of the Earth at Wikimedia Commons

- Down To The Earth's Core (HD)[281] on YouTube
- The Earth's Core[282] on *In Our Time* at the BBC

Heat

Earth's internal heat budget

Earth's internal heat budget is fundamental to the thermal history of the Earth. The **flow of heat from Earth's interior to the surface** is estimated at 47 ± 2 terawatts (TW)[283] and comes from two main sources in roughly equal amounts: the *radiogenic heat* produced by the radioactive decay of isotopes in the mantle and crust, and the *primordial heat* left over from the formation of the Earth.

Earth's internal heat powers most geological processes[284] and drives plate tectonics. Despite its geological significance, this heat energy coming from Earth's interior is actually only 0.03% of Earth's total energy budget at the surface, which is dominated by 173,000 TW of incoming solar radiation. The insolation that eventually, after reflection, reaches the surface penetrates only several tens of centimeters on the daily cycle and only several tens of meters on the annual cycle. This renders solar radiation minimally relevant for internal processes.[285]

Heat and early estimate of Earth's age

Based on calculations of Earth's cooling rate, which assumed constant conductivity in the Earth's interior, in 1862 William Thomson (later made Lord Kelvin) estimated the age of the Earth at 98 million years,[286] which contrasts with the age of 4.5 billion years obtained in the 20th century by radiometric dating. As pointed out by John Perry in 1895 a variable conductivity in the Earth's interior could expand the computed age of the Earth to billions of years, as later confirmed by radiometric dating. Contrary to the usual representation of Kelvin's argument, the observed thermal gradient of the Earth's crust would not be explained by the addition of radioactivity as a heat source. More significantly, mantle convection alters how heat is transported within the Earth, invalidating Kelvin's assumption of purely conductive cooling.

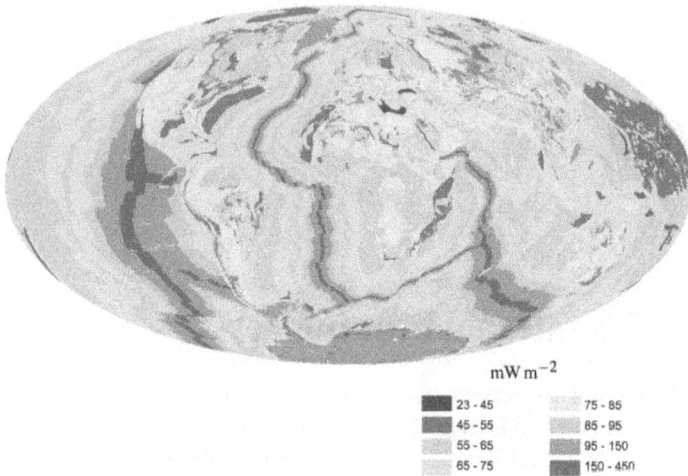

$$mW\,m^{-2}$$

▮	23 - 45	▦	75 - 85
▮	45 - 55	▦	85 - 95
▦	55 - 65	▦	95 - 150
▦	65 - 75	▮	150 - 450

Figure 78: *Global map of the flux of heat, in mW/m², from Earth's interior to the surface. The largest values of heat flux coincide with mid ocean ridges, and the smallest values of heat flux occur in stable continental interiors.*

Global internal heat flow

Estimates of the total heat flow from Earth's interior to surface span a range of 43 to 49 terawatts (TW) (a terawatt is 10^{12} watts).[287] One recent estimate is 47 TW, equivalent to an average heat flux of 91.6 mW/m², and is based on more than 38,000 measurements. The respective mean heat flows of continental and oceanic crust are 70.9 and 105.4 mW/m².

While the total internal Earth heat flow to the surface is well constrained, the relative contribution of the two main sources of Earth's heat, radiogenic and primordial heat, are highly uncertain because their direct measurement is difficult. Chemical and physical models give estimated ranges of 15–41 TW and 12–30 TW for radiogenic heat and primordial heat, respectively.

The structure of the Earth is a rigid outer crust that is composed of thicker continental crust and thinner oceanic crust, solid but plastically flowing mantle, a liquid outer core, and a solid inner core. The fluidity of a material is proportional to temperature; thus, the solid mantle can still flow on long time scales, as a function of its temperature and therefore as a function of the flow of Earth's internal heat. The mantle convects in response to heat escaping from Earth's interior, with hotter and more buoyant mantle rising and cooler, and

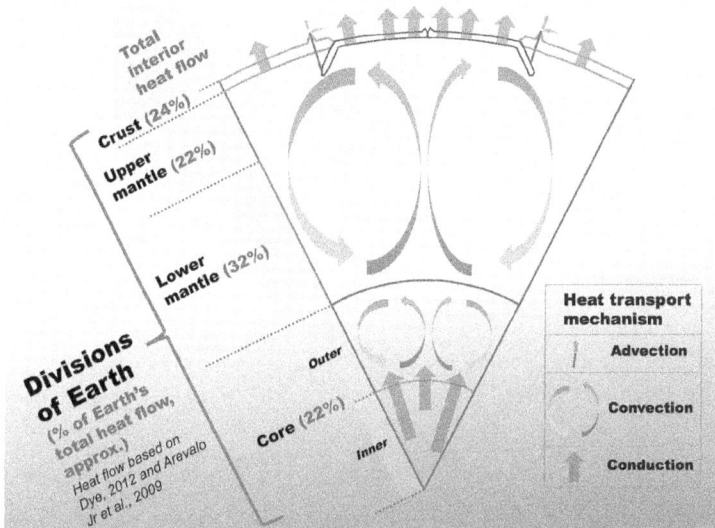

Figure 79: *Cross section of the Earth showing its main divisions and their approximate contributions to Earth's total internal heat flow to the surface, and the dominant heat transport mechanisms within the Earth.*

therefore denser, mantle sinking. This convective flow of the mantle drives the movement of Earth's lithospheric plates; thus, an additional reservoir of heat in the lower mantle is critical for the operation of plate tectonics and one possible source is an enrichment of radioactive elements in the lower mantle.[288]

Earth heat transport occurs by conduction, mantle convection, hydrothermal convection, and volcanic advection.[289] Earth's internal heat flow to the surface is thought to be 80% due to mantle convection, with the remaining heat mostly originating in the Earth's crust,[290] with about 1% due to volcanic activity, earthquakes, and mountain building. Thus, about 99% of Earth's internal heat loss at the surface is by conduction through the crust, and mantle convection is the dominant control on heat transport from deep within the Earth. Most of the heat flow from the thicker continental crust is attributed to internal radiogenic sources, in contrast the thinner oceanic crust has only 2% internal radiogenic heat. The remaining heat flow at the surface would be due to basal heating of the crust from mantle convection. Heat fluxes are negatively correlated with rock age, with the highest heat fluxes from the youngest rock at mid-ocean ridge spreading centers (zones of mantle upwelling), as observed in the global map of Earth heat flow.

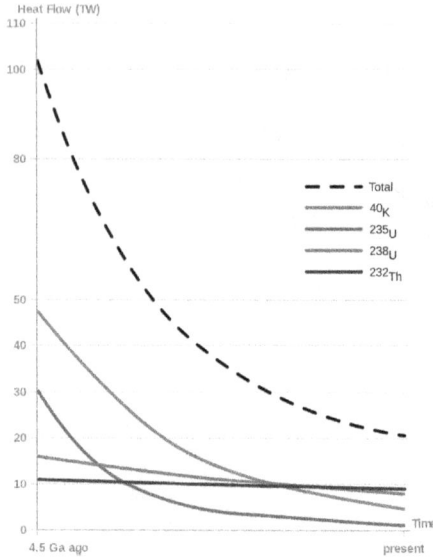

Figure 80: *The evolution of Earth's radiogenic heat flow over time.*

Radiogenic heat

The radioactive decay of elements in the Earth's mantle and crust results in production of daughter isotopes and release of geoneutrinos and heat energy, or
radiogenic heat. Four radioactive isotopes are responsible for the majority of
radiogenic heat because of their enrichment relative to other radioactive isotopes: uranium-238 (^{238}U), uranium-235 (^{235}U), thorium-232 (^{232}Th), and
potassium-40 (^{40}K).[291] Due to a lack of rock samples from below 200 km
depth, it is difficult to determine precisely the radiogenic heat throughout the
whole mantle, although some estimates are available. For the Earth's core,
geochemical studies indicate that it is unlikely to be a significant source of
radiogenic heat due to an expected low concentration of radioactive elements
partitioning into iron. Radiogenic heat production in the mantle is linked to
the structure of mantle convection, a topic of much debate, and it is thought
that the mantle may either have a layered structure with a higher concentration
of radioactive heat-producing elements in the lower mantle, or small reservoirs
enriched in radioactive elements dispersed throughout the whole mantle.

An estimate of the present-day major heat-producing isotopes

Isotope	Heat release W/kg isotope	Half-life years	Mean mantle concentration kg isotope/kg mantle	Heat release W/kg mantle
^{238}U	94.6×10^{-6}	4.47×10^9	30.8×10^{-9}	2.91×10^{-12}
^{235}U	569×10^{-6}	0.704×10^9	0.22×10^{-9}	0.125×10^{-12}
^{232}Th	26.4×10^{-6}	14.0×10^9	124×10^{-9}	3.27×10^{-12}
^{40}K	29.2×10^{-6}	1.25×10^9	36.9×10^{-9}	1.08×10^{-12}

Geoneutrino detectors can detect the decay of ^{238}U and ^{232}Th and thus allow estimation of their contribution to the present radiogenic heat budget, while ^{235}U and ^{40}K is not detectable. Regardless, ^{40}K is estimated to contribute 4 TW of heating.[292] However, due to the short half-lives the decay of ^{235}U and ^{40}K contributed a large fraction of radiogenic heat flux to the early Earth, which was also much hotter than at present. Initial results from measuring the geoneutrino products of radioactive decay from within the Earth, a proxy for radiogenic heat, yielded a new estimate of half of the total Earth internal heat source being radiogenic, and this is consistent with previous estimates.[293]

Primordial heat

Primordial heat is the heat lost by the Earth as it continues to cool from its original formation, and this is in contrast to its still actively-produced radiogenic heat. The Earth core's heat flow—heat leaving the core and flowing into the overlying mantle—is thought to be due to primordial heat, and is estimated at 5–15 TW.[294] Estimates of mantle primordial heat loss range between 7 and 15 TW, which is calculated as the remainder of heat after removal of core heat flow and bulk-Earth radiogenic heat production from the observed surface heat flow.

The early formation of the Earth's dense core could have caused superheating and rapid heat loss, and the heat loss rate would slow once the mantle solidified. Heat flow from the core is necessary for maintaining the convecting outer core and the geodynamo and Earth's magnetic field, therefore primordial heat from the core enabled Earth's atmosphere and thus helped retain Earth's liquid water.

Heat flow and plate tectonics

Controversy over the exact nature of mantle convection makes the linked evolution of Earth's heat budget and the dynamics and structure of the mantle difficult to unravel. There is evidence that the processes of plate tectonics were not

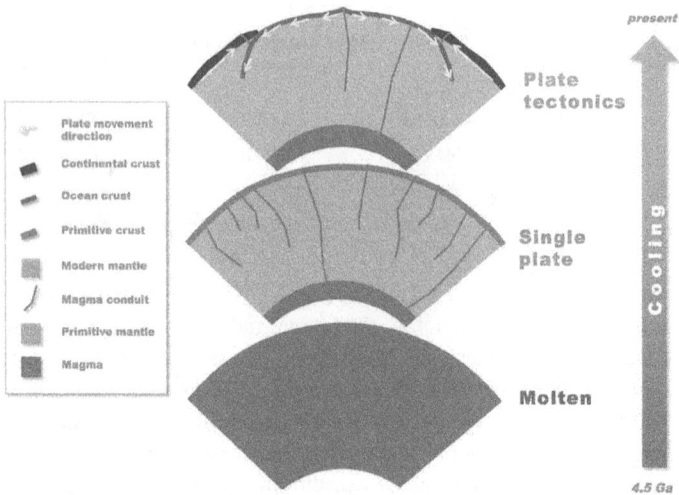

Figure 81: *Earth's tectonic evolution over time from a molten state at 4.5 Ga, to a single-plate lithosphere, to modern plate tectonics sometime between 3.2 Ga[295] and 1.0 Ga.[296]*

active in the Earth before 3.2 billion years ago, and that early Earth's internal heat loss could have been dominated by advection via heat-pipe volcanism.[297] Terrestrial bodies with lower heat flows, such as the Moon and Mars, conduct their internal heat through a single lithospheric plate, and higher heat flows, such as on Jupiter's moon Io, result in advective heat transport via enhanced volcanism, while the active plate tectonics of Earth occur with an intermediate heat flow and a convecting mantle.

Tectonic plates

Plate tectonics

<indicator name="pp-default"> 🔒 </indicator>

Plate tectonics (from the Late Latin *tectonicus*, from the Greek: τεκτονικός "pertaining to building")[298] is a scientific theory describing the large-scale motion of seven large plates and the movements of a larger number of smaller plates of the Earth's lithosphere, since tectonic processes began on Earth between 3 and 3.5 billion years ago. The model builds on the concept of continental drift, an idea developed during the first decades of the 20th century. The geoscientific community accepted plate-tectonic theory after seafloor spreading was validated in the late 1950s and early 1960s.

The lithosphere, which is the rigid outermost shell of a planet (the crust and upper mantle), is broken into tectonic plates. The Earth's lithosphere is composed of seven or eight major plates (depending on how they are defined) and many minor plates. Where the plates meet, their relative motion determines the type of boundary: convergent, divergent, or transform. Earthquakes, volcanic activity, mountain-building, and oceanic trench formation occur along these plate boundaries (or faults). The relative movement of the plates typically ranges from zero to 100 mm annually.[299]

Tectonic plates are composed of oceanic lithosphere and thicker continental lithosphere, each topped by its own kind of crust. Along convergent boundaries, subduction, or one plate moving under another, carries the lower one down into the mantle; the material lost is roughly balanced by the formation of new (oceanic) crust along divergent margins by seafloor spreading. In this way, the total surface of the lithosphere remains the same. This prediction of plate tectonics is also referred to as the conveyor belt principle. Earlier theories, since disproven, proposed gradual shrinking (contraction) or gradual expansion of the globe.[300]

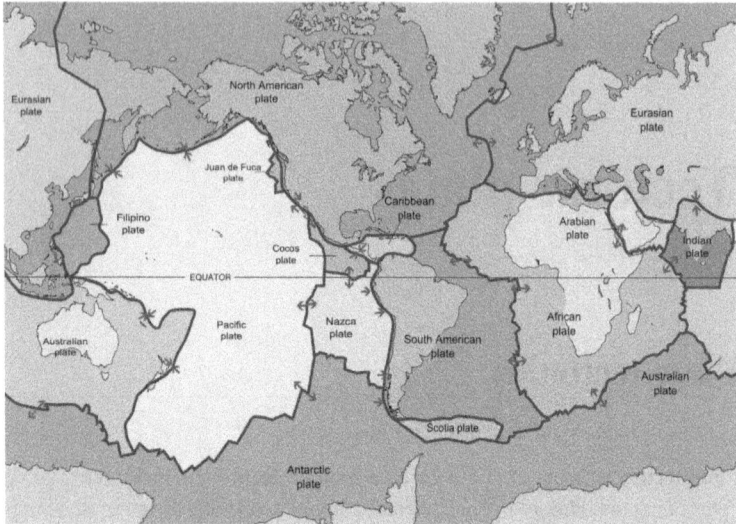

Figure 82: *The tectonic plates of the world were mapped in the second half of the 20th century.*

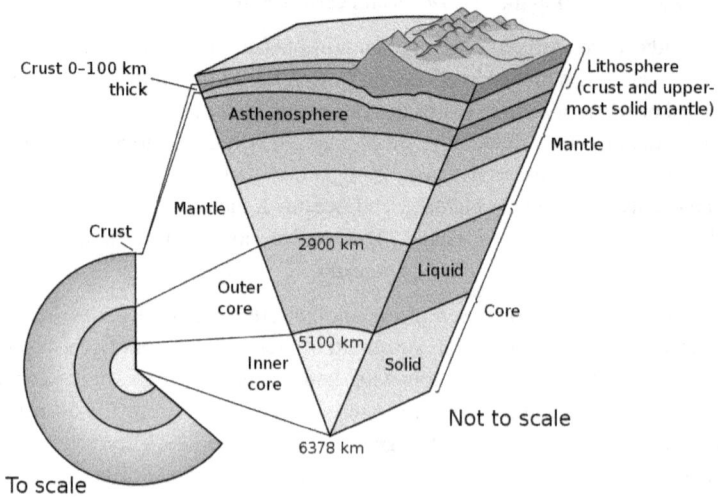

Figure 83: *Diagram of the internal layering of the Earth showing the lithosphere above the asthenosphere (not to scale)*

Tectonic plates are able to move because the Earth's lithosphere has greater mechanical strength than the underlying asthenosphere. Lateral density variations in the mantle result in convection; that is, the slow creeping motion of Earth's solid mantle. Plate movement is thought to be driven by a combination of the motion of the seafloor away from spreading ridges due to variations in topography (the ridge is a topographic high) and density changes in the crust (density increases as newly formed crust cools and moves away from the ridge). At subduction zones the relatively cold, dense crust is "pulled" or sinks down into the mantle over the downward convecting limb of a mantle cell. Another explanation lies in the different forces generated by tidal forces of the Sun and Moon. The relative importance of each of these factors and their relationship to each other is unclear, and still the subject of much debate.

Key principles

The outer layers of the Earth are divided into the lithosphere and asthenosphere. The division is based on differences in mechanical properties and in the method for the transfer of heat. The lithosphere is cooler and more rigid, while the asthenosphere is hotter and flows more easily. In terms of heat transfer, the lithosphere loses heat by conduction, whereas the asthenosphere also transfers heat by convection and has a nearly adiabatic temperature gradient. This division should not be confused with the *chemical* subdivision of these same layers into the mantle (comprising both the asthenosphere and the mantle portion of the lithosphere) and the crust: a given piece of mantle may be part of the lithosphere or the asthenosphere at different times depending on its temperature and pressure.

The key principle of plate tectonics is that the lithosphere exists as separate and distinct *tectonic plates*, which ride on the fluid-like (visco-elastic solid) asthenosphere. Plate motions range up to a typical 10–40 mm/year (Mid-Atlantic Ridge; about as fast as fingernails grow), to about 160 mm/year (Nazca Plate; about as fast as hair grows).[301] The driving mechanism behind this movement is described below.

Tectonic lithosphere plates consist of lithospheric mantle overlain by one or two types of crustal material: oceanic crust (in older texts called *sima* from silicon and magnesium) and continental crust (*sial* from silicon and aluminium). Average oceanic lithosphere is typically 100 km (62 mi) thick;[302] its thickness is a function of its age: as time passes, it conductively cools and subjacent cooling mantle is added to its base. Because it is formed at mid-ocean ridges and spreads outwards, its thickness is therefore a function of its distance from the mid-ocean ridge where it was formed. For a typical distance that oceanic lithosphere must travel before being subducted, the thickness varies from about 6 km (4 mi) thick at mid-ocean ridges to greater than 100 km (62 mi) at subduction zones; for shorter or longer distances, the subduction zone (and therefore also the mean) thickness becomes smaller or larger, respectively.[303] Continental lithosphere is typically about 200 km thick, though this varies considerably between basins, mountain ranges, and stable cratonic interiors of continents.

The location where two plates meet is called a *plate boundary*. Plate boundaries are commonly associated with geological events such as earthquakes and the creation of topographic features such as mountains, volcanoes, mid-ocean ridges, and oceanic trenches. The majority of the world's active volcanoes occur along plate boundaries, with the Pacific Plate's Ring of Fire being the most active and widely known today. These boundaries are discussed in further detail below. Some volcanoes occur in the interiors of plates, and these have been variously attributed to internal plate deformation[304] and to mantle plumes.

As explained above, tectonic plates may include continental crust or oceanic crust, and most plates contain both. For example, the African Plate includes the continent and parts of the floor of the Atlantic and Indian Oceans. The distinction between oceanic crust and continental crust is based on their modes of formation. Oceanic crust is formed at sea-floor spreading centers, and continental crust is formed through arc volcanism and accretion of terranes through tectonic processes, though some of these terranes may contain ophiolite sequences, which are pieces of oceanic crust considered to be part of the continent when they exit the standard cycle of formation and spreading centers and subduction beneath continents. Oceanic crust is also denser than continental crust owing to their different compositions. Oceanic crust is denser because it has less silicon and more heavier elements ("mafic") than continental crust ("felsic").[305] As a result of this density stratification, oceanic crust generally lies below sea level (for example most of the Pacific Plate), while continental crust buoyantly projects above sea level (see the page isostasy for explanation of this principle).

Types of plate boundaries

Three types of plate boundaries exist,[306] with a fourth, mixed type, characterized by the way the plates move relative to each other. They are associated with different types of surface phenomena. The different types of plate boundaries are:

1. *Transform boundaries (Conservative)* occur where two lithospheric plates slide, or perhaps more accurately, grind past each other along transform faults, where plates are neither created nor destroyed. The relative motion

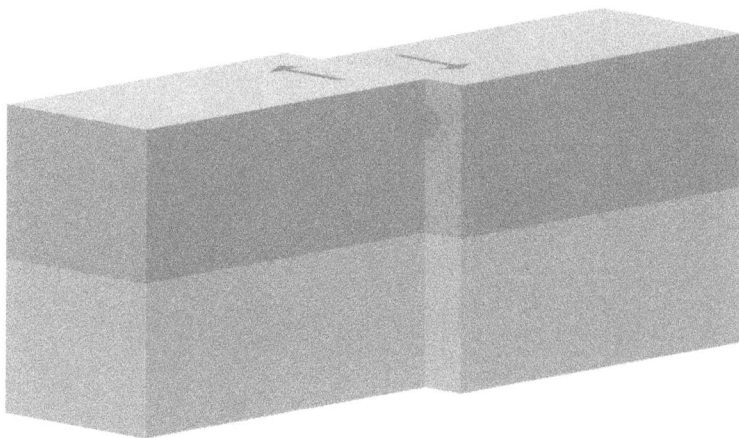

Figure 84: *Transform boundary*

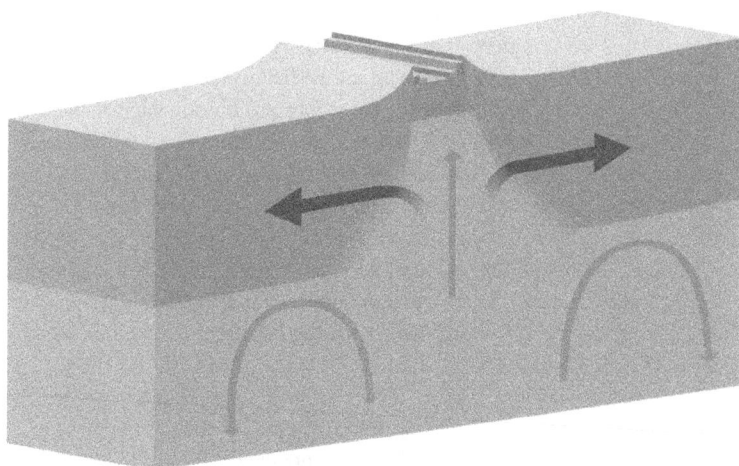

Figure 85: *Divergent boundary*

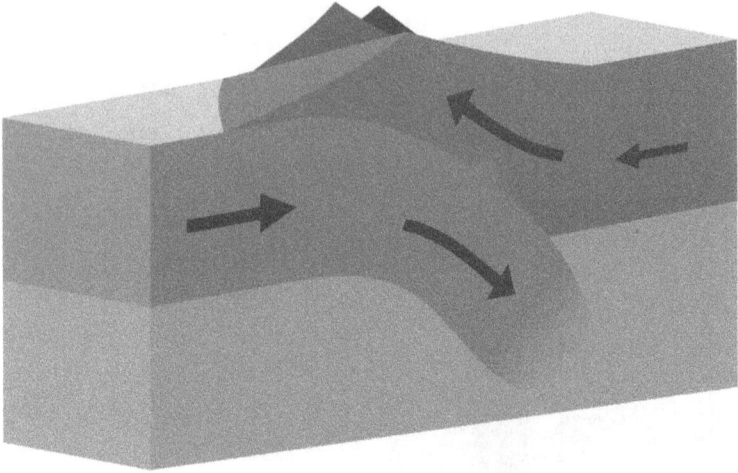

Figure 86: *Convergent boundary*

of the two plates is either sinistral (left side toward the observer) or dextral (right side toward the observer). Transform faults occur across a spreading center. Strong earthquakes can occur along a fault. The San Andreas Fault in California is an example of a transform boundary exhibiting dextral motion.

2. *Divergent boundaries (Constructive)* occur where two plates slide apart from each other. At zones of ocean-to-ocean rifting, divergent boundaries form by seafloor spreading, allowing for the formation of new ocean basin. As the ocean plate splits, the ridge forms at the spreading center, the ocean basin expands, and finally, the plate area increases causing many small volcanoes and/or shallow earthquakes. At zones of continent-to-continent rifting, divergent boundaries may cause new ocean basin to form as the continent splits, spreads, the central rift collapses, and ocean fills the basin. Active zones of mid-ocean ridges (e.g., the Mid-Atlantic Ridge and East Pacific Rise), and continent-to-continent rifting (such as Africa's East African Rift and Valley and the Red Sea), are examples of divergent boundaries.

3. *Convergent boundaries (Destructive)* (or *active margins*) occur where two plates slide toward each other to form either a subduction zone (one plate moving underneath the other) or a continental collision. At zones of ocean-to-continent subduction (e.g. the Andes mountain range in South America, and the Cascade Mountains in Western United States), the dense oceanic lithosphere plunges beneath the less dense continent. Earthquakes trace the path of the downward-moving plate as it descends

into asthenosphere, a trench forms, and as the subducted plate is heated it releases volatiles, mostly water from hydrous minerals, into the surrounding mantle. The addition of water lowers the melting point of the mantle material above the subducting slab, causing it to melt. The magma that results typically leads to volcanism. At zones of ocean-to-ocean subduction (e.g. Aleutian islands, Mariana Islands, and the Japanese island arc), older, cooler, denser crust slips beneath less dense crust. This motion causes earthquakes and a deep trench to form in an arc shape. The upper mantle of the subducted plate then heats and magma rises to form curving chains of volcanic islands. Deep marine trenches are typically associated with subduction zones, and the basins that develop along the active boundary are often called "foreland basins". Closure of ocean basins can occur at continent-to-continent boundaries (e.g., Himalayas and Alps): collision between masses of granitic continental lithosphere; neither mass is subducted; plate edges are compressed, folded, uplifted.

4. *Plate boundary zones* occur where the effects of the interactions are unclear, and the boundaries, usually occurring along a broad belt, are not well defined and may show various types of movements in different episodes.

Driving forces of plate motion

It has generally been accepted that tectonic plates are able to move because of the relative density of oceanic lithosphere and the relative weakness of the asthenosphere. Dissipation of heat from the mantle is acknowledged to be the original source of the energy required to drive plate tectonics through convection or large scale upwelling and doming. The current view, though still a matter of some debate, asserts that as a consequence, a powerful source of plate motion is generated due to the excess density of the oceanic lithosphere sinking in subduction zones. When the new crust forms at mid-ocean ridges, this oceanic lithosphere is initially less dense than the underlying asthenosphere, but it becomes denser with age as it conductively cools and thickens. The greater density of old lithosphere relative to the underlying asthenosphere allows it to sink into the deep mantle at subduction zones, providing most of the driving force for plate movement. The weakness of the asthenosphere allows the tectonic plates to move easily towards a subduction zone. Although subduction is thought to be the strongest force driving plate motions, it cannot be the only force since there are plates such as the North American Plate which are moving, yet are nowhere being subducted. The same is true for the enormous Eurasian Plate. The sources of plate motion are a matter of intensive research and discussion among scientists. One of the main points is that the kinematic pattern of the movement itself should be separated clearly

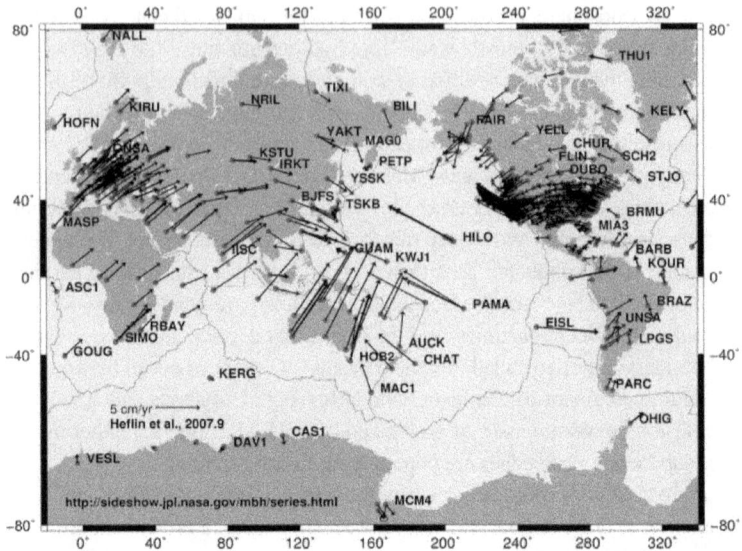

Figure 87: *Plate motion based on Global Positioning System (GPS) satellite data from NASA JPL*[307]. *The vectors show direction and magnitude of motion.*

from the possible geodynamic mechanism that is invoked as the driving force of the observed movement, as some patterns may be explained by more than one mechanism.[308] In short, the driving forces advocated at the moment can be divided into three categories based on the relationship to the movement: mantle dynamics related, gravity related (mostly secondary forces), and earth rotation related.

Driving forces related to mantle dynamics

For much of the last quarter century, the leading theory of the driving force behind tectonic plate motions envisaged large scale convection currents in the upper mantle, which can be transmitted through the asthenosphere. This theory was launched by Arthur Holmes and some forerunners in the 1930s and was immediately recognized as the solution for the acceptance of the theory as originally discussed in the papers of Alfred Wegener in the early years of the century. However, despite its acceptance, it was long debated in the scientific community because the leading theory still envisaged a static Earth without moving continents up until the major breakthroughs of the early sixties.

Two- and three-dimensional imaging of Earth's interior (seismic tomography) shows a varying lateral density distribution throughout the mantle. Such density variations can be material (from rock chemistry), mineral (from variations in mineral structures), or thermal (through thermal expansion and contraction from heat energy). The manifestation of this varying lateral density is mantle convection from buoyancy forces.[309]

How mantle convection directly and indirectly relates to plate motion is a matter of ongoing study and discussion in geodynamics. Somehow, this energy must be transferred to the lithosphere for tectonic plates to move. There are essentially two main types of forces that are thought to influence plate motion: friction and gravity.

- Basal drag (friction): Plate motion driven by friction between the convection currents in the asthenosphere and the more rigid overlying lithosphere.
- Slab suction (gravity): Plate motion driven by local convection currents that exert a downward pull on plates in subduction zones at ocean trenches. Slab suction may occur in a geodynamic setting where basal tractions continue to act on the plate as it dives into the mantle (although perhaps to a greater extent acting on both the under and upper side of the slab).

Lately, the convection theory has been much debated, as modern techniques based on 3D seismic tomography still fail to recognize these predicted large scale convection cells; therefore, alternative views have been proposed:

In the theory of plume tectonics developed during the 1990s, a modified concept of mantle convection currents is used. It asserts that super plumes rise from the deeper mantle and are the drivers or substitutes of the major convection cells. These ideas, which find their roots in the early 1930s with the so-called "fixistic" ideas of the European and Russian Earth Science Schools, find resonance in the modern theories which envisage hot spots or mantle plumes which remain fixed and are overridden by oceanic and continental lithosphere plates over time and leave their traces in the geological record (though these phenomena are not invoked as real driving mechanisms, but rather as modulators). Modern theories that continue building on the older mantle doming concepts and see plate movements as a secondary phenomena are beyond the scope of this article and are discussed elsewhere (for example on the Plume tectonics article).

Another theory is that the mantle flows neither in cells nor large plumes but rather as a series of channels just below the Earth's crust, which then provide basal friction to the lithosphere. This theory, called "surge tectonics", became quite popular in geophysics and geodynamics during the 1980s and 1990s.[310] Recent research, based on three-dimensional computer modeling, suggests that plate geometry is governed by a feedback between mantle convection patterns and the strength of the lithosphere.

Driving forces related to gravity

Forces related to gravity are usually invoked as secondary phenomena within the framework of a more general driving mechanism such as the various forms of mantle dynamics described above.

Gravitational sliding away from a spreading ridge: According to many authors, plate motion is driven by the higher elevation of plates at ocean ridges.[311] As oceanic lithosphere is formed at spreading ridges from hot mantle material, it gradually cools and thickens with age (and thus adds distance from the ridge). Cool oceanic lithosphere is significantly denser than the hot mantle material from which it is derived and so with increasing thickness it gradually subsides into the mantle to compensate the greater load. The result is a slight lateral incline with increased distance from the ridge axis.

This force is regarded as a secondary force and is often referred to as "ridge push". This is a misnomer as nothing is "pushing" horizontally and tensional features are dominant along ridges. It is more accurate to refer to this mechanism as gravitational sliding as variable topography across the totality of the plate can vary considerably and the topography of spreading ridges is only the most prominent feature. Other mechanisms generating this gravitational secondary force include flexural bulging of the lithosphere before it dives underneath an adjacent plate which produces a clear topographical feature that can offset, or at least affect, the influence of topographical ocean ridges, and mantle plumes and hot spots, which are postulated to impinge on the underside of tectonic plates.

Slab-pull: Current scientific opinion is that the asthenosphere is insufficiently competent or rigid to directly cause motion by friction along the base of the lithosphere. Slab pull is therefore most widely thought to be the greatest force acting on the plates. In this current understanding, plate motion is mostly driven by the weight of cold, dense plates sinking into the mantle at trenches.[312] Recent models indicate that trench suction plays an important role as well. However, the fact that the North American Plate is nowhere being subducted, although it is in motion, presents a problem. The same holds for the African, Eurasian, and Antarctic plates.

Gravitational sliding away from mantle doming: According to older theories, one of the driving mechanisms of the plates is the existence of large scale asthenosphere/mantle domes which cause the gravitational sliding of lithosphere plates away from them. This gravitational sliding represents a secondary phenomenon of this basically vertically oriented mechanism. This can act on various scales, from the small scale of one island arc up to the larger scale of an entire ocean basin.[313]

Driving forces related to Earth rotation

Alfred Wegener, being a meteorologist, had proposed tidal forces and centrifugal forces as the main driving mechanisms behind continental drift; however, these forces were considered far too small to cause continental motion as the concept was of continents plowing through oceanic crust. Therefore, Wegener later changed his position and asserted that convection currents are the main driving force of plate tectonics in the last edition of his book in 1929.

However, in the plate tectonics context (accepted since the seafloor spreading proposals of Heezen, Hess, Dietz, Morley, Vine, and Matthews (see below) during the early 1960s), the oceanic crust is suggested to be in motion *with* the continents which caused the proposals related to Earth rotation to be reconsidered. In more recent literature, these driving forces are:

1. Tidal drag due to the gravitational force the Moon (and the Sun) exerts on the crust of the Earth
2. Global deformation of the geoid due to small displacements of the rotational pole with respect to the Earth's crust;
3. Other smaller deformation effects of the crust due to wobbles and spin movements of the Earth rotation on a smaller time scale.

Forces that are small and generally negligible are:

1. The Coriolis force
2. The centrifugal force, which is treated as a slight modification of gravity:[249]

For these mechanisms to be overall valid, systematic relationships should exist all over the globe between the orientation and kinematics of deformation and the geographical latitudinal and longitudinal grid of the Earth itself. Ironically, these systematic relations studies in the second half of the nineteenth century and the first half of the twentieth century underline exactly the opposite: that the plates had not moved in time, that the deformation grid was fixed with respect to the Earth equator and axis, and that gravitational driving forces were generally acting vertically and caused only local horizontal movements (the so-called pre-plate tectonic, "fixist theories"). Later studies (discussed below on this page), therefore, invoked many of the relationships recognized during this pre-plate tectonics period to support their theories (see the anticipations and reviews in the work of van Dijk and collaborators).[314]

Of the many forces discussed in this paragraph, tidal force is still highly debated and defended as a possible principal driving force of plate tectonics. The other forces are only used in global geodynamic models not using plate tectonics concepts (therefore beyond the discussions treated in this section) or proposed as minor modulations within the overall plate tectonics model.

In 1973, George W. Moore[315] of the USGS and R. C. Bostrom[316] presented evidence for a general westward drift of the Earth's lithosphere with respect to the mantle. He concluded that tidal forces (the tidal lag or "friction") caused by the Earth's rotation and the forces acting upon it by the Moon are a driving force for plate tectonics. As the Earth spins eastward beneath the moon, the moon's gravity ever so slightly pulls the Earth's surface layer back westward, just as proposed by Alfred Wegener (see above). In a more recent 2006 study,[317] scientists reviewed and advocated these earlier proposed ideas. It has also been suggested recently in Lovett (2006) that this observation may also explain why Venus and Mars have no plate tectonics, as Venus has no moon and Mars' moons are too small to have significant tidal effects on the planet. In a recent paper,[318] it was suggested that, on the other hand, it can easily be observed that many plates are moving north and eastward, and that the dominantly westward motion of the Pacific Ocean basins derives simply from the eastward bias of the Pacific spreading center (which is not a predicted manifestation of such lunar forces). In the same paper the authors admit, however, that relative to the lower mantle, there is a slight westward component in the motions of all the plates. They demonstrated though that the westward drift, seen only for the past 30 Ma, is attributed to the increased dominance of the steadily growing and accelerating Pacific plate. The debate is still open.

Relative significance of each driving force mechanism

The vector of a plate's motion is a function of all the forces acting on the plate; however, therein lies the problem regarding the degree to which each process contributes to the overall motion of each tectonic plate.

The diversity of geodynamic settings and the properties of each plate result from the impact of the various processes actively driving each individual plate. One method of dealing with this problem is to consider the relative rate at which each plate is moving as well as the evidence related to the significance of each process to the overall driving force on the plate.

One of the most significant correlations discovered to date is that lithospheric plates attached to downgoing (subducting) plates move much faster than plates not attached to subducting plates. The Pacific plate, for instance, is essentially surrounded by zones of subduction (the so-called Ring of Fire) and moves much faster than the plates of the Atlantic basin, which are attached (perhaps one could say 'welded') to adjacent continents instead of subducting plates. It is thus thought that forces associated with the downgoing plate (slab pull and slab suction) are the driving forces which determine the motion of plates, except for those plates which are not being subducted.[312] This view however has been contradicted by a recent study which found that the actual motions of the Pacific Plate and other plates associated with the East Pacific Rise do

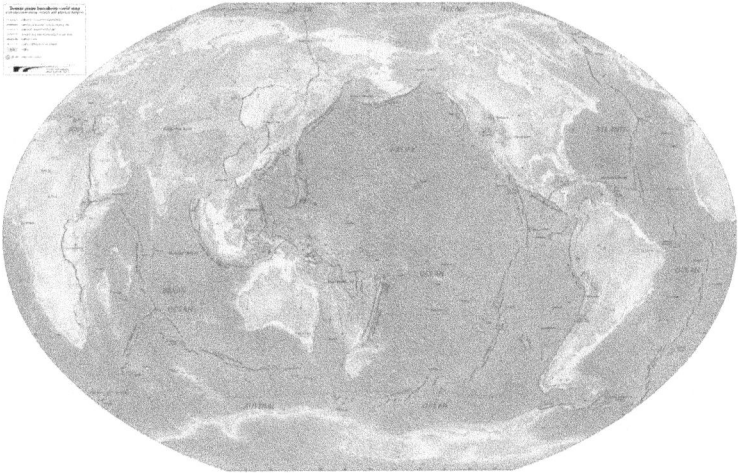

Figure 88: *Detailed map showing the tectonic plates with their movement vectors.*

not correlate mainly with either slab pull or slab push, but rather with a mantle convection upwelling whose horizontal spreading along the bases of the various plates drives them along via viscosity-related traction forces.[319] The driving forces of plate motion continue to be active subjects of on-going research within geophysics and tectonophysics.

Development of the theory

Summary

In line with other previous and contemporaneous proposals, in 1912 the meteorologist Alfred Wegener amply described what he called continental drift, expanded in his 1915 book *The Origin of Continents and Oceans*[320] and the scientific debate started that would end up fifty years later in the theory of plate tectonics.[321] Starting from the idea (also expressed by his forerunners) that the present continents once formed a single land mass (which was called Pangea later on) that drifted apart, thus releasing the continents from the Earth's mantle and likening them to "icebergs" of low density granite floating on a sea of denser basalt.[322] Supporting evidence for the idea came from the dovetailing outlines of South America's east coast and Africa's west coast, and from the matching of the rock formations along these edges. Confirmation of their previous contiguous nature also came from the fossil plants *Glossopteris* and *Gangamopteris*, and the therapsid or mammal-like reptile *Lystrosaurus*, all widely distributed over South America, Africa, Antarctica, India, and Australia. The evidence for such an erstwhile joining of these continents was patent to field geologists working in the southern hemisphere. The South African Alex du Toit put together a mass of such information in his 1937 publication *Our Wandering Continents*, and went further than Wegener in recognising the strong links between the Gondwana fragments.

But without detailed evidence and a force sufficient to drive the movement, the theory was not generally accepted: the Earth might have a solid crust and mantle and a liquid core, but there seemed to be no way that portions of the crust could move around. Distinguished scientists, such as Harold Jeffreys and Charles Schuchert, were outspoken critics of continental drift.

Despite much opposition, the view of continental drift gained support and a lively debate started between "drifters" or "mobilists" (proponents of the theory) and "fixists" (opponents). During the 1920s, 1930s and 1940s, the former reached important milestones proposing that convection currents might have driven the plate movements, and that spreading may have occurred below the sea within the oceanic crust. Concepts close to the elements now incorporated in plate tectonics were proposed by geophysicists and geologists (both fixists and mobilists) like Vening-Meinesz, Holmes, and Umbgrove.

One of the first pieces of geophysical evidence that was used to support the movement of lithospheric plates came from paleomagnetism. This is based on the fact that rocks of different ages show a variable magnetic field direction, evidenced by studies since the mid–nineteenth century. The magnetic north and south poles reverse through time, and, especially important in paleotectonic studies, the relative position of the magnetic north pole varies through time. Initially, during the first half of the twentieth century, the latter phenomenon was explained by introducing what was called "polar wander" (see apparent polar wander), i.e., it was assumed that the north pole location had been shifting through time. An alternative explanation, though, was that the continents had moved (shifted and rotated) relative to the north pole, and each continent, in fact, shows its own "polar wander path". During the late 1950s it was successfully shown on two occasions that these data could show the validity of continental drift: by Keith Runcorn in a paper in 1956,[323] and by Warren Carey in a symposium held in March 1956.[324]

The second piece of evidence in support of continental drift came during the late 1950s and early 60s from data on the bathymetry of the deep ocean floors and the nature of the oceanic crust such as magnetic properties and, more generally, with the development of marine geology[325] which gave evidence for the association of seafloor spreading along the mid-oceanic ridges and magnetic field reversals, published between 1959 and 1963 by Heezen, Dietz, Hess, Mason, Vine & Matthews, and Morley.[326]

Simultaneous advances in early seismic imaging techniques in and around Wadati–Benioff zones along the trenches bounding many continental margins, together with many other geophysical (e.g. gravimetric) and geological observations, showed how the oceanic crust could disappear into the mantle, providing the mechanism to balance the extension of the ocean basins with shortening along its margins.

All this evidence, both from the ocean floor and from the continental margins, made it clear around 1965 that continental drift was feasible and the theory of plate tectonics, which was defined in a series of papers between 1965 and 1967, was born, with all its extraordinary explanatory and predictive power. The theory revolutionized the Earth sciences, explaining a diverse range of geological phenomena and their implications in other studies such as paleogeography and paleobiology.

Continental drift

In the late 19th and early 20th centuries, geologists assumed that the Earth's major features were fixed, and that most geologic features such as basin development and mountain ranges could be explained by vertical crustal movement, described in what is called the geosynclinal theory. Generally, this was placed in the context of a contracting planet Earth due to heat loss in the course of a relatively short geological time.

It was observed as early as 1596 that the opposite coasts of the Atlantic Ocean—or, more precisely, the edges of the continental shelves—have similar shapes and seem to have once fitted together.[327]

Since that time many theories were proposed to explain this apparent complementarity, but the assumption of a solid Earth made these various proposals difficult to accept.[328]

The discovery of radioactivity and its associated heating properties in 1895 prompted a re-examination of the apparent age of the Earth.[329] This had previously been estimated by its cooling rate under the assumption that the Earth's surface radiated like a black body.[330] Those calculations had implied that, even if it started at red heat, the Earth would have dropped to its present temperature in a few tens of millions of years. Armed with the knowledge of a new heat source, scientists realized that the Earth would be much older, and that its core was still sufficiently hot to be liquid.

By 1915, after having published a first article in 1912,[331] Alfred Wegener was making serious arguments for the idea of continental drift in the first edition of *The Origin of Continents and Oceans*.[320] In that book (re-issued in four successive editions up to the final one in 1936), he noted how the east coast of South America and the west coast of Africa looked as if they were once

Figure 89: *Alfred Wegener in Greenland in the winter of 1912–13.*

attached. Wegener was not the first to note this (Abraham Ortelius, Antonio Snider-Pellegrini, Eduard Suess, Roberto Mantovani and Frank Bursley Taylor preceded him just to mention a few), but he was the first to marshal significant fossil and paleo-topographical and climatological evidence to support this simple observation (and was supported in this by researchers such as Alex du Toit). Furthermore, when the rock strata of the margins of separate continents are very similar it suggests that these rocks were formed in the same way, implying that they joined initially. For instance, parts of Scotland and Ireland contain rocks very similar to those found in Newfoundland and New Brunswick. Furthermore, the Caledonian Mountains of Europe and parts of the Appalachian Mountains of North America are very similar in structure and lithology.

However, his ideas were not taken seriously by many geologists, who pointed out that there was no apparent mechanism for continental drift. Specifically, they did not see how continental rock could plow through the much denser rock that makes up oceanic crust. Wegener could not explain the force that drove continental drift, and his vindication did not come until after his death in 1930.

Preliminary Determination of Epicenters
358,214 Events, 1963 - 1998

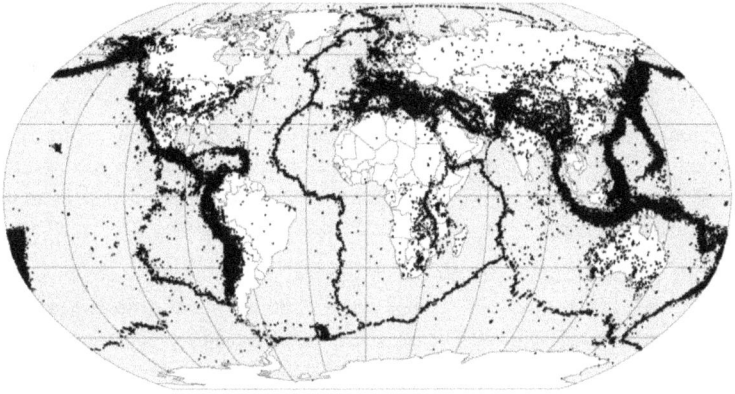

Figure 90: *Global earthquake epicenters, 1963–1998. Most earthquakes occur in narrow belts that correspond to the locations of lithospheric plate boundaries.*

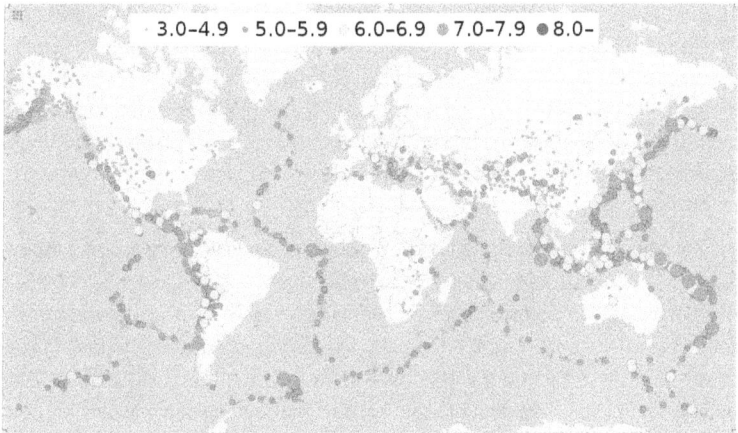

Figure 91: *Map of earthquakes in 2016*

Floating continents, paleomagnetism, and seismicity zones

As it was observed early that although granite existed on continents, seafloor seemed to be composed of denser basalt, the prevailing concept during the first half of the twentieth century was that there were two types of crust, named "sial" (continental type crust) and "sima" (oceanic type crust). Furthermore, it was supposed that a static shell of strata was present under the continents. It therefore looked apparent that a layer of basalt (sial) underlies the continental rocks.

However, based on abnormalities in plumb line deflection by the Andes in Peru, Pierre Bouguer had deduced that less-dense mountains must have a downward projection into the denser layer underneath. The concept that mountains had "roots" was confirmed by George B. Airy a hundred years later, during study of Himalayan gravitation, and seismic studies detected corresponding density variations. Therefore, by the mid-1950s, the question remained unresolved as to whether mountain roots were clenched in surrounding basalt or were floating on it like an iceberg.

During the 20th century, improvements in and greater use of seismic instruments such as seismographs enabled scientists to learn that earthquakes tend to be concentrated in specific areas, most notably along the oceanic trenches and spreading ridges. By the late 1920s, seismologists were beginning to identify several prominent earthquake zones parallel to the trenches that typically were inclined 40–60° from the horizontal and extended several hundred kilometers into the Earth. These zones later became known as Wadati–Benioff zones, or simply Benioff zones, in honor of the seismologists who first recognized them, Kiyoo Wadati of Japan and Hugo Benioff of the United States. The study of global seismicity greatly advanced in the 1960s with the establishment of the Worldwide Standardized Seismograph Network (WWSSN) to monitor the compliance of the 1963 treaty banning above-ground testing of nuclear weapons. The much improved data from the WWSSN instruments allowed seismologists to map precisely the zones of earthquake concentration worldwide.

Meanwhile, debates developed around the phenomena of polar wander. Since the early debates of continental drift, scientists had discussed and used evidence that polar drift had occurred because continents seemed to have moved through different climatic zones during the past. Furthermore, paleomagnetic data had shown that the magnetic pole had also shifted during time. Reasoning in an opposite way, the continents might have shifted and rotated, while the pole remained relatively fixed. The first time the evidence of magnetic polar wander was used to support the movements of continents was in a paper by Keith Runcorn in 1956,[323] and successive papers by him and his students Ted Irving (who was actually the first to be convinced of the fact that paleomagnetism supported continental drift) and Ken Creer.

This was immediately followed by a symposium in Tasmania in March 1956.[332] In this symposium, the evidence was used in the theory of an expansion of the global crust. In this hypothesis the shifting of the continents can be simply explained by a large increase in size of the Earth since its formation. However, this was unsatisfactory because its supporters could offer no convincing mechanism to produce a significant expansion of the Earth. Certainly there is no evidence that the moon has expanded in the past 3 billion years; other work would soon show that the evidence was equally in support of continental drift on a globe with a stable radius.

During the thirties up to the late fifties, works by Vening-Meinesz, Holmes, Umbgrove, and numerous others outlined concepts that were close or nearly identical to modern plate tectonics theory. In particular, the English geologist Arthur Holmes proposed in 1920 that plate junctions might lie beneath the sea, and in 1928 that convection currents within the mantle might be the driving force.[333] Often, these contributions are forgotten because:

- At the time, continental drift was not accepted.
- Some of these ideas were discussed in the context of abandoned fixistic ideas of a deforming globe without continental drift or an expanding Earth.
- They were published during an episode of extreme political and economic instability that hampered scientific communication.
- Many were published by European scientists and at first not mentioned or given little credit in the papers on sea floor spreading published by the American researchers in the 1960s.

Mid-oceanic ridge spreading and convection

In 1947, a team of scientists led by Maurice Ewing utilizing the Woods Hole Oceanographic Institution's research vessel *Atlantis* and an array of instruments, confirmed the existence of a rise in the central Atlantic Ocean, and found that the floor of the seabed beneath the layer of sediments consisted of basalt, not the granite which is the main constituent of continents. They also found that the oceanic crust was much thinner than continental crust. All these new findings raised important and intriguing questions.[334]

The new data that had been collected on the ocean basins also showed particular characteristics regarding the bathymetry. One of the major outcomes of these datasets was that all along the globe, a system of mid-oceanic ridges was detected. An important conclusion was that along this system, new ocean floor was being created, which led to the concept of the "Great Global Rift". This was described in the crucial paper of Bruce Heezen (1960),[335] which would trigger a real revolution in thinking. A profound consequence of seafloor spreading is that new crust was, and still is, being continually created along the oceanic ridges. Therefore, Heezen advocated the so-called "expanding Earth" hypothesis of S. Warren Carey (see above). So, still the question remained: how can new crust be continuously added along the oceanic ridges without increasing the size of the Earth? In reality, this question had been solved already by numerous scientists during the forties and the fifties, like Arthur Holmes, Vening-Meinesz, Coates and many others: The crust in excess disappeared along what were called the oceanic trenches, where so-called "subduction" occurred. Therefore, when various scientists during the early sixties started to reason on the data at their disposal regarding the ocean floor, the pieces of the theory quickly fell into place.

The question particularly intrigued Harry Hammond Hess, a Princeton University geologist and a Naval Reserve Rear Admiral, and Robert S. Dietz, a scientist with the U.S. Coast and Geodetic Survey who first coined the term *seafloor spreading*. Dietz and Hess (the former published the same idea one year earlier in *Nature*,[336] but priority belongs to Hess who had already distributed an unpublished manuscript of his 1962 article by 1960)[337] were among the small handful who really understood the broad implications of sea floor spreading and how it would eventually agree with the, at that time, unconventional and unaccepted ideas of continental drift and the elegant and mobilistic models proposed by previous workers like Holmes.

In the same year, Robert R. Coats of the U.S. Geological Survey described the main features of island arc subduction in the Aleutian Islands. His paper, though little noted (and even ridiculed) at the time, has since been called "seminal" and "prescient". In reality, it actually shows that the work by the European scientists on island arcs and mountain belts performed and published during the 1930s up until the 1950s was applied and appreciated also in the United States.

If the Earth's crust was expanding along the oceanic ridges, Hess and Dietz reasoned like Holmes and others before them, it must be shrinking elsewhere. Hess followed Heezen, suggesting that new oceanic crust continuously spreads away from the ridges in a conveyor belt–like motion. And, using the mobilistic concepts developed before, he correctly concluded that many millions of years later, the oceanic crust eventually descends along the continental margins where oceanic trenches – very deep, narrow canyons – are formed, e.g. along the rim of the Pacific Ocean basin. The important step Hess made was that convection currents would be the driving force in this process, arriving at the same conclusions as Holmes had decades before with the only difference that the thinning of the ocean crust was performed using Heezen's mechanism of spreading along the ridges. Hess therefore concluded that the Atlantic Ocean was expanding while the Pacific Ocean was shrinking. As old oceanic crust is "consumed" in the trenches (like Holmes and others, he thought this was done by thickening of the continental lithosphere, not, as now understood, by underthrusting at a larger scale of the oceanic crust itself into the mantle), new magma rises and erupts along the spreading ridges to form new crust. In effect, the ocean basins are perpetually being "recycled," with the creation of

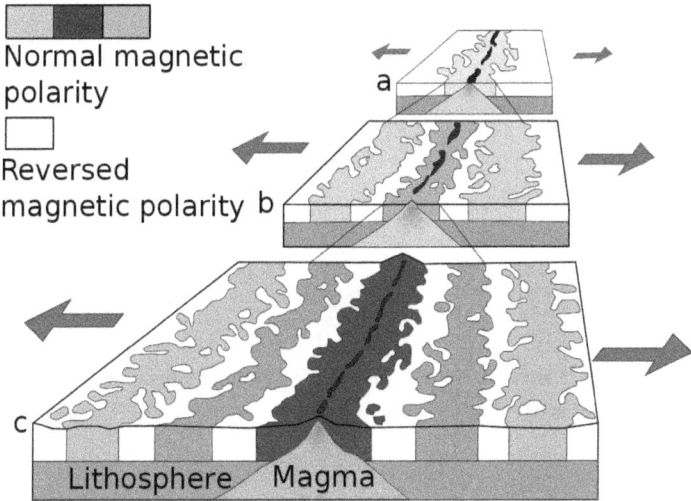

Figure 92: *Seafloor magnetic striping.*

new crust and the destruction of old oceanic lithosphere occurring simultane-ously. Thus, the new mobilistic concepts neatly explained why the Earth does not get bigger with sea floor spreading, why there is so little sediment accu-mulation on the ocean floor, and why oceanic rocks are much younger than continental rocks.

Magnetic striping

Beginning in the 1950s, scientists like Victor Vacquier, using magnetic in-struments (magnetometers) adapted from airborne devices developed during World War II to detect submarines, began recognizing odd magnetic varia-tions across the ocean floor. This finding, though unexpected, was not entirely surprising because it was known that basalt—the iron-rich, volcanic rock mak-ing up the ocean floor—contains a strongly magnetic mineral (magnetite) and can locally distort compass readings. This distortion was recognized by Ice-landic mariners as early as the late 18th century. More important, because the presence of magnetite gives the basalt measurable magnetic properties, these newly discovered magnetic variations provided another means to study the deep ocean floor. When newly formed rock cools, such magnetic materials recorded the Earth's magnetic field at the time.

As more and more of the seafloor was mapped during the 1950s, the magnetic variations turned out not to be random or isolated occurrences, but instead

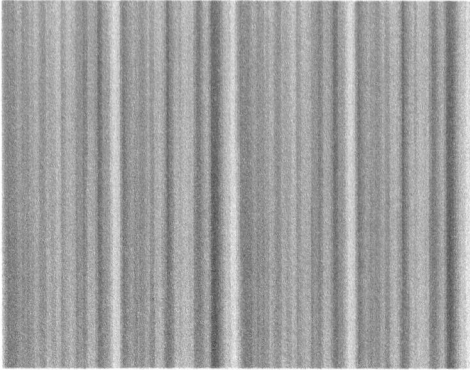

Figure 93: *A demonstration of magnetic striping. (The darker the color is, the closer it is to normal polarity)*

revealed recognizable patterns. When these magnetic patterns were mapped over a wide region, the ocean floor showed a zebra-like pattern: one stripe with normal polarity and the adjoining stripe with reversed polarity. The overall pattern, defined by these alternating bands of normally and reversely polarized rock, became known as magnetic striping, and was published by Ron G. Mason and co-workers in 1961, who did not find, though, an explanation for these data in terms of sea floor spreading, like Vine, Matthews and Morley a few years later.[338]

The discovery of magnetic striping called for an explanation. In the early 1960s scientists such as Heezen, Hess and Dietz had begun to theorise that mid-ocean ridges mark structurally weak zones where the ocean floor was being ripped in two lengthwise along the ridge crest (see the previous paragraph). New magma from deep within the Earth rises easily through these weak zones and eventually erupts along the crest of the ridges to create new oceanic crust. This process, at first denominated the "conveyer belt hypothesis" and later called seafloor spreading, operating over many millions of years continues to form new ocean floor all across the 50,000 km-long system of mid-ocean ridges.

Only four years after the maps with the "zebra pattern" of magnetic stripes were published, the link between sea floor spreading and these patterns was correctly placed, independently by Lawrence Morley, and by Fred Vine and Drummond Matthews, in 1963,[339] now called the Vine-Matthews-Morley hypothesis. This hypothesis linked these patterns to geomagnetic reversals and was supported by several lines of evidence:[340]

1. the stripes are symmetrical around the crests of the mid-ocean ridges; at or near the crest of the ridge, the rocks are very young, and they become progressively older away from the ridge crest;
2. the youngest rocks at the ridge crest always have present-day (normal) polarity;
3. stripes of rock parallel to the ridge crest alternate in magnetic polarity (normal-reversed-normal, etc.), suggesting that they were formed during different epochs documenting the (already known from independent studies) normal and reversal episodes of the Earth's magnetic field.

By explaining both the zebra-like magnetic striping and the construction of the mid-ocean ridge system, the seafloor spreading hypothesis (SFS) quickly gained converts and represented another major advance in the development of the plate-tectonics theory. Furthermore, the oceanic crust now came to be appreciated as a natural "tape recording" of the history of the geomagnetic field reversals (GMFR) of the Earth's magnetic field. Today, extensive studies are dedicated to the calibration of the normal-reversal patterns in the oceanic crust on one hand and known timescales derived from the dating of basalt layers in sedimentary sequences (magnetostratigraphy) on the other, to arrive at estimates of past spreading rates and plate reconstructions.

Definition and refining of the theory

After all these considerations, Plate Tectonics (or, as it was initially called "New Global Tectonics") became quickly accepted in the scientific world, and numerous papers followed that defined the concepts:

- In 1965, Tuzo Wilson who had been a promotor of the sea floor spreading hypothesis and continental drift from the very beginning[341] added the concept of transform faults to the model, completing the classes of fault types necessary to make the mobility of the plates on the globe work out.[342]
- A symposium on continental drift was held at the Royal Society of London in 1965 which must be regarded as the official start of the acceptance of plate tectonics by the scientific community, and which abstracts are issued as Blacket, Bullard & Runcorn (1965). In this symposium, Edward Bullard and co-workers showed with a computer calculation how the continents along both sides of the Atlantic would best fit to close the ocean, which became known as the famous "Bullard's Fit".
- In 1966 Wilson published the paper that referred to previous plate tectonic reconstructions, introducing the concept of what is now known as the "Wilson Cycle".[343]

- In 1967, at the American Geophysical Union's meeting, W. Jason Morgan proposed that the Earth's surface consists of 12 rigid plates that move relative to each other.[344]
- Two months later, Xavier Le Pichon published a complete model based on 6 major plates with their relative motions, which marked the final acceptance by the scientific community of plate tectonics.[345]
- In the same year, McKenzie and Parker independently presented a model similar to Morgan's using translations and rotations on a sphere to define the plate motions.[346]

Plate Tectonics Revolution

The Plate Tectonics Revolution was the scientific and cultural change which developed from the acceptance of the plate tectonics theory. The event was a paradigm shift and scientific revolution.

Implications for biogeography

Continental drift theory helps biogeographers to explain the disjunct biogeographic distribution of present-day life found on different continents but having similar ancestors.[347] In particular, it explains the Gondwanan distribution of ratites and the Antarctic flora.

Plate reconstruction

Reconstruction is used to establish past (and future) plate configurations, helping determine the shape and make-up of ancient supercontinents and providing a basis for paleogeography.

Defining plate boundaries

Current plate boundaries are defined by their seismicity.[348] Past plate boundaries within existing plates are identified from a variety of evidence, such as the presence of ophiolites that are indicative of vanished oceans.[349]

Past plate motions

Tectonic motion is believed to have begun around 3 to 3.5 billion years ago.[350]Wikipedia:Please clarify

Various types of quantitative and semi-quantitative information are available to constrain past plate motions. The geometric fit between continents, such as between west Africa and South America is still an important part of plate reconstruction. Magnetic stripe patterns provide a reliable guide to relative plate motions going back into the Jurassic period. The tracks of hotspots give absolute reconstructions, but these are only available back to the Cretaceous.[351] Older reconstructions rely mainly on paleomagnetic pole data, although these only constrain the latitude and rotation, but not the longitude. Combining poles of different ages in a particular plate to produce apparent polar wander paths provides a method for comparing the motions of different plates through time.[352] Additional evidence comes from the distribution of certain sedimentary rock types, faunal provinces shown by particular fossil groups, and the position of orogenic belts.[351]

Formation and break-up of continents

The movement of plates has caused the formation and break-up of continents over time, including occasional formation of a supercontinent that contains most or all of the continents. The supercontinent Columbia or Nuna formed during a period of 2,000 to 1,800[353] million years ago and broke up about 1,500 to 1,300[354] million years ago. The supercontinent Rodinia is thought to have formed about 1 billion years ago and to have embodied most or all of Earth's continents, and broken up into eight continents around 600[355] million years ago. The eight continents later re-assembled into another supercontinent called Pangaea; Pangaea broke up into Laurasia (which became North America and Eurasia) and Gondwana (which became the remaining continents).

The Himalayas, the world's tallest mountain range, are assumed to have been formed by the collision of two major plates. Before uplift, they were covered by the Tethys Ocean.

Current plates

Depending on how they are defined, there are usually seven or eight "major" plates: African, Antarctic, Eurasian, North American, South American, Pacific, and Indo-Australian. The latter is sometimes subdivided into the Indian and Australian plates.

There are dozens of smaller plates, the seven largest of which are the Arabian, Caribbean, Juan de Fuca, Cocos, Nazca, Philippine Sea, and Scotia.

DIGITAL TECTONIC ACTIVITY MAP OF THE EARTH
Tectonism and Volcanism of the Last One Million Years

The current motion of the tectonic plates is today determined by remote sensing satellite data sets, calibrated with ground station measurements.

Other celestial bodies (planets, moons)

The appearance of plate tectonics on terrestrial planets is related to planetary mass, with more massive planets than Earth expected to exhibit plate tectonics. Earth may be a borderline case, owing its tectonic activity to abundant water[356] (silica and water form a deep eutectic).

Venus

Venus shows no evidence of active plate tectonics. There is debatable evidence of active tectonics in the planet's distant past; however, events taking place since then (such as the plausible and generally accepted hypothesis that the Venusian lithosphere has thickened greatly over the course of several hundred million years) has made constraining the course of its geologic record difficult. However, the numerous well-preserved impact craters have been utilized as a dating method to approximately date the Venusian surface (since there are thus far no known samples of Venusian rock to be dated by more reliable methods). Dates derived are dominantly in the range 500 to 750[357] million years ago, although ages of up to 1,200[358] million years ago have been calculated. This research has led to the fairly well accepted hypothesis that Venus has undergone an essentially complete volcanic resurfacing at least once in its distant past, with the last event taking place approximately within the range of estimated surface ages. While the mechanism of such an impressive thermal event remains a debated issue in Venusian geosciences, some scientists are advocates of processes involving plate motion to some extent.

One explanation for Venus's lack of plate tectonics is that on Venus temperatures are too high for significant water to be present.[359] The Earth's crust is soaked with water, and water plays an important role in the development of shear zones. Plate tectonics requires weak surfaces in the crust along which crustal slices can move, and it may well be that such weakening never took place on Venus because of the absence of water. However, some researchersWikipedia:Manual of Style/Words to watch#Unsupported attributions remain convinced that plate tectonics is or was once active on this planet.

Mars

Mars is considerably smaller than Earth and Venus, and there is evidence for ice on its surface and in its crust.

In the 1990s, it was proposed that Martian Crustal Dichotomy was created by plate tectonic processes.[360] Scientists today disagree, and think that it was created either by upwelling within the Martian mantle that thickened the crust of the Southern Highlands and formed Tharsis[361] or by a giant impact that excavated the Northern Lowlands.[362]

Valles Marineris may be a tectonic boundary.

Observations made of the magnetic field of Mars by the *Mars Global Surveyor* spacecraft in 1999 showed patterns of magnetic striping discovered on this planet. Some scientists interpreted these as requiring plate tectonic processes, such as seafloor spreading. However, their data fail a "magnetic reversal test", which is used to see if they were formed by flipping polarities of a global magnetic field.[363]

Icy satellites

Some of the satellites of Jupiter have features that may be related to plate-tectonic style deformation, although the materials and specific mechanisms may be different from plate-tectonic activity on Earth. On 8 September 2014, NASA reported finding evidence of plate tectonics on Europa, a satellite of Jupiter—the first sign of subduction activity on another world other than Earth.

Titan, the largest moon of Saturn, was reported to show tectonic activity in images taken by the *Huygens* probe, which landed on Titan on January 14, 2005.[364]

Exoplanets

On Earth-sized planets, plate tectonics is more likely if there are oceans of water. However, in 2007, two independent teams of researchers came to opposing conclusions about the likelihood of plate tectonics on larger super-Earths with one team saying that plate tectonics would be episodic or stagnant[365] and the other team saying that plate tectonics is very likely on super-earths even if the planet is dry.[356]

Consideration of plate tectonics is a part of the search for extraterrestrial intelligence and extraterrestrial life.

References

Cited books

<templatestyles src="Template:Refbegin/styles.css" />

- Butler, Robert F. (1992). "Applications to paleogeography". *Paleomagnetism: Magnetic domains to geologic terranes*[366] (PDF). Blackwell. ISBN 0-86542-070-X. Archived from the original[367] (PDF) on 17 August 2010. Retrieved 18 June 2010.
- Carey, S. W. (1958). "The tectonic approach to continental drift". In Carey, S.W. *Continental Drift – A symposium, held in March 1956*. Hobart: Univ. of Tasmania. pp. 177–363. Expanding Earth from pp. 311–49.
- Condie, K.C. (1997). *Plate tectonics and crustal evolution*[368] (4th ed.). Butterworth-Heinemann. p. 282. ISBN 978-0-7506-3386-4. Retrieved 2010-06-18.
- Foulger, Gillian R. (2010). *Plates vs Plumes: A Geological Controversy*. Wiley-Blackwell. ISBN 978-1-4051-6148-0.
- Frankel, H. (1987). "The Continental Drift Debate". In H.T. Engelhardt Jr; A.L. Caplan. *Scientific Controversies: Case Studies in the Resolution and Closure of Disputes in Science and Technology*[369]. Cambridge University Press. ISBN 978-0-521-27560-6.
- Hancock, Paul L.; Skinner, Brian J.; Dineley, David L. (2000). *The Oxford Companion to The Earth*. Oxford University Press. ISBN 0-19-854039-6.
- Hess, H. H. (November 1962). "History of Ocean Basins"[370] (PDF). In A. E. J. Engel; Harold L. James; B. F. Leonard. *Petrologic studies: a volume to honor of A. F. Buddington*. Boulder, CO: Geological Society of America. pp. 599–620.
- Holmes, Arthur (1978). *Principles of Physical Geology* (3 ed.). Wiley. pp. 640–41. ISBN 0-471-07251-6.

- Joly, John (1909). *Radioactivity and Geology: An Account of the Influence of Radioactive Energy on Terrestrial History*. London: Archibald Constable. p. 36. ISBN 1-4021-3577-7.
- Kious, W. Jacquelyne; Tilling, Robert I. (February 2001) [1996]. "Historical perspective"[371]. *This Dynamic Earth: the Story of Plate Tectonics*[372] (Online ed.). U.S. Geological Survey. ISBN 0-16-048220-8. Retrieved 2008-01-29. <q>Abraham Ortelius in his work Thesaurus Geographicus... suggested that the Americas were 'torn away from Europe and Africa... by earthquakes and floods... The vestiges of the rupture reveal themselves, if someone brings forward a map of the world and considers carefully the coasts of the three [continents].'</q>
- Lippsett, Laurence (2006). "Maurice Ewing and the Lamont-Doherty Earth Observatory". In William Theodore De Bary; Jerry Kisslinger; Tom Mathewson. *Living Legacies at Columbia*[373]. Columbia University Press. pp. 277–97. ISBN 0-231-13884-9. Retrieved 2010-06-22.
- Little, W.; Fowler, H.W.; Coulson, J. (1990). Onions C.T., ed. *The Shorter Oxford English Dictionary: on historical principles*. **II** (3 ed.). Clarendon Press. ISBN 978-0-19-861126-4.
- Lliboutry, L. (2000). *Quantitative geophysics and geology*[374]. Springer. p. 480. ISBN 978-1-85233-115-3. Retrieved 2010-06-18.
- McKnight, Tom (2004). *Geographica: The complete illustrated Atlas of the world*. New York: Barnes and Noble Books. ISBN 0-7607-5974-X.
- Meissner, Rolf (2002). *The Little Book of Planet Earth*. New York: Copernicus Books. p. 202. ISBN 978-0-387-95258-1.
- Meyerhoff, Arthur Augustus; Taner, I.; Morris, A. E. L.; Agocs, W. B.; Kamen-Kaye, M.; Bhat, Mohammad I.; Smoot, N. Christian; Choi, Dong R. (1996). Donna Meyerhoff Hull, ed. *Surge tectonics: a new hypothesis of global geodynamics*[375]. Solid Earth Sciences Library. **9**. Springer Netherlands. p. 348. ISBN 978-0-7923-4156-7.
- Moss, S.J.; Wilson, M.E.J. (1998). "Biogeographic implications from the Tertiary palaeogeographic evolution of Sulawesi and Borneo"[376] (PDF). In Hall R; Holloway JD. *Biogeography and Geological Evolution of SE Asia* (PDF). Leiden, The Netherlands: Backhuys. pp. 133–63. ISBN 90-73348-97-8. Archived from the original on 2008-02-16.
- Oreskes, Naomi, ed. (2003). *Plate Tectonics: An Insider's History of the Modern Theory of the Earth*. Westview. ISBN 0-8133-4132-9.
- Read, Herbert Harold; Watson, Janet (1975). *Introduction to Geology*. New York: Halsted. pp. 13–15. ISBN 978-0-470-71165-1. OCLC 317775677[377].
- Schmidt, Victor A.; Harbert, William (1998). "The Living Machine: Plate Tectonics"[378]. *Planet Earth and the New Geosciences*[379] (3 ed.). p. 442. ISBN 0-7872-4296-9. Archived from the original on 2010-01-24.

Retrieved 2008-01-28.

- Schubert, Gerald; Turcotte, Donald L.; Olson, Peter (2001). *Mantle Convection in the Earth and Planets*. Cambridge: Cambridge University Press. ISBN 0-521-35367-X.
- Stanley, Steven M. (1999). *Earth System History*. W.H. Freeman. pp. 211–28. ISBN 0-7167-2882-6.
- Stein, Seth; Wysession, Michael (2009). *An Introduction to Seismology, Earthquakes, and Earth Structure*. Chichester: John Wiley & Sons. ISBN 978-1-4443-1131-0.
- Sverdrup, H. U., Johnson, M. W. and Fleming, R. H. (1942). *The Oceans: Their physics, chemistry and general biology*. Englewood Cliffs: Prentice-Hall. p. 1087.
- Thompson, Graham R. & Turk, Jonathan (1991). *Modern Physical Geology*. Saunders College Publishing. ISBN 0-03-025398-5.
- Torsvik, Trond Helge; Steinberger, Bernhard (December 2006). "Fra kontinentaldrift til manteldynamikk"[380] [From Continental Drift to Mantle Dynamics]. *Geo* (in Norwegian). **8**: 20–30. Archived from the original[381] on 23 July 2011. Retrieved 22 June 2010., translation: Torsvik, Trond Helge; Steinberger, Bernhard (2008). "From Continental Drift to Mantle Dynamics"[382] (PDF). In Trond Slagstad; Rolv Dahl Gråsteinen. *Geology for Society for 150 years – The Legacy after Kjerulf*. **12**. Trondheim: Norges Geologiske Undersokelse. pp. 24–38. Archived from the original[383] (PDF) on 2011-07-23 [Norwegian Geological Survey, Popular Science].
- Turcotte, D.L.; Schubert, G. (2002). "Plate Tectonics". *Geodynamics* (2 ed.). Cambridge University Press. pp. 1–21. ISBN 0-521-66186-2.
- Wegener, Alfred (1929). *Die Entstehung der Kontinente und Ozeane* (4 ed.). Braunschweig: Friedrich Vieweg & Sohn Akt. Ges. ISBN 3-443-01056-3.
- Wegener, Alfred (1966). *The origin of continents and oceans*. Biram John (translator). Courier Dover. p. 246. ISBN 0-486-61708-4.
- Winchester, Simon (2003). *Krakatoa: The Day the World Exploded: August 27, 1883*. HarperCollins. ISBN 0-06-621285-5.

Cited articles

<templatestyles src="Template:Refbegin/styles.css" />

- Andrews-Hanna, Jeffrey C.; Zuber, Maria T.; Banerdt, W. Bruce (2008). "The Borealis basin and the origin of the martian crustal dichotomy". *Nature*. **453** (7199): 1212–15. Bibcode: 2008Natur.453.1212A[384]. doi: 10.1038/nature07011[385]. PMID 18580944[386].

- Blacket, P.M.S.; Bullard, E.; Runcorn, S.K., eds. (1965). *A Symposium on Continental Drift, held in 28 October 1965*. Philosophical Transactions of the Royal Society A. **258**. The Royal Society of London. p. 323.
- Bostrom, R.C. (31 December 1971). "Westward displacement of the lithosphere". *Nature*. **234** (5331): 536–38. Bibcode: 1971Natur.234..536B[387]. doi: 10.1038/234536a0[388].
- Connerney, J.E.P.; Acuña, M.H.; Wasilewski, P.J.; Ness, N.F.; Rème H.; Mazelle C.; Vignes D.; Lin R.P.; Mitchell D.L.; Cloutier P.A. (1999). "Magnetic Lineations in the Ancient Crust of Mars". *Science*. **284** (5415): 794–98. Bibcode: 1999Sci...284..794C[389]. doi: 10.1126/science.284.5415.794[390]. PMID 10221909[391].
- Connerney, J.E.P.; Acuña, M.H.; Ness, N.F.; Kletetschka, G.; Mitchell D.L.; Lin R.P.; Rème H. (2005). "Tectonic implications of Mars crustal magnetism"[392]. *Proceedings of the National Academy of Sciences*. **102** (42): 14970–175. Bibcode: 2005PNAS..10214970C[393]. doi: 10.1073/pnas.0507469102[394]. PMC 1250232[392] ☐. PMID 16217034[395].
- Conrad, Clinton P.; Lithgow-Bertelloni, Carolina (2002). "How Mantle Slabs Drive Plate Tectonics"[396]. *Science*. **298** (5591): 207–09. Bibcode: 2002Sci...298..207C[397]. doi: 10.1126/science.1074161[398]. PMID 12364804[399]. Archived from the original[400] on September 20, 2009.
- Dietz, Robert S. (June 1961). "Continent and Ocean Basin Evolution by Spreading of the Sea Floor". *Nature*. **190** (4779): 854–57. Bibcode: 1961Natur.190..854D[401]. doi: 10.1038/190854a0[402].
- van Dijk, Janpieter; Okkes, F.W. Mark (1990). "The analysis of shear zones in Calabria; implications for the geodynamics of the Central Mediterranean". *Rivista Italiana di Paleontologia e Stratigrafia*. **96** (2–3): 241–70.
- van Dijk, J.P.; Okkes, F.W.M. (1991). "Neogene tectonostratigraphy and kinematics of Calabrian Basins: implications for the geodynamics of the Central Mediterranean". *Tectonophysics*. **196**: 23–60. Bibcode: 1991Tect.196...23V[403]. doi: 10.1016/0040-1951(91)90288-4[404].
- van Dijk, Janpieter (1992). "Late Neogene fore-arc basin evolution in the Calabrian Arc (Central Mediterranean). Tectonic sequence stratigraphy and dynamic geohistory. With special reference to the geology of Central Calabria"[405]. *Geologica Ultraiectina*. **92**: 288. Archived from the original[406] on 2013-04-20.
- Frankel, Henry (July 1978). "Arthur Holmes and continental drift". *The British Journal for the History of Science*. **11** (2): 130–50. doi: 10.1017/S0007087400016551[407]. JSTOR 4025726[408].
- Harrison, C.G.A. (2000). "Questions About Magnetic Lineations in the Ancient Crust of Mars". *Science*. **287** (5453): 547a. doi: 10.1126/sci-

ence.287.5453.547a[409].

- Heezen, B. (1960). "The rift in the ocean floor". *Scientific American*. **203** (4): 98–110. Bibcode: 1960SciAm.203d..98H[410]. doi: 10.1038/scientificamerican1060-98[411].
- Heirtzler, James R.; Le Pichon, Xavier; Baron, J. Gregory (1966). "Magnetic anomalies over the Reykjanes Ridge". *Deep-Sea Research*. **13** (3): 427–32. Bibcode: 1966DSROA..13..427H[412]. doi: 10.1016/0011-7471(66)91078-3[413].
- Holmes, Arthur (1928). "Radioactivity and Earth movements". *Transactions of the Geological Society of Glasgow*. **18**: 559–606.
- Hughes, Patrick (8 February 2001). "Alfred Wegener (1880–1930): A Geographic Jigsaw Puzzle"[414]. *On the shoulders of giants*. Earth Observatory, NASA. Retrieved 2007-12-26. <q>... on January 6, 1912, Wegener... proposed instead a grand vision of drifting continents and widening seas to explain the evolution of Earth's geography.</q>
- Hughes, Patrick (8 February 2001). "Alfred Wegener (1880–1930): The origin of continents and oceans"[415]. *On the Shoulders of Giants*. Earth Observatory, NASA. Retrieved 2007-12-26. <q>By his third edition (1922), Wegener was citing geological evidence that some 300 million years ago all the continents had been joined in a supercontinent stretching from pole to pole. He called it Pangaea (all lands),...</q>
- Kasting, James F. (1988). "Runaway and moist greenhouse atmospheres and the evolution of Earth and Venus". *Icarus*. **74** (3): 472–94. Bibcode: 1988Icar...74..472K[416]. doi: 10.1016/0019-1035(88)90116-9[417]. PMID 11538226[418].
- Korgen, Ben J. (1995). "A voice from the past: John Lyman and the plate tectonics story"[419] (PDF). *Oceanography*. The Oceanography Society. **8** (1): 19–20. doi: 10.5670/oceanog.1995.29[420]. Archived from the original[421] (PDF) on 2007-09-26.
- Lippsett, Laurence (2001). "Maurice Ewing and the Lamont-Doherty Earth Observatory"[422]. *Living Legacies*. Retrieved 2008-03-04.
- Lovett, Richard A (24 January 2006). "Moon Is Dragging Continents West, Scientist Says"[423]. *National Geographic News*.
- Lyman, J.; Fleming, R.H. (1940). "Composition of Seawater". *Journal of Marine Research*. **3**: 134–46.
- Mason, Ronald G.; Raff, Arthur D. (1961). "Magnetic survey off the west coast of the United States between 32°N latitude and 42°N latitude". *Bulletin of the Geological Society of America*. **72** (8): 1259–66. Bibcode: 1961GSAB...72.1259M[424]. doi: 10.1130/0016-7606(1961)72[1259:MSOTWC]2.0.CO;2[425]. ISSN 0016-7606[426].
- Mc Kenzie, D.; Parker, R.L. (1967). "The North Pacific: an example of tectonics on a sphere". *Nature*. **216** (5122): 1276–1280. Bibcode:

1967Natur.216.1276M[427]. doi: 10.1038/2161276a0[428].

- Moore, George W. (1973). "Westward Tidal Lag as the Driving Force of Plate Tectonics". *Geology*. **1** (3): 99–100. Bibcode: 1973Geo.....1...99M[429]. doi: 10.1130/0091-7613(1973)1<99:WTLATD>2.0.CO;2[430]. ISSN 0091-7613[431].

- Morgan, W. Jason (1968). "Rises, Trenches, Great Faults, and Crustal Blocks"[432] (PDF). *Journal of Geophysical Research*. **73** (6): 1959–182. Bibcode: 1968JGR....73.1959M[433]. doi: 10.1029/JB073i006p01959[434].

- Le Pichon, Xavier (15 June 1968). "Sea-floor spreading and continental drift". *Journal of Geophysical Research*. **73** (12): 3661–97. Bibcode: 1968JGR....73.3661L[435]. doi: 10.1029/JB073i012p03661[436].

- Quilty, Patrick G.; Banks, Maxwell R. (2003). "Samuel Warren Carey, 1911–2002"[437]. *Biographical memoirs*. Australian Academy of Science. Archived from the original[438] on 2010-12-21. Retrieved 2010-06-19. <q>This memoir was originally published in *Historical Records of Australian Science* (2003) **14** (3).</q>

- Raff, Arthur D.; Mason, Roland G. (1961). "Magnetic survey off the west coast of the United States between 40°N latitude and 52°N latitude". *Bulletin of the Geological Society of America*. **72** (8): 1267–70. Bibcode: 1961GSAB...72.1267R[439]. doi: 10.1130/0016-7606(1961)72[1267:MSOTWC]2.0.CO;2[440]. ISSN 0016-7606[426].

- Runcorn, S.K. (1956). "Paleomagnetic comparisons between Europe and North America". *Proceedings, Geological Association of Canada*. **8** (1088): 7785. Bibcode: 1965RSPTA.258....1R[441]. doi: 10.1098/rsta.1965.0016[442].

- Scalera, G. & Lavecchia, G. (2006). "Frontiers in earth sciences: new ideas and interpretation". *Annals of Geophysics*. **49** (1). doi: 10.4401/ag-4406[443] (inactive 2017-01-16).

- Scoppola, B.; Boccaletti, D.; Bevis, M.; Carminati, E.; Doglioni, C. (2006). "The westward drift of the lithosphere: A rotational drag?". *Geological Society of America Bulletin*. **118**: 199–209. Bibcode: 2006GSAB..118..199S[444]. doi: 10.1130/B25734.1[445].

- Segev, A (2002). "Flood basalts, continental breakup and the dispersal of Gondwana: evidence for periodic migration of upwelling mantle flows (plumes)"[446] (PDF). *EGU Stephan Mueller Special Publication Series*. **2**: 171–91. doi: 10.5194/smsps-2-171-2002[447]. Retrieved 5 August 2010.

- Sleep, Norman H. (1994). "Martian plate tectonics"[448] (PDF). *Journal of Geophysical Research*. **99**: 5639. Bibcode: 1994JGR....99.5639S[449]. doi: 10.1029/94JE00216[450].

- Soderblom, Laurence A.; Tomasko, Martin G.; Archinal, Brent A.; Becker, Tammy L.; Bushroe, Michael W.; Cook, Debbie A.; Doose, Lyn R.; Galuszka, Donna M.; Hare, Trent M.; Howington-Kraus, Elpitha;

Karkoschka, Erich; Kirk, Randolph L.; Lunine, Jonathan I.; McFar-
lane, Elisabeth A.; Redding, Bonnie L.; Rizk, Bashar; Rosiek, Mark
R.; See, Charles; Smith, Peter H. (2007). "Topography and geomorphology of the Huygens landing site on Titan". *Planetary and Space
Science*. **55** (13): 2015–24. Bibcode: 2007P&SS...55.2015S[451]. doi:
10.1016/j.pss.2007.04.015[452].

- Spence, William (1987). "Slab pull and the seismotectonics of subducting
  lithosphere"[453] (PDF). *Reviews of Geophysics*. **25** (1): 55–69. Bibcode:
  1987RvGeo..25...55S[454]. doi: 10.1029/RG025i001p00055[455].

- Spiess, Fred; Kuperman, William (2003). "The Marine Physical Laboratory at Scripps"[456] (PDF). *Oceanography*. The Oceanography Society.
  16 (3): 45–54. doi: 10.5670/oceanog.2003.30[457]. Archived from the
  original[458] (PDF) on 2007-09-26.

- Tanimoto, Toshiro; Lay, Thorne (7 November 2000). "Mantle dynamics
  and seismic tomography"[459]. *Proceedings of the National Academy of
  Sciences*. **97** (23): 12409–110. Bibcode: 2000PNAS...9712409T[460]. doi:
  10.1073/pnas.210382197[461]. PMC 34063[459] ∂. PMID 11035784[462].

- Thomson, W (1863). "On the secular cooling of the earth"[463]. *Philosophical Magazine*. **4** (25): 1–14. doi: 10.1080/14786446308643410[464].

- Torsvik, Trond H.; Steinberger, Bernhard; Gurnis, Michael; Gaina, Car-
  men (2010). "Plate tectonics and net lithosphere rotation over the past
  150 My"[465] (PDF). *Earth and Planetary Science Letters*. **291**: 106–12.
  Bibcode: 2010E&PSL.291..106T[466]. doi: 10.1016/j.epsl.2009.12.055[467].
  Archived from the original[468] (PDF) on 16 May 2011. Retrieved 18 June
  2010.

- Valencia, Diana; O'Connell, Richard J.; Sasselov, Dimitar D (November
  2007). "Inevitability of Plate Tectonics on Super-Earths". *Astrophysical
  Journal Letters*. **670** (1): L45–L48. arXiv: 0710.0699[469] ∂. Bibcode:
  2007ApJ...670L..45V[470]. doi: 10.1086/524012[471].

- Vine, F.J.; Matthews, D.H. (1963). "Magnetic anomalies over
  oceanic ridges". *Nature*. **199** (4897): 947–949. Bibcode:
  1963Natur.199..947V[472]. doi: 10.1038/199947a0[473].

- Wegener, Alfred (6 January 1912). "Die Herausbildung der Grossformen
  der Erdrinde (Kontinente und Ozeane), auf geophysikalischer Grund-
  lage"[474] (PDF). *Petermanns Geographische Mitteilungen*. **63**: 185–95,
  253–56, 305–09. Archived from the original[475] (PDF) on 5 July 2010.

- White, R.; McKenzie, D. (1989). "Magmatism at rift zones: The generation of volcanic continental margins and flood basalts". *Journal of Geophysical Research*. **94**: 7685–729. Bibcode: 1989JGR....94.7685W[476].
  doi: 10.1029/JB094iB06p07685[477].

- Wilson, J.T. (8 June 1963). "Hypothesis on the Earth's behaviour".
  Nature. **198** (4884): 849–65. Bibcode: 1963Natur.198..925T[478]. doi:

10.1038/198925a0[479].

- Wilson, J. Tuzo (July 1965). "A new class of faults and their bearing on continental drift"[480] (PDF). *Nature*. **207** (4995): 343–47. Bibcode: 1965Natur.207..343W[481]. doi: 10.1038/207343a0[482]. Archived from the original[483] (PDF) on August 6, 2010.
- Wilson, J. Tuzo (13 August 1966). "Did the Atlantic close and then re-open?"[484] (PDF). *Nature*. **211** (5050): 676–81. Bibcode: 1966Natur.211..676W[485]. doi: 10.1038/211676a0[486].Wikipedia:Link rot
- Zhen Shao, Huang (1997). "Speed of the Continental Plates"[487]. *The Physics Factbook*. Archived from the original[488] on 2012-02-05.
- Zhao, Guochun, Cawood, Peter A., Wilde, Simon A., and Sun, M. (2002). "Review of global 2.1–1.8 Ga orogens: implications for a pre-Rodinia supercontinent". *Earth-Science Reviews*. **59**: 125–62. Bibcode: 2002ESRv...59..125Z[489]. doi: 10.1016/S0012-8252(02)00073-9[490].
- Zhao, Guochun, Sun, M., Wilde, Simon A., and Li, S.Z. (2004). "A Paleo-Mesoproterozoic supercontinent: assembly, growth and breakup". *Earth-Science Reviews*. **67**: 91–123. Bibcode: 2004ESRv...67...91Z[491]. doi: 10.1016/j.earscirev.2004.02.003[492].
- Zhong, Shijie; Zuber, Maria T. (2001). "Degree-1 mantle convection and the crustal dichotomy on Mars"[493] (PDF). *Earth and Planetary Science Letters*. **189**: 75–84. Bibcode: 2001E&PSL.189...75Z[494]. doi: 10.1016/S0012-821X(01)00345-4[495].

External links

	The Wikibook *Historical Geology* has a page on the topic of: ***Plate tectonics: overview***

	Wikimedia Commons has media related to ***Plate tectonics***.

- This Dynamic Earth: The Story of Plate Tectonics[372]. USGS.
- Understanding Plate Tectonics[496]. USGS.
- An explanation of tectonic forces[497]. Example of calculations to show that Earth Rotation could be a driving force.
- Bird, P. (2003); An updated digital model of plate boundaries[498].
- Map of tectonic plates[499].
- MORVEL plate velocity estimates and information[500]. C. DeMets, D. Argus, & R. Gordon.
- Plate Tectonics[501] on *In Our Time* at the BBC

Videos

- Khan Academy Explanation of evidence[502]
- 750 million years of global tectonic activity[503]. Movie.
- Multiple videos of plate tectonic movements[504] Quartz December 31, 2015

Surface

Lithosphere

<indicator name="pp-default"> 🔒 </indicator>

A **lithosphere** (Ancient Greek: λίθος [*lithos*] for "rocky", and σφαίρα [*sphaira*] for "sphere") is the rigid,[505] outermost shell of a terrestrial-type planet, or natural satellite, that is defined by its rigid mechanical properties. On Earth, it is composed of the crust and the portion of the upper mantle that behaves elastically on time scales of thousands of years or greater. The outermost shell of a rocky planet, the crust, is defined on the basis of its chemistry and mineralogy.

The study of past and current formations of landscapes is called geomorphology.

Earth's lithosphere

Earth's lithosphere includes the crust and the uppermost mantle, which constitute the hard and rigid outer layer of the Earth. The lithosphere is subdivided into tectonic plates. The uppermost part of the lithosphere that chemically reacts to the atmosphere, hydrosphere and biosphere through the soil forming process is called the pedosphere. The lithosphere is underlain by the asthenosphere which is the weaker, hotter, and deeper part of the upper mantle. The Lithosphere-Asthenosphere boundary is defined by a difference in response to stress: the lithosphere remains rigid for very long periods of geologic time in which it deforms elastically and through brittle failure, while the asthenosphere deforms viscously and accommodates strain through plastic deformation.

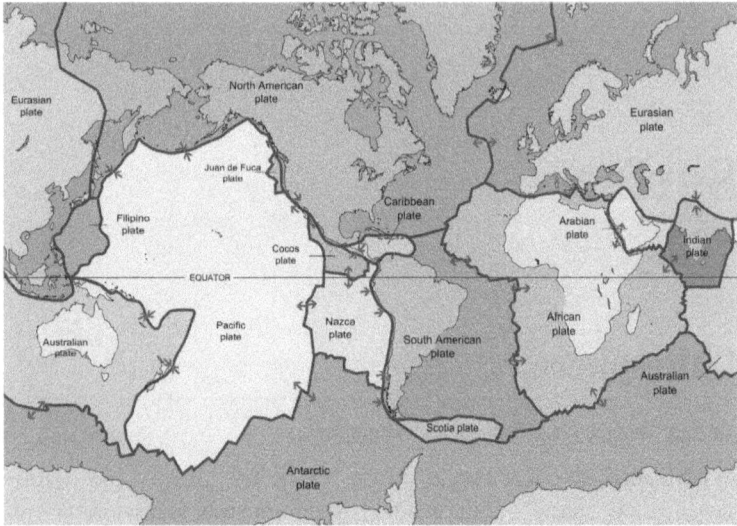

Figure 94: *The tectonic plates of the lithosphere on Earth*

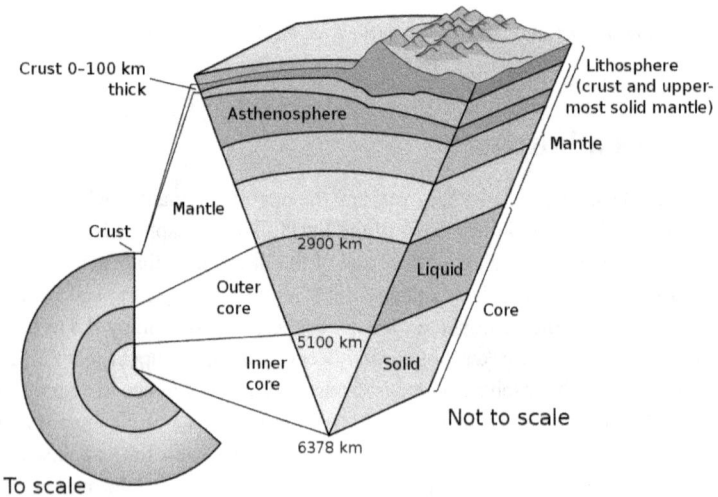

Figure 95: *Earth cutaway from core to crust, the lithosphere comprising the crust and lithospheric mantle (detail not to scale)*

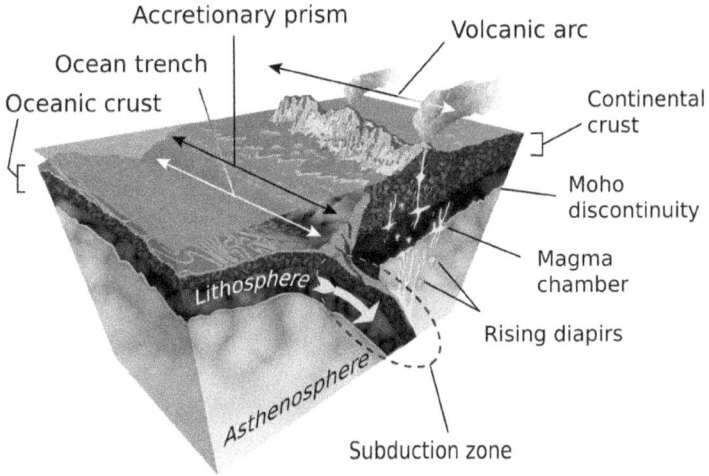

Figure 96: *Different types of lithosphere*

History of the concept

The concept of the lithosphere as Earth's strong outer layer was described by A.E.H. Love in his 1911 monograph "Some problems of Geodynamics" and further developed by Joseph Barrell, who wrote a series of papers about the concept and introduced the term "lithosphere". The concept was based on the presence of significant gravity anomalies over continental crust, from which he inferred that there must exist a strong, solid upper layer (which he called the lithosphere) above a weaker layer which could flow (which he called the asthenosphere). These ideas were expanded by Reginald Aldworth Daly in 1940 with his seminal work "Strength and Structure of the Earth."[506] They have been broadly accepted by geologists and geophysicists. These concepts of a strong lithosphere resting on a weak asthenosphere are essential to the theory of plate tectonics.

Types

There are two types of lithosphere:

- Oceanic lithosphere, which is associated with oceanic crust and exists in the ocean basins (mean density of about 2.9 grams per cubic centimeter)
- Continental lithosphere, which is associated with continental crust (mean density of about 2.7 grams per cubic centimeter)

The thickness of the lithosphere is considered to be the depth to the isotherm associated with the transition between brittle and viscous behavior. The temperature at which olivine begins to deform viscously ($\sim$1000 °C) is often used to set this isotherm because olivine is generally the weakest mineral in the upper mantle. Oceanic lithosphere is typically about 50–140 km thick (but beneath the mid-ocean ridges is no thicker than the crust), while continental lithosphere has a range in thickness from about 40 km to perhaps 280 km; the upper $\sim$30 to $\sim$50 km of typical continental lithosphere is crust. The mantle part of the lithosphere consists largely of peridotite. The crust is distinguished from the upper mantle by the change in chemical composition that takes place at the Moho discontinuity.

Oceanic lithosphere

Oceanic lithosphere consists mainly of mafic crust and ultramafic mantle (peridotite) and is denser than continental lithosphere, for which the mantle is associated with crust made of felsic rocks. Oceanic lithosphere thickens as it ages and moves away from the mid-ocean ridge. This thickening occurs by conductive cooling, which converts hot asthenosphere into lithospheric mantle and causes the oceanic lithosphere to become increasingly thick and dense with age. In fact, oceanic lithosphere is a thermal boundary layer for the convection[507] in the mantle. The thickness of the mantle part of the oceanic lithosphere can be approximated as a thermal boundary layer that thickens as the square root of time.

$$h \sim 2\sqrt{\kappa t}$$

Here, h is the thickness of the oceanic mantle lithosphere, κ is the thermal diffusivity (approximately 10^{-6} m²/s) for silicate rocks, and t is the age of the given part of the lithosphere. The age is often equal to L/V, where L is the distance from the spreading centre of mid-oceanic ridge, and V is velocity of the lithospheric plate.

Oceanic lithosphere is less dense than asthenosphere for a few tens of millions of years but after this becomes increasingly denser than asthenosphere. This is because the chemically differentiated oceanic crust is lighter than asthenosphere, but thermal contraction of the mantle lithosphere makes it more dense than the asthenosphere. The gravitational instability of mature oceanic lithosphere has the effect that at subduction zones, oceanic lithosphere invariably sinks underneath the overriding lithosphere, which can be oceanic or continental. New oceanic lithosphere is constantly being produced at mid-ocean ridges and is recycled back to the mantle at subduction zones. As a result, oceanic lithosphere is much younger than continental lithosphere: the oldest oceanic lithosphere is about 170 million years old, while parts of the continental lithosphere are billions of years old. The oldest parts of continental lithosphere underlie cratons, and the mantle lithosphere there is thicker and less dense than typical; the relatively low density of such mantle "roots of cratons" helps to stabilize these regions.

Subducted lithosphere

Geophysical studies in the early 21st century posit that large pieces of the lithosphere have been subducted into the mantle as deep as 2900 km to near the core-mantle boundary, while others "float" in the upper mantle, while some stick down into the mantle as far as 400 km but remain "attached" to the continental plate above, similar to the extent of the "tectosphere" proposed by Jordan in 1988.

Mantle xenoliths

Geoscientists can directly study the nature of the subcontinental mantle by examining mantle xenoliths[508] brought up in kimberlite, lamproite, and other volcanic pipes. The histories of these xenoliths have been investigated by many methods, including analyses of abundances of isotopes of osmium and rhenium. Such studies have confirmed that mantle lithospheres below some cratons have persisted for periods in excess of 3 billion years, despite the mantle flow that accompanies plate tectonics.

Further reading

- Chernicoff, Stanley; Whitney, Donna (1990). *Geology. An Introduction to Physical Geology* (4th ed.). Pearson. ISBN 0-13-175124-7.

External links

Wikimedia Commons has media related to *Lithospheres*.

- Earth's Crust, Lithosphere and Asthenosphere[509]
- Crust and Lithosphere[510]

Landform

A **landform** is a natural feature of the solid surface of the Earth or other planetary body. Landforms together make up a given terrain, and their arrangement in the landscape is known as topography. Typical landforms include hills, mountains, plateaus, canyons, and valleys, as well as shoreline features such as bays, peninsulas, and seas,Wikipedia:Citation needed including submerged features such as mid-ocean ridges, volcanoes, and the great ocean basins.

Physical characteristics

Landforms are categorized by characteristic physical attributes such as elevation, slope, orientation, stratification, rock exposure, and soil type. Gross physical features or landforms include intuitive elements such as berms, mounds, hills, ridges, cliffs, valleys, rivers, peninsulas, volcanoes, and numerous other structural and size-scaled (i.e. ponds vs. lakes, hills vs. mountains) elements including various kinds of inland and oceanic waterbodies and subsurface features.

File:Cades Cove Panorama.JPG

This panorama in Great Smoky Mountains National Park has the readily identifiable **physical features** of a rolling plain, actually part of a broad valley, distant foothills, and a backdrop of the old, much weathered Appalachian mountain range

Hierarchy of classes

Oceans and continents exemplify the highest-order landforms. Landform elements are parts of a high-order landforms that can be further identified and systematically given a cohesive definition such as hill-tops, shoulders, saddles, foreslopes and backslopes.

Some generic landform elements including: pits, peaks, channels, ridges, passes, pools and plains.

Figure 97: *This conical hill in Salar de Arizaro, Salta, Argentina called Cono de Arita constitutes a landform.*

Figure 98: *Karst towers landforms along Lijiang River, Guilin, China*

Terrain (or *relief*) is the third or vertical dimension of *land surface*. Topography is the study of terrain, although the word is often used as a synonym for relief itself. When relief is described underwater, the term bathymetry is used. In cartography, many different techniques are used to describe relief, including contour lines and TIN (Triangulated irregular network).

Elementary landforms (segments, facets, relief units) are the smallest homogeneous divisions of the land surface, at the given scale/resolution. These are areas with relatively homogeneous morphometric properties, bounded by lines of discontinuity. A plateau or a hill can be observed at various scales ranging from few hundred meters to hundreds of kilometers. Hence, the spatial distribution of landforms is often scale-dependent as is the case for soils and geological strata.

A number of factors, ranging from plate tectonics to erosion and deposition, can generate and affect landforms. Biological factors can also influence landforms— for example, note the role of vegetation in the development of dune systems and salt marshes, and the work of corals and algae in the formation of coral reefs.

Landforms do not include man-made features, such as canals, ports and many harbors; and geographic features, such as deserts, forests, and grasslands. Many of the terms are not restricted to refer to features of the planet Earth, and can be used to describe surface features of other planets and similar objects in the Universe. Examples are mountains, hills, polar caps, and valleys, which are found on all of the terrestrial planets.

The scientific study of landforms is known as geomorphology.

Recent developments

Landforms may be extracted from a digital elevation model using some automated techniques where the data has been gathered by modern satellites and stereoscopic aerial surveillance cameras. Until recently, compiling the data found in such data sets required time consuming and expensive techniques involving many man-hours. The most detailed DEMs available are measured directly using LIDAR techniques.

External links

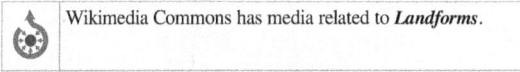

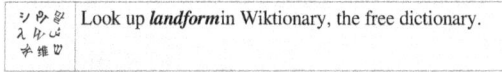

| | Wikimedia Commons has media related to *Landforms*. |

| シ ウ 岁
ス ㅐ ﹙ù
夲 推 ♡ | Look up *landform* in Wiktionary, the free dictionary. |

- Open-Geomorphometry Project[511]

Extreme points of Earth

This is a list of **extreme points of Earth**, the geographical locations that are farther north or south than, higher or lower in elevation than, or farthest inland or out to sea from, any other locations on the landmasses, continents or countries.

For other lists of extreme points on Earth, including places that hold temperature and weather records, see Extremes on Earth, Lists of extreme points, and List of weather records.

Earth

Latitude and longitude

- The **northernmost point on Earth** is the Geographic North Pole, in the Arctic Ocean.
  - The **northernmost point on land** is the northern tip of Kaffeklubben Island, north of Greenland (83°40′N 29°50′W[512]), which lies slightly north of Cape Morris Jesup, Greenland (83°38′N 32°40′W[513]). Various shifting gravel bars lie farther north, the most famous being Oodaaq.
- The **southernmost point on Earth** and the **southernmost point on land** is the geographic South Pole, which is on the continent of Antarctica.
  - The **southernmost point of water** is a bay on the Filchner-Ronne Ice Shelf along the coast of Antarctica (83°S 59°W[514]) about 100 kilometres (62 mi) south of Berkner Island, the southernmost island in the world. The **southernmost point of ocean** is located on the Gould Coast (84°30′S 150°0′W[515]);[516] the **southernmost point of open ocean** is also part of the Ross Sea, namely the Bay of Whales at 78°30′S, at the edge of the Ross Ice Shelf.

Figure 99: *Chimborazo in Ecuador is the farthest point from Earth's center.*

- The **westernmost and easternmost points on Earth**, based on the east-west standard for describing longitude, can be found anywhere along the 180th meridian in Siberia (including Wrangel Island), Antarctica, or the three islands of Fiji through which the 180th meridian passes (Vanua Levu's eastern peninsula, the middle of Taveuni, and the western part of Rabi Island).
  - Using the path of the International Date Line, the **westernmost point on land** is Attu Island, Alaska, and the **easternmost point on land** is Caroline Island, Kiribati.[517]

Elevation

Highest points

- The **highest point on Earth's surface** measured from sea level is the summit of Mount Everest on the border of Nepal and China. While measurements of its height vary slightly, the elevation of its peak is usually given as 8,848 m (29,029 ft) above sea level. It was first reached by Sir Edmund Hillary of New Zealand and Sherpa of Nepal Tenzing Norgay in 1953 (with speculation that it may have been reached in 1924).
- The **point farthest from Earth's center** is the summit of Chimborazo in Ecuador, at 6,384.4 km (3,967.1 mi) from Earth's center; the peak's elevation relative to sea level is 6,263.47 m (20,549 ft).[518]</ref> This is because Earth is an oblate spheroid rather than a perfect sphere; it is wider

at the Equator and narrower between the poles. Therefore, the summit of Chimborazo, which is near the Equator, is farther away from Earth's center than the summit of Mount Everest is; the latter is 2,168 m (7,112.9 ft) closer, at 6,382.3 km (3,965.8 mi) from Earth's center. Peru's Huascarán (at 6,768 m (22,205 ft)) contends closely with Chimborazo, the difference in the mountains' heights being just 23 metres (75 ft).

- The **fastest point on Earth** or, in other words, **the point furthest from the axis of Earth** is the summit of Cayambe in Ecuador, at 1,675.89 km/h (1,041 mph) and 6,383.95 km (3,967 mi) from the axis. Like Chimborazo, which is the fourth fastest peak at 1,675.47 km/h (1,041 mph), it is close to the Equator and takes advantage of the oblate spheroid figure of Earth. More importantly, however, it being so near the Equator means that the majority of its distance from Earth's center goes into it being away from the axis. The importance of latitude becomes most apparent when one looks at the Challenger Deep (speed of 1,639.15 km/h (1,019 mph)) compared to Mount Everest (speed of 1,481.67 km/h (921 mph)).

Highest points attainable by transportation

- The **highest point accessible...**
  - **...by land vehicle** is an elevation of 6,688 m (21,942 ft) on Ojos del Salado in Chile, which was reached by the Chilean duo of Gonzalo Bravo G. and Eduardo Canales Moya on 21 April 2007 with a modified Suzuki Samurai, setting the high-altitude record for a four-wheeled vehicle.
  - **...by road (dead end)** is on a mining road to the summit of Aucanquilcha in Chile, which reaches an elevation of 6,176 m (20,262 ft). It was once usable by 20-tonne mining trucks.[519] The road is no longer usable. 21.214°S 68.475°W[520]
  - **...by road (mountain pass)** is disputed; there are a number of competing claims for this title due to the definition of "motorable pass" (i.e. a surfaced road or one simply passable by a vehicle):
    - The **highest asphalted road** crosses Tibet's Semo La pass at 5,565 m (18,258 feet). It is used by trucks and buses regularly. The Ticlio pass, on the Central Road of Peru, is the highest surfaced road in the Americas, at an elevation of 4,818 m (15,807 feet).
    - The **highest unsurfaced road** is claimed by several different roads. All are unsurfaced or gravel roads including the barely passable road to Umling La, 17 kilometres (11 mi) west of Demchok in Ladakh, India, which reaches 5,800 m (19,029 feet) ("19,300 feet" according to a Border Roads Organisation sign there that recognizes it as

Figure 100: *La Rinconada, Peru*

the "World's Highest Motorable Pass"), and Mana Pass, between India and Tibet, which is crossed by a gravel road reaching 5,610 m
(18,406 feet). The heavily trafficked Khardung La in Ladakh lies
at 5,359 m (17,582 feet). A possibly motorable gravel road crosses
Marsimik La in Ladakh at 5,582 m (18,314 feet).

- **...by train** is Tanggula Pass, located on the Qinghai–Tibet (Qingzang)
Railway in the Tanggula Mountains of Qinghai/Tibet, China, at
5,072 m (16,640 feet). The Tanggula railway station is the world's
highest railway station at 5,068 m (16,627 feet). Before the Qingzang
Railway was built, the highest railway ran between Lima and Huancayo
in Peru, reaching 4,829 m (15,843 feet) at Ticlio.

- **...by oceangoing vessel** is a segment of the Rhine–Main–Danube Canal
between the Hilpoltstein and Bachhausen locks in Bavaria, Germany.
The locks artificially raise the surface level of the water in the canal to
406 m (1,332 feet) above mean sea level, higher than any other lock
system in the world, making it the highest point currently accessible by
oceangoing commercial watercraft.

- The **highest commercial airport** is Daocheng Yading Airport, Sichuan,
China, at 4,411 m (14,472 feet). The proposed Nagqu Dagring Airport
in Tibet, China, if built, will be 25 m (82 feet) higher at 4,436 m (14,554
feet).

- The **highest helipad** is Sonam, Siachen Glacier, India, at a height of 6,400 m (20,997 feet) above sea level.
- The **highest permanent human settlement** is La Rinconada, Peru, 5,100 m (16,732 feet), in the Peruvian Andes.
- The **farthest road from the Earth's center** is the Road to Carrel Hut in the Ecuadorian Andes, at an elevation of 4,850 m (15,912 feet) above sea level and a distance of 6,382.9 km (3,966 miles) from the center of the Earth.

Highest geographical features

- The **highest volcano** is Ojos del Salado on the Argentina–Chile border. It has the highest summit, 6,893 m (22,615 feet), of any volcano on Earth.
- The **highest natural lake** is an unnamed crater lake on Ojos del Salado at 6,390 m (20,965 feet), on the Argentina side. Another candidate was Lhagba Pool on the northeast slopes of Mount Everest, Tibet, China, at an elevation of 6,368 m (20,892 feet), which has since dried up.
- The **highest navigable lake** is Lake Titicaca, on the border of Bolivia and Peru in the Andes, at 3,812 m (12,507 feet).
- The **highest glacier** is the Khumbu Glacier on the southwest slopes of Mount Everest in Nepal, beginning on the west side of Lhotse at an elevation of 7,600 to 8,000 m (24,900 to 26,200 feet).
- The **highest river** is disputed; one candidate from many possibilities is the Ating Ho, which flows into the Aong Tso (Hagung Tso), a large lake in Tibet, China, and has an elevation of about 6,100 m (20,013 feet) at its source at 32°49′30"N 81°03′45"E[521]. A very large and high river is the Yarlung Tsangpo or upper Brahmaputra River in Tibet, China, whose main stem, the Maquan River, has its source at about 6,020 m (19,751 feet) above sea level at 30°48′59"N 82°42′45"E[522]. Above these elevations, there are no constantly flowing rivers since the temperature is almost always below freezing.
- The **highest island** is one of a number of islands in the Orba Co lake in Tibet, China, at an elevation of 5,209 m (17,090 feet).

Lowest points

Lowest artificial points

- The **lowest point underground** ever reached was 12,262 m (40,230 feet) deep (SG-3 at Kola superdeep borehole).
- The **lowest human-sized point underground** is 3,900 m (12,800 feet) below ground at the TauTona Mine, Carletonville, South Africa.
- The **lowest (from sea level) artificially made point with open sky** may be the Hambach surface mine, Germany, which reaches a depth of 293 m (961 feet) below sea level.
- The **lowest (from surface) artificially made point with open sky** may be the Bingham Canyon open-pit mine, Salt Lake City, United States, at a depth of 1,200 m (3,900 feet) below surface level.
- The **lowest point underwater** is the 10,685 m (35,056 feet)-deep (as measured from the subsea wellhead) oil and gas well drilled on the Tiber Oil Field in the Gulf of Mexico. The wellhead of this well was an additional 1,259 m (4,131 feet) underwater for a total distance of 11,944 m (39,186 feet) as measured from sea level. 28.736667°N 88.386944°W[523]

Lowest natural points

- The **lowest known point** is Challenger Deep, at the bottom of the Mariana Trench, 11,034 m (36,201 feet) below sea level. Only three humans have reached the bottom of the trench: Jacques Piccard and U.S. Navy Lieutenant Don Walsh in 1960 aboard the bathyscaphe *Trieste*, and filmmaker James Cameron in 2012 aboard *Deepsea Challenger*.
- The **lowest point underground** is more than 2,000 m (6,600 feet) under the Earth's surface. For example, the altitude difference between the entrance and the deepest explored point (the maximum depth) of the Krubera Cave in Georgia is 2,191 ± 20 m (7,188 ± 66 feet). In 2012, Ukrainian cave diver Gennadiy Samokhin reached the lowest point, breaking the world record.
- The **lowest point on land not covered by liquid water** is the valley under the Byrd Glacier in Antarctica, which reaches 2,780 m (9,121 feet) below sea level. It is, however, covered by a thick layer of ice.
- The **lowest point on dry land** is the shore of the Dead Sea, shared by Jordan, Palestine, and Israel, 432.65 m (1,419 feet) below sea level. As the Dead Sea waters are receding, it loses some 100-120 cm every year.
- The **point closest to the Earth's center** on the Earth's surface (interpreted as a natural surface of the land or sea that is accessible by a person) is the surface of the Arctic Ocean at the Geographic North Pole (6,356.77 km or 3,950 miles).

Figure 101: *The shore of the Dead Sea in Israel*

- The **closest point on the ground** (interpreted as a land surface or sea floor) is the bottom of the Litke Deep, the deepest point of the Arctic Ocean, which is 6,351.61 km (3,947 miles) from the center of the Earth. By comparison, the bottom of the deepest oceanic trench in the world, the Mariana Trench in the Pacific Ocean, is 14.7 km (9 miles) farther from the center of the Earth.

Lowest points attainable by transportation

- The **lowest point accessible...**
  - **...by road**, excluding roads in mines, is any of the roads alongside the Dead Sea in Israel and Jordan, which are the lowest on Earth at 418 m (1,371 feet) below sea level.
    - The **lowest undersea highway tunnel** is the Eiksund Tunnel, in Norway, at 287 m (942 feet) below sea level.
  - **...by train**, excluding the tracks inside some South African gold mines, which can be several thousand meters below sea level, is located in the Seikan Tunnel of Japan railroad, at 240 m (787 feet) below sea level. By way of comparison, the undersea Channel Tunnel between England and France reaches a depth of 75 m (246 feet) below sea level.

The lowest railroad station was the Japanese Yoshioka-Kaitei Station, at 150 m (492 feet) below sea level, but it was closed in 2014. The lowest railroad not inside a tunnel is 71 m (233 feet) below sea level, in the Mojave Desert between Yuma, Arizona, and Palm Springs, California, in the United States of America.

- The **lowest airfield** is the Bar Yehuda Airfield (MTZ), near Masada, Israel, at 378 m (1,240 feet) below sea level.
- The **lowest commercial airport** is Atyrau Airport (GUW), near Atyrau, Kazakhstan, at 22 m (72 feet) below sea level, in the basin of the Caspian Sea.

Lowest cities

Baku is located 28 metres (92 ft) below sea level, which makes it the lowest lying national capital in the world and also the largest city in the world located below sea level.

Remoteness

Poles of inaccessibility

Each continent has its own continental pole of inaccessibility, defined as the place on the continent that is farthest from any ocean. Similarly, each ocean has its own oceanic pole of inaccessibility, defined as the place in the ocean that is farthest from any land.

Continental

- The **most distant point from an ocean** is the Eurasian Pole of Inaccessibility (or "EPIA") 46°17′N 86°40′E[524], in China's Xinjiang region near the border with Kazakhstan. Calculations have shown that this point, located in the Dzoosotoyn Elisen Desert, is 2,645 km (1,644 miles) from the nearest coastline. The nearest settlement to the EPIA is Suluk at 46°15′N 86°50′E[525], about 11 km (6.8 miles) to the east.Wikipedia:Citation needed A recent study suggests that the historical calculation of the EPIA failed to recognize the point where the Gulf of Ob joins the Arctic Ocean, and proposes instead that varying definitions of coastline could result in other locations for the EPIA:
  - EPIA1, somewhere between 44°17′N 82°11′E[526] and 44°29′N 82°19′E[527], is about 2,510 ± 10 kilometres (1,559.6 ± 6.2 mi) from the nearest ocean.
  - EPIA2, somewhere between 45°17′N 88°08′E[528] and 45°28′N 88°14′E[529], is about 2,514 ± 7 kilometres (1,562.1 ± 4.3 mi) from the nearest ocean.

If adopted, this would place the final EPIA roughly 130 km (81 miles) closer to the ocean than the point that is currently agreed upon. Coincidentally, EPIA1, or EPIA2, and the most remote of the Oceanic Pole of Inaccessibility (specifically, the point in the South Pacific Ocean that is farthest from land) are similarly remote; EPIA1 is less than 200 km (120

miles) closer to the ocean than the Oceanic Pole of Inaccessibility is to
land.

- The continental poles of inaccessibility for the other continents are as fol-
  lows:
  - Africa: 5.65°N 26.17°E[530], close to the tripoint of the Central African
    Republic, South Sudan, and the Democratic Republic of the Congo
  - Australia: either 23°2′S 132°10′E[531],[532] or 23.17°S 132.27°E[533], near
    Papunya, Northern Territory
  - North America: 43.36°N 101.97°W[534], between Kyle, South Dakota
    and Allen, South Dakota.
  - South America: 14.05°S 56.85°W[535], near Arenápolis, Mato Grosso,
    Brazil

Oceanic

- The **most distant point from land** is the Pacific pole of inaccessibility
  (also called "Point Nemo"), which lies in the South Pacific Ocean at
  48°52.6′S 123°23.6′W[536], approximately 2,688 km (1,670 mi) from the
  nearest land (equidistant from Ducie Island in the Pitcairn Islands to the
  north, Motu Nui off Rapa Nui to the northeast, and Maher Island off Siple
  Island near Marie Byrd Land, Antarctica, to the south).

Other places considered the most remote

- The **most remote island** is Bouvet Island, a small, uninhabited island
  in the South Atlantic Ocean that is a dependency of Norway. It lies at
  coordinates 54°26′S 3°24′E[537]. The nearest land is the uninhabited Queen
  Maud Land, Antarctica, over 1,600 km (994 mi) to the south. The nearest
  inhabited lands are Tristan da Cunha, 2,260 km (1,404 mi) away, and the
  coast of South Africa, 2,580 km (1,603 mi) away.
- The title for **most remote inhabited island or archipelago** (the farthest
  away from any other permanently inhabited place) depends on how the
  question is interpreted. If the south Atlantic island Tristan da Cunha
  (population about 300) and its dependency Gough Island (with a small
  staffed research post), which are 399 km (248 mi) from each other, are
  considered part of the same archipelago, or if Gough Island is not counted
  because it has no permanent residents, then Tristan da Cunha is the
  world's most remote inhabited island/archipelago: the main island, also
  called Tristan da Cunha, is 2,434 km (1,512 mi) from the island Saint
  Helena, 2,816 km (1,750 mi) from South Africa, and 3,360 km (2,090
  miles) from South America. It is 1,845 km (1,146 mi) away from uninhabited Bouvet Island. However, if Gough and Tristan da Cunha are
  considered separately, they disqualify each other, and the most remote inhabited island is Easter Island in the South Pacific Ocean, which lies 2,075

ISS017E016161

Figure 102: *Bouvet Island*

kilometres (1,289 mi) from Pitcairn Island (about 50 residents in 2013), 2,606 km (1,619 mi) from Rikitea on the island of Mangareva (the nearest town with a population over 500), and 3,512 kilometres (2,182 mi) from the coast of Chile (the nearest continental point). The Kerguelen Islands in the southern Indian Ocean are another contender, lying 1,340 kilometres (830 mi) from the small Alfred Faure scientific station in Île de la Possession, but otherwise more than 3,300 kilometres (2,100 mi) from the coast of Madagascar (the nearest permanently inhabited place), 450 km (280 mi) northwest of uninhabited Heard Island and McDonald Islands, and 1,440 km (890 mi) from the non-permanent scientific station located in Île Amsterdam.

- The **most remote city...**
  - **...with a population in excess of one million from the nearest city with a population in excess of one million** is Auckland, New Zealand. The nearest city of comparable size or greater is Sydney, Australia, 2,168.9 kilometres (1,347.7 mi) away.[538]
  - **...with a population in excess of one million from the nearest city with a population above 100,000** is Perth, Australia, located 2,138 kilometres (1,328 mi) away from Adelaide, Australia.
  - **...with a population in excess of 100,000 from the nearest city of at least that population** is Honolulu, Hawaii, United States. The nearest city of comparable size or greater is San Francisco, 3,850 km (2,390

miles) away.

- **...that is a national capital from the nearest national capital** is a tie between Wellington, New Zealand, and Canberra, Australia, which are 2,326 km (1,445 mi) apart from each other.
- The **most remote airport in the world** from another airport is Mataveri International Airport (IPC) on Easter Island, which has a single runway for military and public use. It is located 2,603 km (1,617 mi) from Totegegie Airport (GMR; very few flights) in the Gambier Islands, French Polynesia and 3,759 km (2,336 mi) from Santiago, Chile (SCL; a fairly large airport). In comparison, the airport at the Amundsen–Scott South Pole Station (NZSP) is not very remote at all, being located only 1,355 kilometres (842 mi) from Williams Field (NZWD) near Ross Island.

Farthest-apart cities

The pairs of cities (with a population over 100,000) with the greatest distance between them are:

1. Rosario, Argentina to Xinghua, China: 19,996 km (12,425 mi)
2. Lu'an, China to Río Cuarto, Argentina: 19,994 km (12,424 mi)
3. Cuenca, Ecuador to Subang Jaya, Malaysia: 19,989 km (12,421 mi)
4. Rancagua, Chile to Xi'an, China: 19,972 km (12,410 mi)
5. Salamanca, Spain to Lower Hutt, New Zealand: 19,961 km (12,403 mi)
6. Marbella, Spain to Auckland, New Zealand: 19,960 km (12,403 mi)

Centre

- Since the Earth is a spheroid, its centre (the core) is thousands of kilometres beneath its crust. On the surface, the **center of the standard geographic model** as viewed on a traditional world map is the point 0°, 0° (the coordinates of zero degrees latitude by zero degrees longitude), which is located in the Atlantic Ocean approximately 614 km (382 miles) south of Accra, Ghana, in the Gulf of Guinea, at the intersection of the Equator and the Prime Meridian. However, the selection for the Prime Meridian as the 0° longitude meridian is culturally and historically dependent and therefore arbitrary.
- The **center of population**, the place to which there is the shortest average route for every individual human being in the world, could also be considered a "center of the world". This point is located in the north of the Indian subcontinent, although the precise location has never been calculated and is constantly shifting due to changes in the distribution of the human population across the planet.

Longest lines between two points

Along constant latitude

- The **longest continuous east-west distance on land** is 10,726 km (6,665 miles) along the latitude 48°24'53"N, from the west coast of France (48°24'53"N 4°47'44"W[539]) through Central Europe, Ukraine, Russia, Kazakhstan, Mongolia and China, to a point on the east coast of Russia (48°24'53"N 140°6'3"E[540]).
  - The **longest continuous east-west distance on land including permanent ice shelf** is 7,958 km (4,945 miles) along the latitude 78°35'S; this is the minimum extent of the Ross Ice Shelf, Antarctica and is subject to change.
- The **longest continuous east-west distance at sea** is 22,471 km (13,963 miles) along the latitude 55°59'S, south of Cape Horn, South America. The longest in the northern hemisphere is 4,435 km (2,756 miles) along the latitude 83°40'N, north of Kaffeklubben Island, Greenland.
  - The **longest continuous east-west distance at sea between two continents** is 15,409 km (9,575 miles) along the latitude 18°39'12"N, from the coast of Hainan, China (18°39'12"N 110°15'9"E[541]) across the Pacific Ocean to the coast of Michoacán, Mexico (18°39'12"N 103°42'6"W[542]).

Along constant longitude

- The **longest continuous north-south distance on land** is 7,590 km (4,720 miles) along the meridian 99°1'30"E, from the northern tip of Siberia in the Russian Federation (76°13'6"N 99°1'30"E[543]), through Mongolia, China, and Myanmar, to a point on the south coast of Thailand (7°53'24"N 99°1'30"E[544]).
  - The longest in Africa is 7,417 km (4,609 miles) along the meridian 20°12'E, from the north coast of Libya (32°19'0"N 20°12'0"E[545]), through Chad, Central African Republic, Democratic Republic of the Congo, Angola, Namibia, and Botswana, to the south coast of South Africa (34°41'30"S 20°12'0"E[546]).
  - The longest in South America is 7,098 km (4,410 miles) along the meridian 70°2'W, from the north coast of Venezuela (11°30'30"N 70°2'0"W[547]), through Colombia, Ecuador, Peru, and Chile, to the southern tip of Argentina (52°33'30"S 70°2'0"W[548]).
  - The longest in North America is 5,813 km (3,612 miles) along the meridian 97°52'30"W, from northern Canada (68°21'0"N 97°52'30"W[549]), through the United States, to southern Mexico (16°1'0"N 97°52'30"W[550]).

- The **longest continuous north-south distance at sea** is 15,986 km (9,933 miles) along the meridian 34°45'45"W, from the coast of Eastern Greenland (66°23'45"N 34°45'45"W[551]) across the Atlantic Ocean to the Filchner-Ronne Ice Shelf, on the coast of Antarctica (77°37'0"S 34°45'45"W[552]). The longest in the Pacific Ocean is 15,883 km (9,869 miles) along the meridian 172°8'30"W, from the coast of Siberia (64°45'0"N 172°8'30"W[553]) to the Ross Ice Shelf in Antarctica (78°20'0"S 172°8'30"W[554]).

- The **meridian that crosses the greatest total distance on land** (disregarding intervening bodies of water) is still to be determined. It is likely located in the vicinity of 22°E, which is the longest integer meridian that fits that criterion, crossing a total of 13,035 km (8,100 miles) of land through Europe (3,370 km (2,090 miles)), Africa (7,458 km (4,634 miles)), and Antarctica (2,207 km (1,371 miles)). More than 65% of the meridian's length is located on land. The meridian that crosses Giza Great Pyramid (31°08'3.69"E) is 855 km (531 miles) shorter.

  - The next six longest integer meridians by total distance over land are, in order:
    - 23°E: 12,953 km (8,049 miles) through Europe (3,325 km (2,066 miles)), Africa (7,415 km (4,607 miles)), and Antarctica (2,214 km (1,376 miles))
    - 27°E: 12,943 km (8,042 miles) through Europe (3,254 km (2,022 miles)), Asia (246 km (153 miles)), Africa (7,223 km (4,488 miles)), and Antarctica (2,221 km (1,380 miles))
    - 25°E: 12,875 km (8,000 miles) through Europe (3,344 km (2,078 miles)), Africa (7,327 km (4,553 miles)), and Antarctica (2,204 km (1,370 miles))
    - 26°E: 12,858 km (7,990 miles) through Europe (3,404 km (2,115 miles)), Africa (7,258 km (4,510 miles)), and Antarctica (2,196 km (1,365 miles))
    - 24°E: 12,794 km (7,950 miles) through Europe (3,263 km (2,028 miles)), Africa (7,346 km (4,565 miles)), and Antarctica (2,185 km (1,358 miles))
    - 28°E: 12,778 km (7,940 miles) through Europe (3,039 km (1,888 miles)), Asia (388 km (241 miles)), and Africa (7,117 km (4,422 miles))

Along any geodesic

These are the longest straight lines that can be drawn between any two points on the surface of the Earth and remain exclusively over land or water; the points need not lie on the same latitude or longitude.

- The **longest continuous straight-line distance in any direction on
  land** is 13,573 km (8,434 miles), along a line that begins on the West
  African coast near Greenville, Liberia (5°2'51.59"N 9°7'23.26"W[555]),
  goes across the Suez Canal, and ends at the top of a peninsula approximately 100 km (62 miles) northeast of Wenzhou, China 28°17'7.68"N
  121°38'17.31"E[556].[557]
  - The longest continuous straight-line land distance solely within conti-
    nental Africa is 8,402 km (5,221 miles), along a line that begins just
    east of Tangier, Morocco and ends 100 km (62 miles) east of Port
    Elizabeth, South Africa. This line passes through Morocco, Algeria,
    Mali, Niger, Nigeria, Cameroon, Equatorial Guinea, Gabon, Republic
    of the Congo, Democratic Republic of the Congo, Angola, Namibia,
    Botswana and South Africa.
  - The longest continuous straight-line land distance solely within con-
    tinental Asia is 10,152 km (6,308 miles), along a line that begins on
    the Indian coast near Kanyakumari and ends at the Bering Sea coast of
    the Chukchi Peninsula in Russia. This line passes through India, Nepal,
    China, Mongolia and Russia.
  - The longest continuous straight-line land distance solely within con-
    tinental Europe (defining the Ural Mountains as the border between
    Europe and Asia) is 5,325 km (3,309 miles), along a line that begins
    at Cape St. Vincent, Portugal and ends at the Urals, near the town of
    Perm, Russia. This line passes through Portugal, Spain, France, Ger-
    many, Poland, Lithuania, Belarus and Russia.
  - The longest continuous straight-line land distance solely within conti-
    nental Australia is 4,053 km (2,518 miles), along a line that begins at
    the southern end of Cape Range National Park in Western Australia
    and ends at the town of Byron Bay in New South Wales.
- There are several possible candidates for the **longest continuous straight-
  line distance in any direction at sea**, as there are many possible ways to
  travel along a great circle for more than the antipodic length of 19,840 km
  (12,330 miles). Some good examples of such routes would be:
  - From the south coast of Balochistan province somewhere near Port
    of Karachi, Pakistan (25°25'N 66°25'E[558]) across the Arabian Sea,
    southwest through the Indian Ocean, near Comoros, passing Namaete
    Canyon, near the South African coast, across the South Atlantic Ocean,
    then west across Cape Horn, then northwest across the Pacific Ocean,
    near Easter Island, passing the antipodal point near Amlia island,
    through the South Bering Sea and ending somewhere on the north-
    east coast of Kamchatka, near Ossora (59°38'N 163°24'E[559]). This
    route is 32,040 km (19,910 miles) long.[560] This route was confirmed to
    be the longest (at about 32090 km) given map data at a 1.8 km level of

resolution.

- From the south coast of Hormozgan province, Iran (25°35′N 58°22′E[561]) across the Gulf of Oman, southeast across the Arabian Sea, passing south of Australia and New Zealand, near the Antarctic coast, then northeast across the South Pacific Ocean, passing the antipodal point and ending on the southwest coast of Mexico somewhere near Ciudad Lázaro Cárdenas (17°57′N 101°57′W[562]). This route is 25,267 km (15,700 miles) long.[563]
- From Invercargill, New Zealand (46°37′S 168°59′E[564]) across Cape Horn, then off the coast of Brazil close to Recife, passing north of Cape Verde, passing the antipodal point and ending somewhere on the southwest coast of Ireland (52°09′N 6°34′W[565]). This route is 20,701 km (12,863 miles) long.[566]

By region

Afro-Eurasia

- Extreme points of Afro-Eurasia
  - Extreme points of Africa
    - Extreme points of Algeria
    - Extreme points of Angola
    - Extreme points of Benin
    - Extreme points of Botswana
    - Extreme points of Burkina Faso
    - Extreme points of Burundi
    - Extreme points of Cameroon
    - Extreme points of Cape Verde
    - Extreme points of Central African Republic
    - Extreme points of Chad
    - Extreme points of Comoros
    - Extreme points of the Democratic Republic of the Congo
    - Extreme points of the Republic of Congo
    - Extreme points of Côte d'Ivoire
    - Extreme points of Djibouti
    - Extreme points of Egypt
    - Extreme points of Equatorial Guinea
    - Extreme points of Eritrea
    - Extreme points of Ethiopia
    - Extreme points of Gabon
    - Extreme points of the Gambia
    - Extreme points of Ghana

- Extreme points of Georgia
- Extreme points of India
- Extreme points of Indonesia
- Extreme points of Iran
- Extreme points of Israel
- Extreme points of Japan
- Extreme points of Jordan
- Extreme points of Kazakhstan
- Extreme points of Kyrgyzstan
- Extreme points of Laos
- Extreme points of the Maldives
- Extreme points of Mongolia
- Extreme points of Myanmar
- Extreme points of Nepal
- Extreme points of North Korea
- Extreme points of Pakistan
- Extreme points of the Philippines
- Extreme points of Russia
- Extreme points of Singapore
- Extreme points of South Korea
- Extreme points of Sri Lanka
- Extreme points of Taiwan
- Extreme points of Tajikistan
- Extreme points of Thailand
- Extreme points of Turkey
- Extreme points of Turkmenistan
- Extreme points of Uzbekistan
- Extreme points of Vietnam
- Extreme points of Europe
  - Extreme points of the European Union
  - Extreme points of Albania
  - Extreme points of Andorra
  - Extreme points of Austria
  - Extreme points of Belarus
  - Extreme points of Belgium
  - Extreme points of Bosnia and Herzegovina
  - Extreme points of Bulgaria
  - Extreme points of Croatia
  - Extreme points of the Czech Republic
  - Extreme points of Denmark
  - Extreme points of Estonia
  - Extreme points of Finland

- Extreme points of France
- Extreme points of Germany
- Extreme points of Greece
- Extreme points of Hungary
- Extreme points of Iceland
- Extreme points of Ireland
- Extreme points of Italy
- Extreme points of Kosovo
- Extreme points of Latvia
- Extreme points of Liechtenstein
- Extreme points of Lithuania
- Extreme points of Luxembourg
- Extreme points of Macedonia
- Extreme points of Malta
- Extreme points of Moldova
- Extreme points of Monaco
- Extreme points of Montenegro
- Extreme points of the Netherlands
- Extreme points of Norway
- Extreme points of Poland
- Extreme points of Portugal
- Extreme points of Romania
- Extreme points of Russia
- Extreme points of San Marino
- Extreme points of Serbia
- Extreme points of Slovakia
- Extreme points of Slovenia
- Extreme points of Spain
- Extreme points of Sweden
- Extreme points of Switzerland
- Extreme points of Ukraine
- Extreme points of the United Kingdom
- Extreme points of Vatican City

The Americas

- Extreme points of the Americas
  - Extreme points of North America
    - Extreme points of Canada
      - Extreme points of Canadian provinces
      - Extreme communities of Canada
    - Extreme points of Greenland
    - Extreme points of Mexico

- Extreme points of the United States
  - Extreme points of U.S. states
  - Extreme points of New England
- Extreme points of Central America
  - Extreme points of Belize
  - Extreme points of Costa Rica
  - Extreme points of El Salvador
  - Extreme points of Guatemala
  - Extreme points of Honduras
  - Extreme points of Nicaragua
  - Extreme points of Panama
- Extreme points of the Caribbean
  - Extreme points of Cuba
  - Extreme points of Jamaica
- Extreme points of South America
  - Extreme points of Argentina
  - Extreme points of Bolivia
  - Extreme points of Brazil
  - Extreme points of Chile
  - Extreme points of Colombia
  - Extreme points of Ecuador
  - Extreme points of French Guiana
  - Extreme points of Guyana
  - Extreme points of Paraguay
  - Extreme points of Peru
  - Extreme points of Suriname
  - Extreme points of Uruguay
  - Extreme points of Venezuela

Oceania

- Extreme points of Oceania
  - Extreme points of Australia
  - Extreme points of Fiji
  - Extreme points of Guam
  - Extreme points of Indonesia
  - Extreme points of Kiribati
  - Extreme points of the Marshall Islands
  - Extreme points of Micronesia
  - Extreme points of Nauru
  - Extreme points of New Zealand
  - Extreme points of Niue
  - Extreme points of the Northern Mariana Islands

- Extreme points of Palau
- Extreme points of Papua New Guinea
- Extreme points of Tuvalu

Antarctica

- Extreme points of Antarctica
- Extreme points of the Antarctic

Arctic

- Extreme points of the Arctic

Hydrosphere

Hydrosphere

The **hydrosphere** (from Greek ὕδωρ *hydōr*, "water"[567] and σφαῖρα *sphaira*, "sphere"[568]) is the combined mass of water found on, under, and above the surface of a planet, minor planet or natural satellite.

It has been estimated that there are 1,386 million cubic kilometers of water on Earth. This includes water in liquid and frozen forms in groundwater, oceans, lakes and streams. Saltwater accounts for 97.5% of this amount. Fresh water accounts for only 2.5%. Of this fresh water, 68.9% is in the form of ice and permanent snow cover in the Arctic, the Antarctic, and mountain glaciers. 30.8% is in the form of fresh groundwater. Only 0.3% of the fresh water on Earth is in easily accessible lakes, reservoirs and river systems. The total mass of the Earth's hydrosphere is about 1.4×10^{18} tonnes, which is about 0.023% of Earth's total mass. About 20×10^{12} tonnes of this is in Earth's atmosphere (for practical purposes, 1 cubic meter of water weighs one tonne). Approximately 71% of Earth's surface, an area of some 361 million square kilometers (139.5 million square miles), is covered by ocean. The average salinity of Earth's oceans is about 35 grams of salt per kilogram of sea water (3.5%).

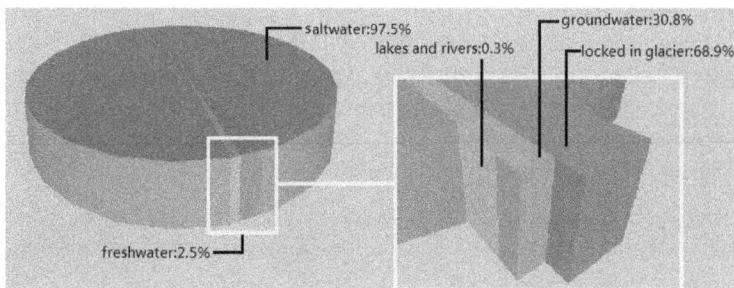

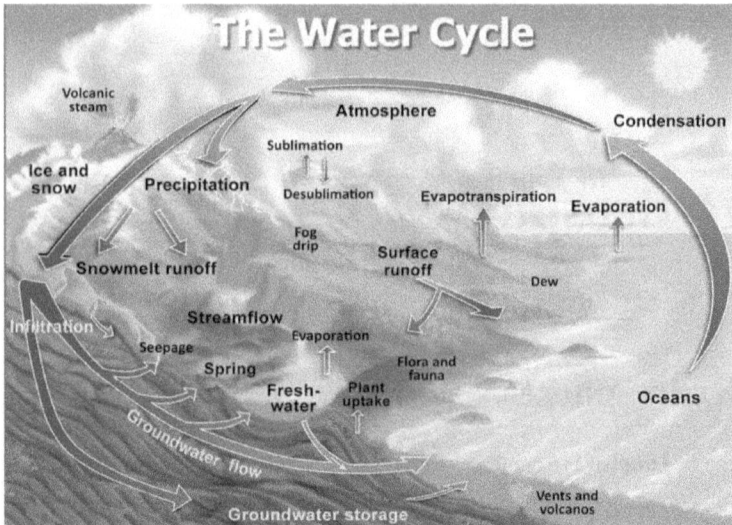

Water cycle

The water cycle refers to the transfer of water from one state or reservoir to another. Reservoirs include atmospheric moisture (snow, rain and clouds), streams, oceans, rivers, lakes, groundwater, subterranean aquifers, polar ice caps and saturated soil. Solar energy, in the form of heat and light (insolation), and gravity cause the transfer from one state to another over periods from hours to thousands of years. Most evaporation comes from the oceans and is returned to the earth as snow or rain.[27] Sublimation refers to evaporation from snow and ice. Transpiration refers to the expiration of water through the minute pores or stomata of trees. Evapotranspiration is the term used by hydrologists in reference to the three processes together, transpiration, sublimation and evaporation.

Marq de Villiers has described the hydrosphere as a closed system in which water exists. The hydrosphere is intricate, complex, interdependent, all-pervading and stable and "seems purpose-built for regulating life."[569]:26 De Villiers claimed that, "On earth, the total amount of water has almost certainly not changed since geological times: what we had then we still have. Water can be polluted, abused, and misused but it is neither created nor destroyed, it only migrates. There is no evidence that water vapor escapes into space.":26

"Every year the turnover of water on Earth involves 577,000 km^3 of water. This is water that evaporates from the oceanic surface (502,800 km^3) and from land (74,200 km^3). The same amount of water falls as atmospheric precipitation, 458,000 km^3 on the ocean and 119,000 km^3 on land. The difference between precipitation and evaporation from the land surface (119,000 - 74,200 = 44,800 km^3/year) represents the total runoff of the Earth's rivers (42,700 km^3/year) and direct groundwater runoff to the ocean (2100 km^3/year). These are the principal sources of fresh water to support life necessities and man's economic activities."

Water is a basic necessity of life. Since 2/3 of the Earth is covered by water, the Earth is also called the blue planet and the watery planet.[570] Hydrosphere plays an important role in the existence of the atmosphere in its present form. Oceans are important in this regard. When the Earth was formed it had only a very thin atmosphere rich in hydrogen and helium similar to the present atmosphere of Mercury. Later the gases hydrogen and helium were expelled from the atmosphere. The gases and water vapor released as the Earth cooled became its present atmosphere. Other gases and water vapor released by volcanoes also entered the atmosphere. As the Earth cooled the water vapor in the atmosphere condensed and fell as rain. The atmosphere cooled further as atmospheric carbon dioxide dissolved in to rain water. In turn this further caused the water vapor to condense and fall as rain. This rain water filled the depressions on the Earth's surface and formed the oceans. It is estimated that this occurred about 4000 million years ago. The first life forms began in the oceans. These organisms did not breathe oxygen. Later, when cyanobacteria evolved, the process of conversion of carbon dioxide into food and oxygen began. As a result, Earth's atmosphere has a distinctly different composition from that of other planets and allowed for life to evolve on Earth.

Recharging reservoirs

According to Igor A. Shiklomanov, it takes 2500 years for the complete recharge and replenishment of oceanic waters, 10,000 years for permafrost and ice, 1500 years for deep groundwater and mountainous glaciers, 17 years in lakes and 16 days in rivers.

Specific fresh water availability

"Specific water availability is the residual (after use) per capita quantity of fresh water." Fresh water resources are unevenly distributed in terms of space and time and can go from floods to water shortages within months in the same area. In 1998 76% of the total population had a specific water availability of less than 5.0 thousand m3 per year per capita. Already by 1998, 35% of the global population suffered "very low or catastrophically low water supplies" and Shiklomanov predicted that the situation would deteriorate in the twenty-first century with "most of the Earth's population will be living under the conditions of low or catastrophically low water supply" by 2025. There is only 2.5% of fresh water in the hydrosphere and only 0.25% of water is accessible for our use.

External links

	Look up *hydrosphere* in Wiktionary, the free dictionary.

- Ground Water - USGS[571]

Atmosphere

Atmosphere of Earth

<indicator name="pp-default"> 🔒 </indicator>

The **atmosphere of Earth** is the layer of gases, commonly known as **air**, that surrounds the planet Earth and is retained by Earth's gravity. The atmosphere of Earth protects life on Earth by creating pressure allowing for liquid water to exist on the Earth's surface, absorbing ultraviolet solar radiation, warming the surface through heat retention (greenhouse effect), and reducing temperature extremes between day and night (the diurnal temperature variation).

By volume, dry air contains 78.09% nitrogen, 20.95% oxygen, 0.93% argon, 0.04% carbon dioxide, and small amounts of other gases. Air also contains a variable amount of water vapor, on average around 1% at sea level, and 0.4% over the entire atmosphere. Air content and atmospheric pressure vary at different layers, and air suitable for use in photosynthesis by terrestrial plants and breathing of terrestrial animals is found only in Earth's troposphere and in artificial atmospheres.

The atmosphere has a mass of about 5.15×10^{18} kg,[572] three quarters of which is within about 11 km (6.8 mi; 36,000 ft) of the surface. The atmosphere becomes thinner and thinner with increasing altitude, with no definite boundary between the atmosphere and outer space. The Kármán line, at 100 km (62 mi), or 1.57% of Earth's radius, is often used as the border between the atmosphere and outer space. Atmospheric effects become noticeable during atmospheric reentry of spacecraft at an altitude of around 120 km (75 mi). Several layers can be distinguished in the atmosphere, based on characteristics such as temperature and composition.

The study of Earth's atmosphere and its processes is called atmospheric science (aerology). Early pioneers in the field include Léon Teisserenc de Bort and Richard Assmann.

Figure 103: *Blue light is scattered more than other wavelengths by the gases in the atmosphere, surrounding Earth in a visibly blue layer when seen from space on board the ISS at an altitude of 335 km (208 mi).*

Composition

The three major constituents of Earth's atmosphere are nitrogen, oxygen, and argon. Water vapor accounts for roughly 0.25% of the atmosphere by mass. The concentration of water vapor (a greenhouse gas) varies significantly from around 10 ppm by volume in the coldest portions of the atmosphere to as much as 5% by volume in hot, humid air masses, and concentrations of other atmospheric gases are typically quoted in terms of dry air (without water vapor).[573] The remaining gases are often referred to as trace gases, among which are the greenhouse gases, principally carbon dioxide, methane, nitrous oxide, and ozone. Filtered air includes trace amounts of many other chemical compounds. Many substances of natural origin may be present in locally and seasonally variable small amounts as aerosols in an unfiltered air sample, including dust of mineral and organic composition, pollen and spores, sea spray, and volcanic ash. Various industrial pollutants also may be present as gases or aerosols, such as chlorine (elemental or in compounds), fluorine compounds and elemental mercury vapor. Sulfur compounds such as hydrogen sulfide and sulfur dioxide (SO_2) may be derived from natural sources or from industrial air pollution.

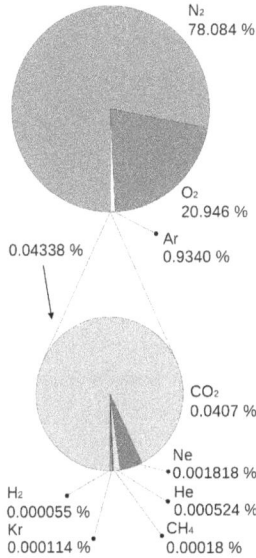

Figure 104: *Composition of Earth's atmosphere by volume. Lower pie represents trace gases that together compose about 0.038% of the atmosphere (0.043% with CO_2 at 2014 concentration). Numbers are mainly from 1987, with CO_2 and methane from 2009, and do not represent any single source.*

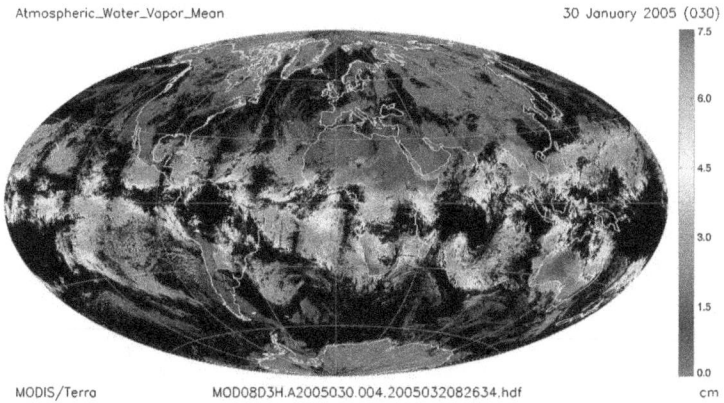

Figure 105: *Mean atmospheric water vapor*

Major constituents of dry air, by volume[574]

Gas		Volume[(A)]	
Name	Formula	in ppmv[(B)]	in %
Nitrogen	N_2	780,840	78.084
Oxygen	O_2	209,460	20.946
Argon	Ar	9,340	0.9340
Carbon dioxide	CO_2	400	0.04
Neon	Ne	18.18	0.001818
Helium	He	5.24	0.000524
Methane	CH_4	1.79	0.000179
Not included in above dry atmosphere:			
Water vapor[(C)]	H_2O	10–50,000[(D)]	0.001%–5%[(D)]

notes:
[(A)] volume fraction is equal to mole fraction for ideal gas only,
  also see volume (thermodynamics)
[(B)] ppmv: parts per million by volume
[(C)] Water vapor is about 0.25% by mass over full atmosphere
[(D)] Water vapor strongly varies locally

The relative concentration of gasses remains constant until about 10,000 m
(33,000 ft).

Structure of the atmosphere

Principal layers

In general, air pressure and density decrease with altitude in the atmosphere.
However, temperature has a more complicated profile with altitude, and may
remain relatively constant or even increase with altitude in some regions (see
the temperature section, below). Because the general pattern of the temper-
ature/altitude profile is constant and measurable by means of instrumented
balloon soundings, the temperature behavior provides a useful metric to distinguish atmospheric layers. In this way, Earth's atmosphere can be divided
(called atmospheric stratification) into five main layers. Excluding the exosphere, the atmosphere has four primary layers, which are the troposphere,
stratosphere, mesosphere, and thermosphere. From highest to lowest, the five
main layers are:

- Exosphere: 700 to 10,000 km (440 to 6,200 miles)
- Thermosphere: 80 to 700 km (50 to 440 miles)
- Mesosphere: 50 to 80 km (31 to 50 miles)

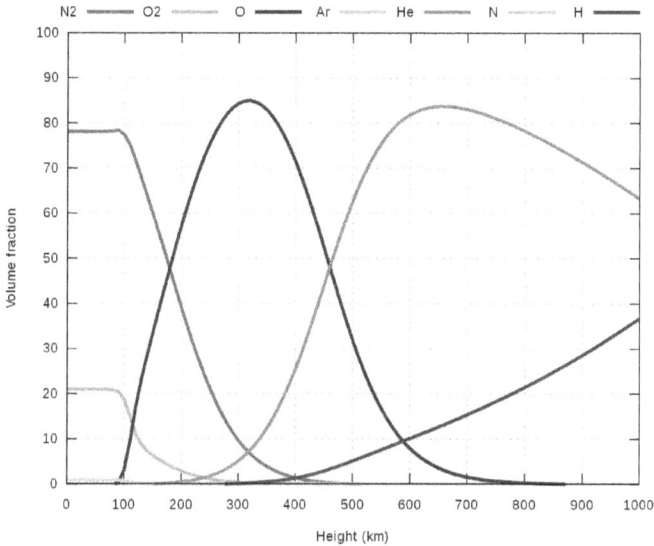

Figure 106: *The volume fraction of the main constituents of the Earth's atmosphere as a function of height according to the MSIS-E-90 atmospheric model.*

- Stratosphere: 12 to 50 km (7 to 31 miles)
- Troposphere: 0 to 12 km (0 to 7 miles)

Exosphere

The exosphere is the outermost layer of Earth's atmosphere (i.e. the upper limit of the atmosphere). It extends from the exobase, which is located at the top of the thermosphere at an altitude of about 700 km above sea level, to about 10,000 km (6,200 mi; 33,000,000 ft) where it merges into the solar wind.

This layer is mainly composed of extremely low densities of hydrogen, helium and several heavier molecules including nitrogen, oxygen and carbon dioxide closer to the exobase. The atoms and molecules are so far apart that they can travel hundreds of kilometers without colliding with one another. Thus, the exosphere no longer behaves like a gas, and the particles constantly escape into space. These free-moving particles follow ballistic trajectories and may migrate in and out of the magnetosphere or the solar wind.

The exosphere is located too far above Earth for any meteorological phenomena to be possible. However, the aurora borealis and aurora australis sometimes occur in the lower part of the exosphere, where they overlap into the thermosphere. The exosphere contains most of the satellites orbiting Earth.

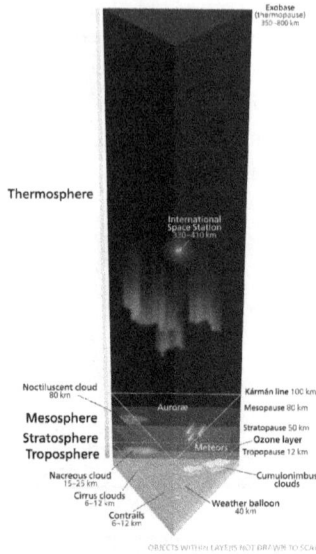

Figure 107: *Earth's atmosphere Lower 4 layers of the atmosphere in 3 dimensions as seen diagonally from above the exobase. Layers drawn to scale, objects within the layers are not to scale. Aurorae shown here at the bottom of the thermosphere can actually form at any altitude in this atmospheric layer.*

Thermosphere

The thermosphere is the second-highest layer of Earth's atmosphere. It extends from the mesopause (which separates it from the mesosphere) at an altitude of about 80 km (50 mi; 260,000 ft) up to the thermopause at an altitude range of 500–1000 km (310–620 mi; 1,600,000–3,300,000 ft). The height of the thermopause varies considerably due to changes in solar activity. Because the thermopause lies at the lower boundary of the exosphere, it is also referred to as the exobase. The lower part of the thermosphere, from 80 to 550 kilometres (50 to 342 mi) above Earth's surface, contains the ionosphere.

The temperature of the thermosphere gradually increases with height. Unlike the stratosphere beneath it, wherein a temperature inversion is due to the absorption of radiation by ozone, the inversion in the thermosphere occurs due to the extremely low density of its molecules. The temperature of this layer can rise as high as 1500 °C (2700 °F), though the gas molecules are so far apart that its temperature in the usual sense is not very meaningful. The air is so rarefied that an individual molecule (of oxygen, for example) travels an average of 1 kilometre (0.62 mi; 3300 ft) between collisions with other molecules.[575]

Although the thermosphere has a high proportion of molecules with high energy, it would not feel hot to a human in direct contact, because its density is too low to conduct a significant amount of energy to or from the skin.

This layer is completely cloudless and free of water vapor. However, non-hydrometeorological phenomena such as the aurora borealis and aurora australis are occasionally seen in the thermosphere. The International Space Station orbits in this layer, between 350 and 420 km (220 and 260 mi).

Mesosphere

The mesosphere is the third highest layer of Earth's atmosphere, occupying the region above the stratosphere and below the thermosphere. It extends from the stratopause at an altitude of about 50 km (31 mi; 160,000 ft) to the mesopause at 80–85 km (50–53 mi; 260,000–280,000 ft) above sea level.

Temperatures drop with increasing altitude to the mesopause that marks the top of this middle layer of the atmosphere. It is the coldest place on Earth and has an average temperature around –85 °C (–120 °F; 190 K).

Just below the mesopause, the air is so cold that even the very scarce water vapor at this altitude can be sublimated into polar-mesospheric noctilucent clouds. These are the highest clouds in the atmosphere and may be visible to the naked eye if sunlight reflects off them about an hour or two after sunset or a similar length of time before sunrise. They are most readily visible when the Sun is around 4 to 16 degrees below the horizon. Lightning-induced discharges known as transient luminous events (TLEs) occasionally form in the mesosphere above tropospheric thunderclouds. The mesosphere is also the layer where most meteors burn up upon atmospheric entrance. It is too high above Earth to be accessible to jet-powered aircraft and balloons, and too low to permit orbital spacecraft. The mesosphere is mainly accessed by sounding rockets and rocket-powered aircraft.

Stratosphere

The stratosphere is the second-lowest layer of Earth's atmosphere. It lies above the troposphere and is separated from it by the tropopause. This layer extends from the top of the troposphere at roughly 12 km (7.5 mi; 39,000 ft) above Earth's surface to the stratopause at an altitude of about 50 to 55 km (31 to 34 mi; 164,000 to 180,000 ft).

The atmospheric pressure at the top of the stratosphere is roughly 1/1000 the pressure at sea level. It contains the ozone layer, which is the part of Earth's atmosphere that contains relatively high concentrations of that gas. The stratosphere defines a layer in which temperatures rise with increasing altitude. This rise in temperature is caused by the absorption of ultraviolet radiation (UV) radiation from the Sun by the ozone layer, which restricts turbulence and mixing. Although the temperature may be –60 °C (–76 °F; 210 K) at the tropopause, the top of the stratosphere is much warmer, and may be near 0 °C.

The stratospheric temperature profile creates very stable atmospheric conditions, so the stratosphere lacks the weather-producing air turbulence that is so prevalent in the troposphere. Consequently, the stratosphere is almost completely free of clouds and other forms of weather. However, polar stratospheric or nacreous clouds are occasionally seen in the lower part of this layer of the atmosphere where the air is coldest. The stratosphere is the highest layer that can be accessed by jet-powered aircraft.

Troposphere

The troposphere is the lowest layer of Earth's atmosphere. It extends from Earth's surface to an average height of about 12 km (7.5 mi; 39,000 ft), although this altitude varies from about 9 km (5.6 mi; 30,000 ft) at the geographic poles to 17 km (11 mi; 56,000 ft) at the Equator, with some variation due to weather. The troposphere is bounded above by the tropopause, a boundary marked in most places by a temperature inversion (i.e. a layer of relatively warm air above a colder one), and in others by a zone which is isothermal with height.

Although variations do occur, the temperature usually declines with increasing altitude in the troposphere because the troposphere is mostly heated through energy transfer from the surface. Thus, the lowest part of the troposphere (i.e. Earth's surface) is typically the warmest section of the troposphere. This promotes vertical mixing (hence, the origin of its name in the Greek word τρόπος, *tropos*, meaning "turn"). The troposphere contains roughly 80% of the mass of Earth's atmosphere. The troposphere is denser than all its overlying atmospheric layers because a larger atmospheric weight sits on top of the troposphere and causes it to be most severely compressed. Fifty percent of the total mass of the atmosphere is located in the lower 5.6 km (3.5 mi; 18,000 ft) of the troposphere.

Nearly all atmospheric water vapor or moisture is found in the troposphere, so it is the layer where most of Earth's weather takes place. It has basically all the weather-associated cloud genus types generated by active wind circulation, although very tall cumulonimbus thunder clouds can penetrate the tropopause from below and rise into the lower part of the stratosphere. Most conventional aviation activity takes place in the troposphere, and it is the only layer that can be accessed by propeller-driven aircraft.

Figure 108: *Space Shuttle Endeavour orbiting in the thermosphere. Because of the angle of the photo, it appears to straddle the stratosphere and mesosphere that actually lie more than 250 km below. The orange layer is the troposphere, which gives way to the whitish stratosphere and then the blue mesosphere.*

Other layers

Within the five principal layers that are largely determined by temperature, several secondary layers may be distinguished by other properties:

- The ozone layer is contained within the stratosphere. In this layer ozone concentrations are about 2 to 8 parts per million, which is much higher than in the lower atmosphere but still very small compared to the main components of the atmosphere. It is mainly located in the lower portion of the stratosphere from about 15–35 km (9.3–21.7 mi; 49,000–115,000 ft), though the thickness varies seasonally and geographically. About 90% of the ozone in Earth's atmosphere is contained in the stratosphere.
- The ionosphere is a region of the atmosphere that is ionized by solar radiation. It is responsible for auroras. During daytime hours, it stretches from 50 to 1,000 km (31 to 621 mi; 160,000 to 3,280,000 ft) and includes the mesosphere, thermosphere, and parts of the exosphere. However, ionization in the mesosphere largely ceases during the night, so auroras are normally seen only in the thermosphere and lower exosphere. The ionosphere forms the inner edge of the magnetosphere. It has practical importance because it influences, for example, radio propagation on Earth.

- The homosphere and heterosphere are defined by whether the atmospheric gases are well mixed. The surface-based homosphere includes the troposphere, stratosphere, mesosphere, and the lowest part of the thermosphere, where the chemical composition of the atmosphere does not depend on molecular weight because the gases are mixed by turbulence. This relatively homogeneous layer ends at the *turbopause* found at about 100 km (62 mi; 330,000 ft), the very edge of space itself as accepted by the FAI, which places it about 20 km (12 mi; 66,000 ft) above the mesopause.

  Above this altitude lies the heterosphere, which includes the exosphere and most of the thermosphere. Here, the chemical composition varies with altitude. This is because the distance that particles can move without colliding with one another is large compared with the size of motions that cause mixing. This allows the gases to stratify by molecular weight, with the heavier ones, such as oxygen and nitrogen, present only near the bottom of the heterosphere. The upper part of the heterosphere is composed almost completely of hydrogen, the lightest element.Wikipedia:Please clarify

- The planetary boundary layer is the part of the troposphere that is closest to Earth's surface and is directly affected by it, mainly through turbulent diffusion. During the day the planetary boundary layer usually is well-mixed, whereas at night it becomes stably stratified with weak or intermittent mixing. The depth of the planetary boundary layer ranges from as little as about 100 metres (330 ft) on clear, calm nights to 3,000 m (9,800 ft) or more during the afternoon in dry regions.

The average temperature of the atmosphere at Earth's surface is 14 °C (57 °F; 287 K) or 15 °C (59 °F; 288 K), depending on the reference.

Physical properties

Pressure and thickness

The average atmospheric pressure at sea level is defined by the International Standard Atmosphere as 101325 pascals (760.00 Torr; 14.6959 psi; 760.00 mmHg). This is sometimes referred to as a unit of standard atmospheres (atm). Total atmospheric mass is 5.1480×10^{18} kg (1.135×10^{19} lb), about 2.5% less than would be inferred from the average sea level pressure and Earth's area of 51007.2 megahectares, this portion being displaced by Earth's mountainous terrain. Atmospheric pressure is the total weight of the air above unit area at the point where the pressure is measured. Thus air pressure varies with location and weather.

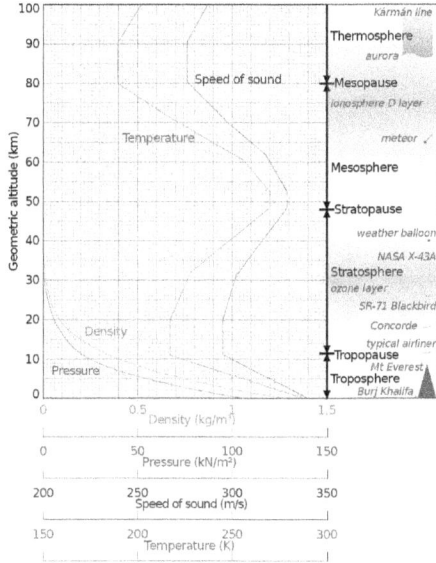

Figure 109: *Comparison of the 1962 US Standard Atmosphere graph of geometric altitude against air density, pressure, the speed of sound and temperature with approximate altitudes of various objects.*[576]

If the entire mass of the atmosphere had a uniform density from sea level, it would terminate abruptly at an altitude of 8.50 km (27,900 ft). It actually decreases exponentially with altitude, dropping by half every 5.6 km (18,000 ft) or by a factor of 1/e every 7.64 km (25,100 ft), the average scale height of the atmosphere below 70 km (43 mi; 230,000 ft). However, the atmosphere is more accurately modeled with a customized equation for each layer that takes gradients of temperature, molecular composition, solar radiation and gravity into account.

In summary, the mass of Earth's atmosphere is distributed approximately as follows:[577]

- 50% is below 5.6 km (18,000 ft).
- 90% is below 16 km (52,000 ft).
- 99.99997% is below 100 km (62 mi; 330,000 ft), the Kármán line. By international convention, this marks the beginning of space where human travelers are considered astronauts.

By comparison, the summit of Mt. Everest is at 8,848 m (29,029 ft); commercial airliners typically cruise between 10 and 13 km (33,000 and 43,000 ft) where the thinner air improves fuel economy; weather balloons reach 30.4 km

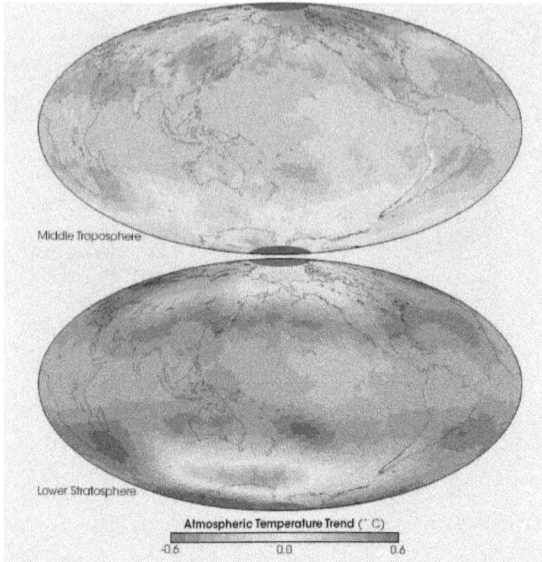

Figure 110: *Temperature trends in two thick layers of the atmosphere as measured between January 1979 and December 2005 by Microwave Sounding Units and Advanced Microwave Sounding Units on NOAA weather satellites. The instruments record microwaves emitted from oxygen molecules in the atmosphere. Source:*

(100,000 ft) and above; and the highest X-15 flight in 1963 reached 108.0 km (354,300 ft).

Even above the Kármán line, significant atmospheric effects such as auroras still occur. Meteors begin to glow in this region, though the larger ones may not burn up until they penetrate more deeply. The various layers of Earth's ionosphere, important to HF radio propagation, begin below 100 km and extend beyond 500 km. By comparison, the International Space Station and Space Shuttle typically orbit at 350–400 km, within the F-layer of the ionosphere where they encounter enough atmospheric drag to require reboosts every few months. Depending on solar activity, satellites can experience noticeable atmospheric drag at altitudes as high as 700–800 km.

Temperature and speed of sound

The division of the atmosphere into layers mostly by reference to temperature is discussed above. Temperature decreases with altitude starting at sea level, but variations in this trend begin above 11 km, where the temperature stabilizes through a large vertical distance through the rest of the troposphere. In

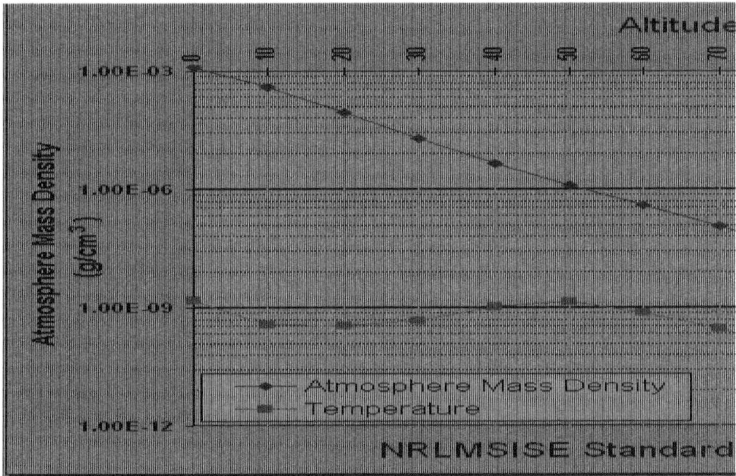

Figure 111: *Temperature and mass density against altitude from the NRLMSISE-00 standard atmosphere model (the eight dotted lines in each "decade" are at the eight cubes 8, 27, 64, ..., 729)*

the stratosphere, starting above about 20 km, the temperature increases with height, due to heating within the ozone layer caused by capture of significant ultraviolet radiation from the Sun by the dioxygen and ozone gas in this region. Still another region of increasing temperature with altitude occurs at very high altitudes, in the aptly-named thermosphere above 90 km.

Because in an ideal gas of constant composition the speed of sound depends only on temperature and not on the gas pressure or density, the speed of sound in the atmosphere with altitude takes on the form of the complicated temperature profile (see illustration to the right), and does not mirror altitudinal changes in density or pressure.

Density and mass

The density of air at sea level is about 1.2 kg/m^3 (1.2 g/L, 0.0012 g/cm^3). Density is not measured directly but is calculated from measurements of temperature, pressure and humidity using the equation of state for air (a form of the ideal gas law). Atmospheric density decreases as the altitude increases. This variation can be approximately modeled using the barometric formula. More sophisticated models are used to predict orbital decay of satellites.

The average mass of the atmosphere is about 5 quadrillion (5×10^{15}) tonnes or 1/1,200,000 the mass of Earth. According to the American National Center for Atmospheric Research, "The total mean mass of the atmosphere is 5.1480×10^{18} kg with an annual range due to water vapor of 1.2 or 1.5×10^{15} kg, depending on whether surface pressure or water vapor data are used; somewhat smaller than the previous estimate. The mean mass of water vapor is estimated as 1.27×10^{16} kg and the dry air mass as $5.1352 \pm 0.0003 \times 10^{18}$ kg."

Optical properties

Solar radiation (or sunlight) is the energy Earth receives from the Sun. Earth also emits radiation back into space, but at longer wavelengths that we cannot see. Part of the incoming and emitted radiation is absorbed or reflected by the atmosphere. In May 2017, glints of light, seen as twinkling from an orbiting satellite a million miles away, were found to be reflected light from ice crystals in the atmosphere.

Scattering

When light passes through Earth's atmosphere, photons interact with it through *scattering*. If the light does not interact with the atmosphere, it is called *direct radiation* and is what you see if you were to look directly at the Sun. *Indirect radiation* is light that has been scattered in the atmosphere. For example, on an overcast day when you cannot see your shadow there is no direct radiation reaching you, it has all been scattered. As another example, due to a phenomenon called Rayleigh scattering, shorter (blue) wavelengths scatter more easily than longer (red) wavelengths. This is why the sky looks blue; you are seeing scattered blue light. This is also why sunsets are red. Because the Sun is close to the horizon, the Sun's rays pass through more atmosphere than normal to reach your eye. Much of the blue light has been scattered out, leaving the red light in a sunset.

Absorption

Different molecules absorb different wavelengths of radiation. For example, O_2 and O_3 absorb almost all wavelengths shorter than 300 nanometers. Water (H_2O) absorbs many wavelengths above 700 nm. When a molecule absorbs a photon, it increases the energy of the molecule. This heats the atmosphere, but the atmosphere also cools by emitting radiation, as discussed below.

The combined absorption spectra of the gases in the atmosphere leave "windows" of low opacity, allowing the transmission of only certain bands of light. The optical window runs from around 300 nm (ultraviolet-C) up into the range humans can see, the visible spectrum (commonly called light), at roughly 400–700 nm and continues to the infrared to around 1100 nm. There

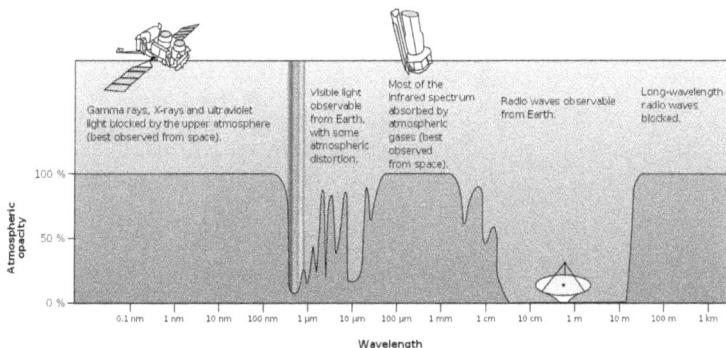

Figure 112: *Rough plot of Earth's atmospheric transmittance (or opacity) to various wavelengths of electromagnetic radiation, including visible light.*

are also infrared and radio windows that transmit some infrared and radio waves at longer wavelengths. For example, the radio window runs from about one centimeter to about eleven-meter waves.

Emission

Emission is the opposite of absorption, it is when an object emits radiation. Objects tend to emit amounts and wavelengths of radiation depending on their "black body" emission curves, therefore hotter objects tend to emit more radiation, with shorter wavelengths. Colder objects emit less radiation, with longer wavelengths. For example, the Sun is approximately 6,000 K (5,730 °C; 10,340 °F), its radiation peaks near 500 nm, and is visible to the human eye. Earth is approximately 290 K (17 °C; 62 °F), so its radiation peaks near 10,000 nm, and is much too long to be visible to humans.

Because of its temperature, the atmosphere emits infrared radiation. For example, on clear nights Earth's surface cools down faster than on cloudy nights. This is because clouds (H_2O) are strong absorbers and emitters of infrared radiation. This is also why it becomes colder at night at higher elevations.

The greenhouse effect is directly related to this absorption and emission effect. Some gases in the atmosphere absorb and emit infrared radiation, but do not interact with sunlight in the visible spectrum. Common examples of these are CO_2 and H_2O.

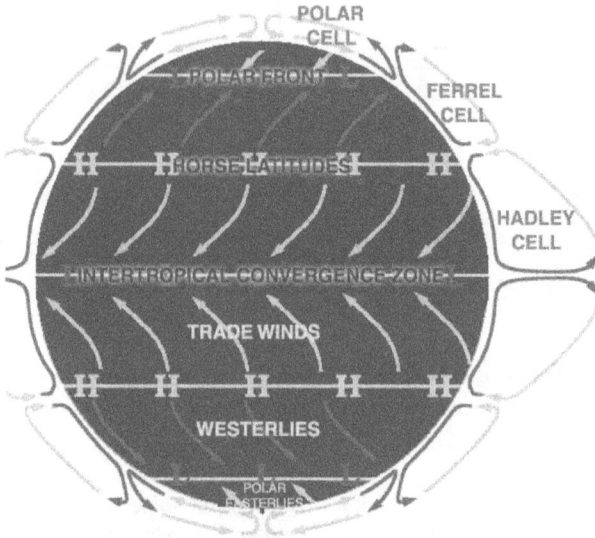

Figure 113: *An idealised view of three large circulation cells.*

Refractive index

The refractive index of air is close to, but just greater than 1. Systematic variations in refractive index can lead to the bending of light rays over long optical paths. One example is that, under some circumstances, observers onboard ships can see other vessels just over the horizon because light is refracted in the same direction as the curvature of Earth's surface.

The refractive index of air depends on temperature, giving rise to refraction effects when the temperature gradient is large. An example of such effects is the mirage.

Circulation

Atmospheric circulation is the large-scale movement of air through the troposphere, and the means (with ocean circulation) by which heat is distributed around Earth. The large-scale structure of the atmospheric circulation varies from year to year, but the basic structure remains fairly constant because it is determined by Earth's rotation rate and the difference in solar radiation between the equator and poles.

Evolution of Earth's atmosphere

Earliest atmosphere

The first atmosphere consisted of gases in the solar nebula, primarily hydrogen. There were probably simple hydrides such as those now found in the gas giants (Jupiter and Saturn), notably water vapor, methane and ammonia.

Second atmosphere

Outgassing from volcanism, supplemented by gases produced during the late heavy bombardment of Earth by huge asteroids, produced the next atmosphere, consisting largely of nitrogen plus carbon dioxide and inert gases. A major part of carbon-dioxide emissions dissolved in water and reacted with metals such as calcium and magnesium during weathering of crustal rocks to form carbonates that were deposited as sediments. Water-related sediments have been found that date from as early as 3.8 billion years ago.[578]

About 3.4 billion years ago, nitrogen formed the major part of the then stable "second atmosphere". The influence of life has to be taken into account rather soon in the history of the atmosphere, because hints of early life-forms appear as early as 3.5 billion years ago.[579] How Earth at that time maintained a climate warm enough for liquid water and life, if the early Sun put out 30% lower solar radiance than today, is a puzzle known as the "faint young Sun paradox".

The geological record however shows a continuous relatively warm surface during the complete early temperature record of Earth – with the exception of one cold glacial phase about 2.4 billion years ago. In the late Archean Eon an oxygen-containing atmosphere began to develop, apparently produced by photosynthesizing cyanobacteria (see Great Oxygenation Event), which have been found as stromatolite fossils from 2.7 billion years ago. The early basic carbon isotopy (isotope ratio proportions) strongly suggests conditions similar to the current, and that the fundamental features of the carbon cycle became established as early as 4 billion years ago.

Ancient sediments in the Gabon dating from between about 2,150 and 2,080 million years ago provide a record of Earth's dynamic oxygenation evolution. These fluctuations in oxygenation were likely driven by the Lomagundi carbon isotope excursion.

Oxygen Content of Earth's Atmosphere
During the Course of the Last Billion Years

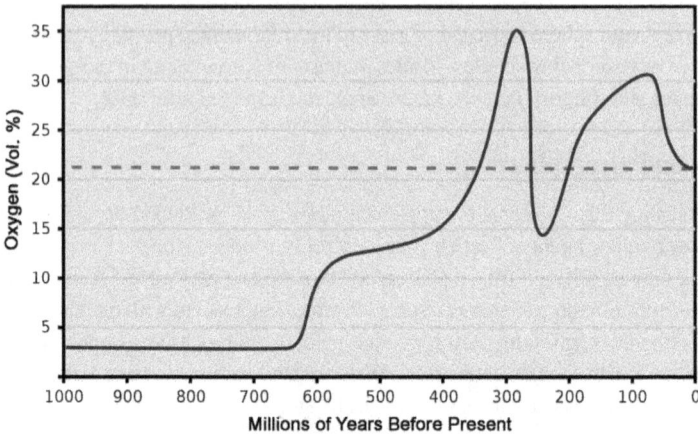

Figure 114: *Oxygen content of the atmosphere over the last billion years*

Third atmosphere

The constant re-arrangement of continents by plate tectonics influences the long-term evolution of the atmosphere by transferring carbon dioxide to and from large continental carbonate stores. Free oxygen did not exist in the atmosphere until about 2.4 billion years ago during the Great Oxygenation Event and its appearance is indicated by the end of the banded iron formations.

Before this time, any oxygen produced by photosynthesis was consumed by oxidation of reduced materials, notably iron. Molecules of free oxygen did not start to accumulate in the atmosphere until the rate of production of oxygen began to exceed the availability of reducing materials that removed oxygen. This point signifies a shift from a reducing atmosphere to an oxidizing atmosphere. O_2 showed major variations until reaching a steady state of more than 15% by the end of the Precambrian.[580] The following time span from 541 million years ago to the present day is the Phanerozoic Eon, during the earliest period of which, the Cambrian, oxygen-requiring metazoan life forms began to appear.

The amount of oxygen in the atmosphere has fluctuated over the last 600 million years, reaching a peak of about 30% around 280 million years ago, significantly higher than today's 21%. Two main processes govern changes in the atmosphere: Plants use carbon dioxide from the atmosphere, releasing

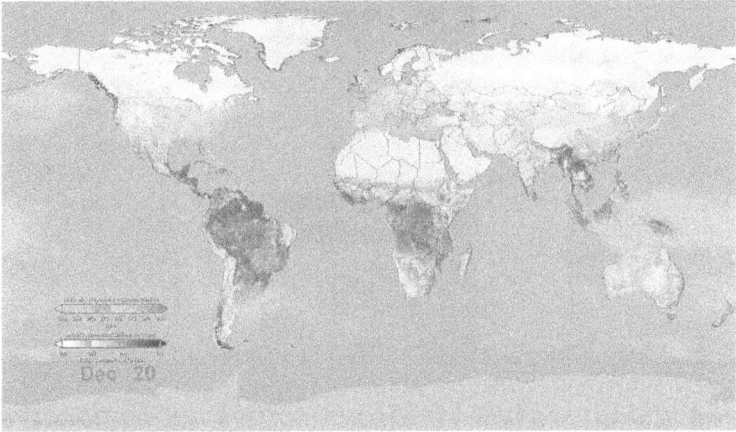

Figure 115: *Animation shows the buildup of tropospheric CO_2 in the Northern Hemisphere with a maximum around May. The maximum in the vegetation cycle follows in the late summer. Following the peak in vegetation, the drawdown of atmospheric CO_2 due to photosynthesis is apparent, particularly over the boreal forests.*

oxygen. Breakdown of pyrite and volcanic eruptions release sulfur into the atmosphere, which oxidizes and hence reduces the amount of oxygen in the atmosphere. However, volcanic eruptions also release carbon dioxide, which plants can convert to oxygen. The exact cause of the variation of the amount of oxygen in the atmosphere is not known. Periods with much oxygen in the atmosphere are associated with rapid development of animals. Today's atmosphere contains 21% oxygen, which is great enough for this rapid development of animals.[581]

Air pollution

Air pollution is the introduction into the atmosphere of chemicals, particulate matter or biological materials that cause harm or discomfort to organisms.[582] Stratospheric ozone depletion is caused by air pollution, chiefly from chlorofluorocarbons and other ozone-depleting substances.

The scientific consensus is that the anthropogenic greenhouse gases currently accumulating in the atmosphere are the main cause of global warming.

Images from space

On October 19, 2015 NASA started a website containing daily images of the full sunlit side of Earth on http://epic.gsfc.nasa.gov/. The images are taken

from the Deep Space Climate Observatory (DSCOVR) and show Earth as it rotates during a day.

<templatestyles src="Gallery/styles.css" />

Blue light is scattered more than other wavelengths by the gases in the atmosphere, giving Earth a blue halo when seen from space.

The geomagnetic storms cause beautiful displays of aurora across the atmosphere.

Limb view, of Earth's atmosphere. Colors roughly denote the layers of the atmosphere.

This image shows the Moon at the centre, with the limb of Earth near the bottom transitioning into the orange-colored troposphere. The troposphere ends abruptly at the tropopause, which appears in the image as the sharp boundary between the orange- and blue-colored atmosphere. The silvery-blue noctilucent clouds extend far above Earth's troposphere.

Earth's atmosphere backlit by the Sun in an eclipse observed from deep space onboard Apollo 12 in 1969.

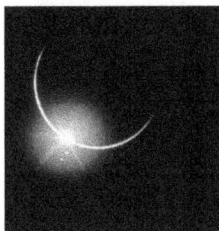

External links

 Wikimedia Commons has media related to *Earth's atmosphere*.

> Wikiquote has quotations related to: *Air*

- Interactive global map of current atmospheric and ocean surface conditions.[583]

Weather

<indicator name="pp-default"> 🔒 </indicator>

Part of the nature series
Weather
Calendar seasons
<ul><li>Winter</li><li>Spring</li><li>Summer</li><li>Autumn</li></ul>
Tropical seasons
<ul><li>Dry season</li><li>Wet season</li></ul>
Storms
<ul><li>Cloud</li><li>Cumulonimbus cloud</li><li>Arcus cloud</li><li>Downburst</li><li>Microburst</li><li>Heat burst</li><li>Dust storm</li><li>Simoom</li><li>Haboob</li><li>Monsoon</li><li>Gale</li><li>Sirocco</li><li>Firestorm</li><li>Lightning</li><li>Supercell</li><li>Thunderstorm</li><li>Severe thunderstorm</li><li>Thundersnow</li><li>Storm surge</li><li>Tornado</li><li>Cyclone</li></ul>

- Mesocyclone
- Anticyclone
- Tropical cyclone (Hurricane)
- Extratropical cyclone
- European windstorm
- Atlantic Hurricane
- Typhoon
- Derecho
- Landspout
- Dust devil
- Fire whirl
- Waterspout
- Winter storm
  - Ice storm
  - Blizzard
  - Ground blizzard
  - Snowsquall

Precipitation

- Drizzle (Freezing drizzle)
- Graupel
- Hail
- Ice pellets (Diamond dust)
- Rain (Freezing rain)
- Cloudburst
- Snow
  - Rain and snow mixed
  - Snow grains
  - Snow roller
  - Slush

Topics

- Air pollution
- Atmosphere

  - Chemistry
  - Convection
  - Physics
  - River
- Climate
- Cloud
  - Physics
- Fog
- Cold wave
- Heat wave
- Jet stream
- Meteorology
- Severe weather
  - List
  - Extreme

Weather is the state of the atmosphere, describing for example the degree to which it is hot or cold, wet or dry, calm or stormy, clear or cloudy.[585] Most weather phenomena occur in the lowest level of the atmosphere, the troposphere,[586,587] just below the stratosphere. Weather refers to day-to-day temperature and precipitation activity, whereas climate is the term for the averaging of atmospheric conditions over longer periods of time. When used without qualification, "weather" is generally understood to mean the weather of Earth.

Weather is driven by air pressure, temperature and moisture differences between one place and another. These differences can occur due to the sun's angle at any particular spot, which varies with latitude. The strong temperature contrast between polar and tropical air gives rise to the largest scale atmospheric circulations: the Hadley Cell, the Ferrel Cell, the Polar Cell, and the jet stream. Weather systems in the mid-latitudes, such as extratropical cyclones, are caused by instabilities of the jet stream flow. Because the Earth's axis is tilted relative to its orbital plane, sunlight is incident at different angles at different times of the year. On Earth's surface, temperatures usually range ±40 °C (–40 °F to 100 °F) annually. Over thousands of years, changes in Earth's orbit can affect the amount and distribution of solar energy received by the Earth, thus influencing long-term climate and global climate change.

Surface temperature differences in turn cause pressure differences. Higher altitudes are cooler than lower altitudes, as most atmospheric heating is due to contact with the Earth's surface while radiative losses to space are mostly constant. Weather forecasting is the application of science and technology to predict the state of the atmosphere for a future time and a given location. The Earth's weather system is a chaotic system; as a result, small changes to one part of the system can grow to have large effects on the system as a whole. Human attempts to control the weather have occurred throughout history, and there is evidence that human activities such as agriculture and industry have modified weather patterns.

Studying how the weather works on other planets has been helpful in understanding how weather works on Earth. A famous landmark in the Solar System, Jupiter's *Great Red Spot*, is an anticyclonic storm known to have existed for at least 300 years. However, weather is not limited to planetary bodies.

Figure 116: *Thunderstorm near Garajau, Madeira*

A star's corona is constantly being lost to space, creating what is essentially a very thin atmosphere throughout the Solar System. The movement of mass ejected from the Sun is known as the solar wind.

Causes

On Earth, the common weather phenomena include wind, cloud, rain, snow, fog and dust storms. Less common events include natural disasters such as tornadoes, hurricanes, typhoons and ice storms. Almost all familiar weather phenomena occur in the troposphere (the lower part of the atmosphere). Weather does occur in the stratosphere and can affect weather lower down in the troposphere, but the exact mechanisms are poorly understood.

Weather occurs primarily due to air pressure, temperature and moisture differences between one place to another. These differences can occur due to the sun angle at any particular spot, which varies by latitude from the tropics. In other words, the farther from the tropics one lies, the lower the sun angle is, which causes those locations to be cooler due the spread of the sunlight over a greater surface.[588] The strong temperature contrast between polar and tropical air gives rise to the large scale atmospheric circulation cells and the jet stream.[589] Weather systems in the mid-latitudes, such as extratropical cyclones, are caused by instabilities of the jet stream flow (see baroclinity).[590]

Figure 117: *Cumulus mediocris cloud surrounded by stratocumulus*

Weather systems in the tropics, such as monsoons or organized thunderstorm systems, are caused by different processes.

Because the Earth's axis is tilted relative to its orbital plane, sunlight is incident at different angles at different times of the year. In June the Northern Hemisphere is tilted towards the sun, so at any given Northern Hemisphere latitude sunlight falls more directly on that spot than in December (see Effect of sun angle on climate).[591] This effect causes seasons. Over thousands to hundreds of thousands of years, changes in Earth's orbital parameters affect the amount and distribution of solar energy received by the Earth and influence long-term climate. (See Milankovitch cycles).[592]

The uneven solar heating (the formation of zones of temperature and moisture gradients, or frontogenesis) can also be due to the weather itself in the form of cloudiness and precipitation.[593] Higher altitudes are typically cooler than lower altitudes, which the result of higher surface temperature and radiational heating, which produces the adiabatic lapse rate. In some situations, the temperature actually increases with height. This phenomenon is known as an inversion and can cause mountaintops to be warmer than the valleys below. Inversions can lead to the formation of fog and often act as a cap that suppresses thunderstorm development. On local scales, temperature differences can occur because different surfaces (such as oceans, forests, ice sheets, or man-made objects) have differing physical characteristics such as reflectivity, roughness, or moisture content.

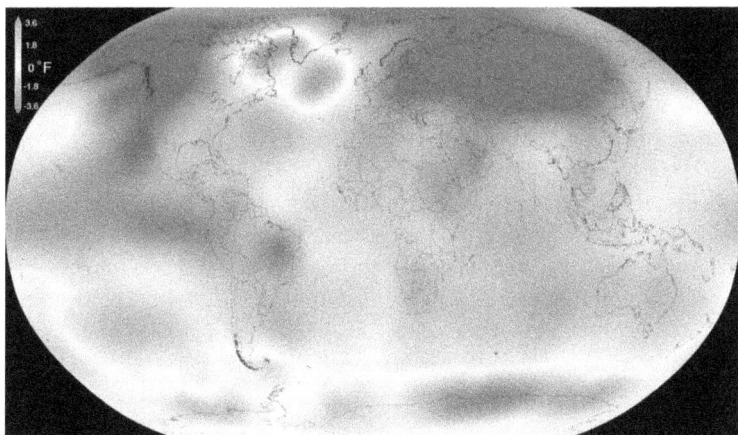

Figure 118:
2015 – Warmest Global Year on Record (since 1880) – Colors indicate temperature anomalies (NASA/NOAA; 20 January 2016).

Surface temperature differences in turn cause pressure differences. A hot surface warms the air above it causing it to expand and lower the density and the resulting surface air pressure.[594] The resulting horizontal pressure gradient moves the air from higher to lower pressure regions, creating a wind, and the Earth's rotation then causes deflection of this air flow due to the Coriolis effect.[595] The simple systems thus formed can then display emergent behaviour to produce more complex systems and thus other weather phenomena. Large scale examples include the Hadley cell while a smaller scale example would be coastal breezes.

The atmosphere is a chaotic system. As a result, small changes to one part of the system can accumulate and magnify to cause large effects on the system as a whole.[596] This atmospheric instability makes weather forecasting less predictable than tides or eclipses. Although it is difficult to accurately predict weather more than a few days in advance, weather forecasters are continually working to extend this limit through meteorological research and refining current methodologies in weather prediction. However, it is theoretically impossible to make useful day-to-day predictions more than about two weeks ahead, imposing an upper limit to potential for improved prediction skill.

Shaping the planet Earth

Weather is one of the fundamental processes that shape the Earth. The process of weathering breaks down the rocks and soils into smaller fragments and then into their constituent substances.[597] During rains precipitation, the water droplets absorb and dissolve carbon dioxide from the surrounding air. This causes the rainwater to be slightly acidic, which aids the erosive properties of water. The released sediment and chemicals are then free to take part in chemical reactions that can affect the surface further (such as acid rain), and sodium and chloride ions (salt) deposited in the seas/oceans. The sediment may reform in time and by geological forces into other rocks and soils. In this way, weather plays a major role in erosion of the surface.[598]

Effect on humans

Weather, seen from an anthropological perspective, is something all humans in the world constantly experience through their senses, at least while being outside. There are socially and scientifically constructed understandings of what weather is, what makes it change, the effect it has on humans in different situations, etc. Therefore, weather is something people often communicate about.

Effects on populations

Weather has played a large and sometimes direct part in human history. Aside from climatic changes that have caused the gradual drift of populations (for example the desertification of the Middle East, and the formation of land bridges during glacial periods), extreme weather events have caused smaller scale population movements and intruded directly in historical events. One such event is the saving of Japan from invasion by the Mongol fleet of Kublai Khan by the Kamikaze winds in 1281.[599] French claims to Florida came to an end in 1565 when a hurricane destroyed the French fleet, allowing Spain to conquer Fort Caroline.[600] More recently, Hurricane Katrina redistributed over one million people from the central Gulf coast elsewhere across the United States, becoming the largest diaspora in the history of the United States.[601]

The Little Ice Age caused crop failures and famines in Europe. The 1690s saw the worst famine in France since the Middle Ages. Finland suffered a severe famine in 1696–1697, during which about one-third of the Finnish population died.[602]

Figure 119: *New Orleans, Louisiana, after being struck by Hurricane Katrina. Katrina was a Category 3 hurricane when it struck although it had been a category 5 hurricane in the Gulf of Mexico.*

Forecasting

Weather forecasting is the application of science and technology to predict the state of the atmosphere for a future time and a given location. Human beings have attempted to predict the weather informally for millennia, and formally since at least the nineteenth century.[603] Weather forecasts are made by collecting quantitative data about the current state of the atmosphere and using scientific understanding of atmospheric processes to project how the atmosphere will evolve.[604]

Once an all-human endeavor based mainly upon changes in barometric pressure, current weather conditions, and sky condition,[605,606] forecast models are now used to determine future conditions. On the other hand, human input is still required to pick the best possible forecast model to base the forecast upon, which involve many disciplines such as pattern recognition skills, teleconnections, knowledge of model performance, and knowledge of model biases.

The chaotic nature of the atmosphere, the massive computational power required to solve the equations that describe the atmosphere, error involved in measuring the initial conditions, and an incomplete understanding of atmospheric processes mean that forecasts become less accurate as the difference in current time and the time for which the forecast is being made (the *range*

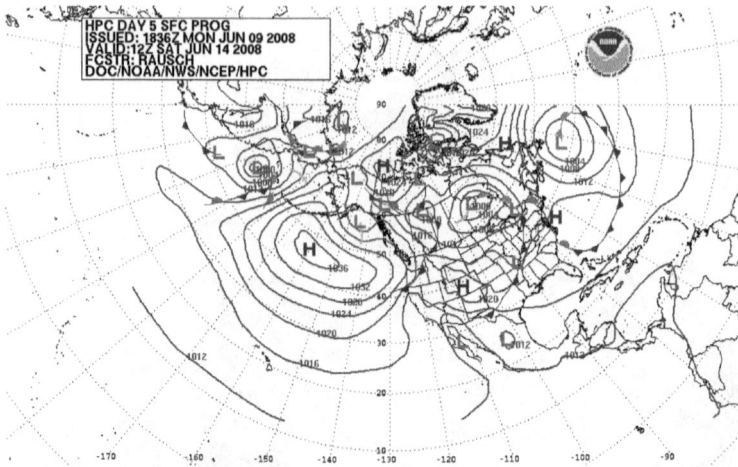

Figure 120: *Forecast of surface pressures five days into the future for the north Pacific, North America, and north Atlantic Ocean as on 9 June 2008*

of the forecast) increases. The use of ensembles and model consensus helps to narrow the error and pick the most likely outcome.[607,608,609]

There are a variety of end users to weather forecasts. Weather warnings are important forecasts because they are used to protect life and property.[610,611] Forecasts based on temperature and precipitation are important to agriculture,[612,613,614,615] and therefore to commodity traders within stock markets. Temperature forecasts are used by utility companies to estimate demand over coming days.[616,617,618]

In some areas, people use weather forecasts to determine what to wear on a given day. Since outdoor activities are severely curtailed by heavy rain, snow and the wind chill, forecasts can be used to plan activities around these events, and to plan ahead to survive through them.

Modification

The aspiration to control the weather is evident throughout human history: from ancient rituals intended to bring rain for crops to the U.S. Military Operation Popeye, an attempt to disrupt supply lines by lengthening the North Vietnamese monsoon. The most successful attempts at influencing weather involve cloud seeding; they include the fog- and low stratus dispersion techniques employed by major airports, techniques used to increase winter precipitation over mountains, and techniques to suppress hail.[619] A recent example of weather control was China's preparation for the 2008 Summer Olympic Games. China shot 1,104 rain dispersal rockets from 21 sites in the city of Beijing in an effort to keep rain away from the opening ceremony of the games on 8 August 2008. Guo Hu, head of the Beijing Municipal Meteorological Bureau (BMB), confirmed the success of the operation with 100 millimeters falling in Baoding City of Hebei Province, to the southwest and Beijing's Fangshan District recording a rainfall of 25 millimeters.

Whereas there is inconclusive evidence for these techniques' efficacy, there is extensive evidence that human activity such as agriculture and industry results in inadvertent weather modification:

- Acid rain, caused by industrial emission of sulfur dioxide and nitrogen oxides into the atmosphere, adversely affects freshwater lakes, vegetation, and structures.
- Anthropogenic pollutants reduce air quality and visibility.
- Climate change caused by human activities that emit greenhouse gases into the air is expected to affect the frequency of extreme weather events such as drought, extreme temperatures, flooding, high winds, and severe storms.[620]
- Heat, generated by large metropolitan areas have been shown to minutely affect nearby weather, even at distances as far as 1,600 kilometres (990 mi).

The effects of inadvertent weather modification may pose serious threats to many aspects of civilization, including ecosystems, natural resources, food and fiber production, economic development, and human health.[621]

Microscale meteorology

Microscale meteorology is the study of short-lived atmospheric phenomena smaller than mesoscale, about 1 km or less. These two branches of meteorology are sometimes grouped together as "mesoscale and microscale meteorology" (MMM) and together study all phenomena smaller than synoptic scale; that is they study features generally too small to be depicted on a weather map. These include small and generally fleeting cloud "puffs" and other small cloud features.

Extremes on Earth

On Earth, temperatures usually range ±40 °C (100 °F to –40 °F) annually. The range of climates and latitudes across the planet can offer extremes of temperature outside this range. The coldest air temperature ever recorded on Earth is –89.2 °C (–128.6 °F), at Vostok Station, Antarctica on 21 July

Figure 121: *Early morning sunshine over Bratislava, Slovakia. February 2008.*

Figure 122: *The same area, just three hours later, after light snowfall*

1983. The hottest air temperature ever recorded was 57.7 °C (135.9 °F) at 'Aziziya, Libya, on 13 September 1922,[622] but that reading is queried. The highest recorded average annual temperature was 34.4 °C (93.9 °F) at Dallol, Ethiopia.[623] The coldest recorded average annual temperature was –55.1 °C (–67.2 °F) at Vostok Station, Antarctica.[624]

The coldest average annual temperature in a permanently inhabited location is at Eureka, Nunavut, in Canada, where the annual average temperature is –19.7 °C (–3.5 °F).[625]

Extraterrestrial within the Solar System

Studying how the weather works on other planets has been seen as helpful in understanding how it works on Earth. Weather on other planets follows many of the same physical principles as weather on Earth, but occurs on different scales and in atmospheres having different chemical composition. The Cassini–Huygens mission to Titan discovered clouds formed from methane or ethane which deposit rain composed of liquid methane and other organic compounds. Earth's atmosphere includes six latitudinal circulation zones, three in each hemisphere.[626] In contrast, Jupiter's banded appearance shows many such zones,[627] Titan has a single jet stream near the 50th parallel north latitude,[628] and Venus has a single jet near the equator.[629]

One of the most famous landmarks in the Solar System, Jupiter's *Great Red Spot*, is an anticyclonic storm known to have existed for at least 300 years.

Figure 123: *Jupiter's Great Red Spot in February 1979, photographed by the unmanned Voyager 1 NASA space probe.*

On other gas giants, the lack of a surface allows the wind to reach enormous speeds: gusts of up to 600 metres per second (about 2,100 km/h or 1,300 mph) have been measured on the planet Neptune. This has created a puzzle for planetary scientists. The weather is ultimately created by solar energy and the amount of energy received by Neptune is only about $1/900$ of that received by Earth, yet the intensity of weather phenomena on Neptune is far greater than on Earth. The strongest planetary winds discovered so far are on the extrasolar planet HD 189733 b, which is thought to have easterly winds moving at more than 9,600 kilometres per hour (6,000 mph).

Space weather

Weather is not limited to planetary bodies. Like all stars, the sun's corona is constantly being lost to space, creating what is essentially a very thin atmosphere throughout the Solar System. The movement of mass ejected from the Sun is known as the solar wind. Inconsistencies in this wind and larger events on the surface of the star, such as coronal mass ejections, form a system that has features analogous to conventional weather systems (such as pressure and wind) and is generally known as space weather. Coronal mass ejections have been tracked as far out in the solar system as Saturn.[630] The activity of this

Figure 124: *Aurora Borealis*

system can affect planetary atmospheres and occasionally surfaces. The inter-action of the solar wind with the terrestrial atmosphere can produce spectac-ular aurorae,[631] and can play havoc with electrically sensitive systems such as electricity grids and radio signals.[632]

External links

> Wikimedia Commons has media related to *Weather*.

|))) | Wikiquote has quotations related to: *Weather* |

<indicator name="good-star"> ⊕ </indicator>

Climate

<indicator name="pp-autoreview"> 🔒 </indicator>

Atmospheric sciences

Atmospheric physics
Atmospheric dynamics (category)
Atmospheric chemistry (category)

Meteorology

Weather (category) · (portal)
Tropical cyclone (category)

Climatology

Climate (category)
Climate change (category)
Global warming (category) · (portal)

Glossaries

Glossary of meteorology

- $\underline{v}$
- $\underline{t}$
- $\underline{e}$[633]

Part of the nature series
Weather
Calendar seasons
<ul><li>Winter</li><li>Spring</li><li>Summer</li><li>Autumn</li></ul>
Tropical seasons

- Dry season
- Wet season

Storms

- Cloud
- Cumulonimbus cloud
- Arcus cloud
- Downburst
- Microburst
- Heat burst
- Dust storm
- Simoom
- Haboob
- Monsoon
- Gale
- Sirocco
- Firestorm
- Lightning
- Supercell
- Thunderstorm
- Severe thunderstorm
- Thundersnow
- Storm surge
- Tornado
- Cyclone
- Mesocyclone
- Anticyclone
- Tropical cyclone (Hurricane)
- Extratropical cyclone
- European windstorm
- Atlantic Hurricane
- Typhoon
- Derecho
- Landspout
- Dust devil
- Fire whirl
- Waterspout
- Winter storm
  - Ice storm
  - Blizzard
  - Ground blizzard
  - Snowsquall

Precipitation

- Drizzle (Freezing drizzle)

- Graupel
- Hail
- Ice pellets (Diamond dust)
- Rain (Freezing rain)
- Cloudburst
- Snow
  - Rain and snow mixed
  - Snow grains
  - Snow roller
  - Slush

Topics

- Air pollution
- Atmosphere

  - Chemistry
  - Convection
  - Physics
  - River
- Climate
- Cloud
  - Physics
- Fog
- Cold wave
- Heat wave
- Jet stream
- Meteorology
- Severe weather
  - List
  - Extreme
- Weather forecasting

Glossaries

- Glossary of meteorology

Weather portal

- $\underline{v}$
- $\underline{t}$
- $\underline{e}^{634}$

Climate is the statistics of weather over long periods of time. It is measured by assessing the patterns of variation in temperature, humidity, atmospheric pressure, wind, precipitation, atmospheric particle count and other meteorological variables in a given region over long periods of time. Climate differs from weather, in that weather only describes the short-term conditions of these variables in a given region.

A region's climate is generated by the **climate system**, which has five components: atmosphere, hydrosphere, cryosphere, lithosphere, and biosphere.[635]

The climate of a location is affected by its latitude, terrain, and altitude, as well as nearby water bodies and their currents. Climates can be classified according to the average and the typical ranges of different variables, most commonly temperature and precipitation. The most commonly used classification scheme was the Köppen climate classification. The Thornthwaite system, in use since 1948, incorporates evapotranspiration along with temperature and precipitation information and is used in studying biological diversity and how climate change affects it. The Bergeron and Spatial Synoptic Classification systems focus on the origin of air masses that define the climate of a region.

Paleoclimatology is the study of ancient climates. Since direct observations of climate are not available before the 19th century, paleoclimates are inferred from proxy variables that include non-biotic evidence such as sediments found in lake beds and ice cores, and biotic evidence such as tree rings and coral. Climate models are mathematical models of past, present and future climates. Climate change may occur over long and short timescales from a variety of factors; recent warming is discussed in global warming. Global warming results in redistributions. For example, "a 3°C change in mean annual temperature corresponds to a shift in isotherms of approximately 300–400 km in latitude (in the temperate zone) or 500 m in elevation. Therefore, species are expected to move upwards in elevation or towards the poles in latitude in response to shifting climate zones".

Definition

Climate (from Ancient Greek *klima*, meaning *inclination*) is commonly defined as the weather averaged over a long period. The standard averaging period is 30 years, but other periods may be used depending on the purpose. Climate also includes statistics other than the average, such as the magnitudes of day-to-day or year-to-year variations. The Intergovernmental Panel on Climate Change (IPCC) 2001 glossary definition is as follows:

<templatestyles src="Template:Quote/styles.css"/>

> *Climate in a narrow sense is usually defined as the "average weather," or more rigorously, as the statistical description in terms of the mean and variability of relevant quantities over a period ranging from months to thousands or millions of years. The classical period is 30 years, as defined by the World Meteorological Organization (WMO). These quantities are most often surface variables such as temperature, precipitation, and wind. Climate in a wider sense is the state, including a statistical description, of the climate system.*[636]

The World Meteorological Organization (WMO) describes climate "normals" as "reference points used by climatologists to compare current climatological trends to that of the past or what is considered 'normal'. A Normal is defined as the arithmetic average of a climate element (e.g. temperature) over a 30-year period. A 30 year period is used, as it is long enough to filter out any interannual variation or anomalies, but also short enough to be able to show longer climatic trends." The WMO originated from the International Meteorological Organization which set up a technical commission for climatology in 1929. At its 1934 Wiesbaden meeting the technical commission designated the thirty-year period from 1901 to 1930 as the reference time frame for climatological standard normals. In 1982 the WMO agreed to update climate normals, and these were subsequently completed on the basis of climate data from 1 January 1961 to 31 December 1990.

The difference between climate and weather is usefully summarized by the popular phrase "Climate is what you expect, weather is what you get."[637] Over historical time spans there are a number of nearly constant variables that determine climate, including latitude, altitude, proportion of land to water, and proximity to oceans and mountains. These change only over periods of millions of years due to processes such as plate tectonics. Other climate determinants are more dynamic: the thermohaline circulation of the ocean leads to a 5 °C (9 °F) warming of the northern Atlantic Ocean compared to other ocean basins.[638] Other ocean currents redistribute heat between land and water on a more regional scale. The density and type of vegetation coverage affects solar heat absorption,[639] water retention, and rainfall on a regional level. Alterations in the quantity of atmospheric greenhouse gases determines the amount of solar energy retained by the planet, leading to global warming or global cooling. The variables which determine climate are numerous and the interactions complex, but there is general agreement that the broad outlines are understood, at least insofar as the determinants of historical climate change are concerned.

Climate classification

There are several ways to classify climates into similar regimes. Originally, climes were defined in Ancient Greece to describe the weather depending upon a location's latitude. Modern climate classification methods can be broadly divided into *genetic* methods, which focus on the causes of climate, and *empiric* methods, which focus on the effects of climate. Examples of genetic classification include methods based on the relative frequency of different air mass types or locations within synoptic weather disturbances. Examples of empiric classifications include climate zones defined by plant hardiness,[640] evapotranspiration, or more generally the Köppen climate classification which was originally designed to identify the climates associated with certain biomes. A common shortcoming of these classification schemes is that they produce distinct

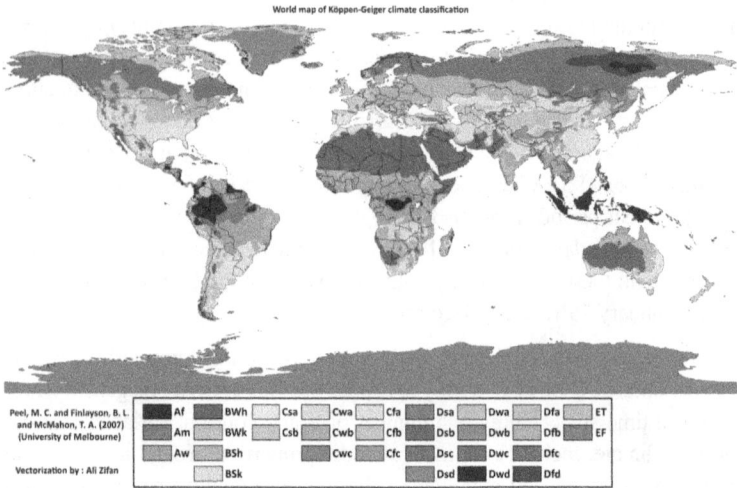

Figure 125:
Worldwide Koppen climate classifications

boundaries between the zones they define, rather than the gradual transition of climate properties more common in nature.

Bergeron and Spatial Synoptic

The simplest classification is that involving air masses. The Bergeron classification is the most widely accepted form of air mass classification. Air mass classification involves three letters. The first letter describes its moisture properties, with c used for continental air masses (dry) and m for maritime air masses (moist). The second letter describes the thermal characteristic of its source region: T for tropical, P for polar, A for Arctic or Antarctic, M for monsoon, E for equatorial, and S for superior air (dry air formed by significant downward motion in the atmosphere). The third letter is used to designate the stability of the atmosphere. If the air mass is colder than the ground below it, it is labeled k. If the air mass is warmer than the ground below it, it is labeled w. While air mass identification was originally used in weather forecasting during the 1950s, climatologists began to establish synoptic climatologies based on this idea in 1973.

Based upon the Bergeron classification scheme is the Spatial Synoptic Classification system (SSC). There are six categories within the SSC scheme: Dry Polar (similar to continental polar), Dry Moderate (similar to maritime superior), Dry Tropical (similar to continental tropical), Moist Polar (similar to maritime

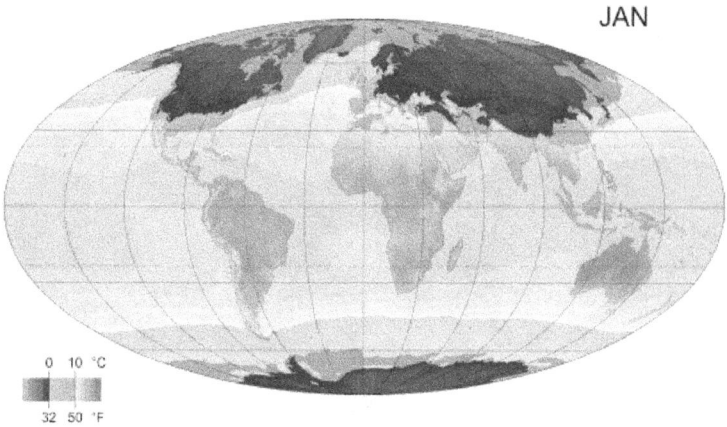

Figure 126: *Monthly average surface temperatures from 1961–1990. This is an example of how climate varies with location and season*

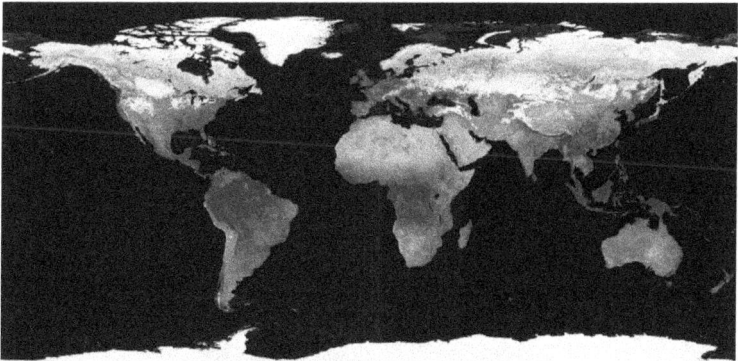

Figure 127: *Monthly global images from NASA Earth Observatory (interactive SVG)*[642]

polar), Moist Moderate (a hybrid between maritime polar and maritime tropical), and Moist Tropical (similar to maritime tropical, maritime monsoon, or maritime equatorial).[641]

Köppen

The Köppen classification depends on average monthly values of temperature and precipitation. The most commonly used form of the Köppen classifica-

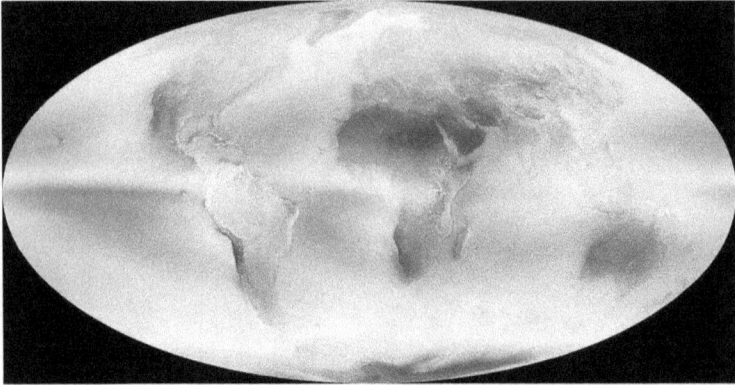

Figure 128: *The world's cloudy and sunny spots. NASA Earth Observatory map using data collected between July 2002 and April 2015.*

tion has five primary types labeled A through E. These primary types are A) tropical, B) dry, C) mild mid-latitude, D) cold mid-latitude, and E) polar. The five primary classifications can be further divided into secondary classifications such as rainforest, monsoon, tropical savanna, humid subtropical, humid continental, oceanic climate, Mediterranean climate, desert, steppe, subarctic climate, tundra, and polar ice cap.

Rainforests are characterized by high rainfall, with definitions setting minimum normal annual rainfall between 1,750 millimetres (69 in) and 2,000 millimetres (79 in). Mean monthly temperatures exceed 18 °C (64 °F) during all months of the year.[643]

A **monsoon** is a seasonal prevailing wind which lasts for several months, ushering in a region's rainy season. Regions within North America, South America, Sub-Saharan Africa, Australia and East Asia are monsoon regimes.[644]

A **tropical savanna** is a grassland biome located in semiarid to semi-humid climate regions of subtropical and tropical latitudes, with average temperatures remain at or above 18 °C (64 °F) year round and rainfall between 750 millimetres (30 in) and 1,270 millimetres (50 in) a year. They are widespread on Africa, and are found in India, the northern parts of South America, Malaysia, and Australia.[645]

The **humid subtropical** climate zone where winter rainfall (and sometimes snowfall) is associated with large storms that the westerlies steer from west to east. Most summer rainfall occurs during thunderstorms and from occasional tropical cyclones. Humid subtropical climates lie on the east side of continents, roughly between latitudes 20° and 40° degrees away from the equator.[646]

Figure 129: *Cloud cover by month for 2014. NASA Earth Observatory*

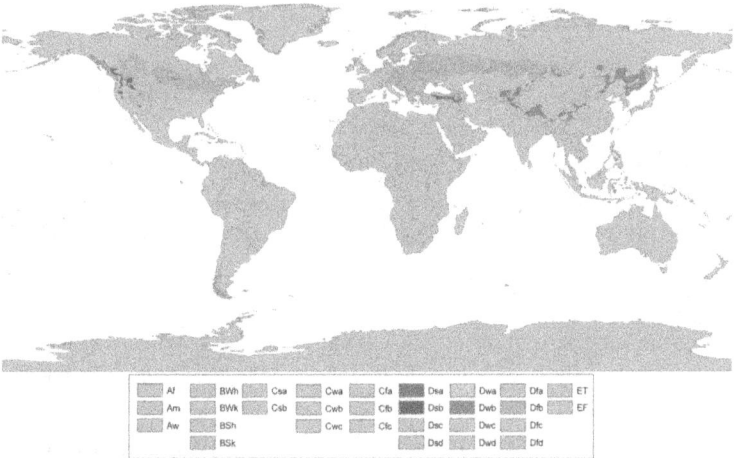

Figure 130:
Humid continental climate, worldwide

Figure 131:
Map of arctic tundra

A **humid continental** climate is marked by variable weather patterns and a large seasonal temperature variance. Places with more than three months of average daily temperatures above 10 °C (50 °F) and a coldest month temperature below –3 °C (27 °F) and which do not meet the criteria for an arid or semiarid climate, are classified as continental.

An **oceanic climate** is typically found along the west coasts at the middle latitudes of all the world's continents, and in southeastern Australia, and is accompanied by plentiful precipitation year-round.[647]

The **Mediterranean climate** regime resembles the climate of the lands in the Mediterranean Basin, parts of western North America, parts of Western and South Australia, in southwestern South Africa and in parts of central Chile. The climate is characterized by hot, dry summers and cool, wet winters.[648]

A **steppe** is a dry grassland with an annual temperature range in the summer of up to 40 °C (104 °F) and during the winter down to –40 °C (–40 °F).[649]

A **subarctic climate** has little precipitation,[650] and monthly temperatures which are above 10 °C (50 °F) for one to three months of the year, with permafrost in large parts of the area due to the cold winters. Winters within subarctic climates usually include up to six months of temperatures averaging below 0 °C (32 °F).[651]

Tundra occurs in the far Northern Hemisphere, north of the taiga belt, including vast areas of northern Russia and Canada.

A **polar ice cap**, or polar ice sheet, is a high-latitude region of a planet or moon that is covered in ice. Ice caps form because high-latitude regions receive less energy as solar radiation from the sun than equatorial regions, resulting in lower surface temperatures.[652]

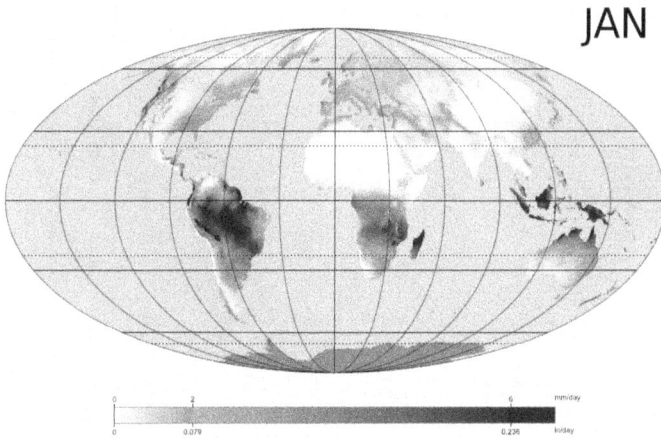

Figure 132: *Precipitation by month*

A **desert** is a landscape form or region that receives very little precipitation. Deserts usually have a large diurnal and seasonal temperature range, with high or low, depending on location daytime temperatures (in summer up to 45 °C or 113 °F), and low nighttime temperatures (in winter down to 0 °C or 32 °F) due to extremely low humidity. Many deserts are formed by rain shadows, as mountains block the path of moisture and precipitation to the desert.[653]

Thornthwaite

Devised by the American climatologist and geographer C. W. Thornthwaite, this climate classification method monitors the soil water budget using evapotranspiration.[654] It monitors the portion of total precipitation used to nourish vegetation over a certain area. It uses indices such as a humidity index and an aridity index to determine an area's moisture regime based upon its average temperature, average rainfall, and average vegetation type.[655] The lower the value of the index in any given area, the drier the area is.

The moisture classification includes climatic classes with descriptors such as hyperhumid, humid, subhumid, subarid, semi-arid (values of –20 to –40), and arid (values below –40).[656] Humid regions experience more precipitation than evaporation each year, while arid regions experience greater evaporation than precipitation on an annual basis. A total of 33 percent of the Earth's land-mass is considered either arid or semi-arid, including southwest North America, southwest South America, most of northern and a small part of southern Africa, southwest and portions of eastern Asia, as well as much of Australia.

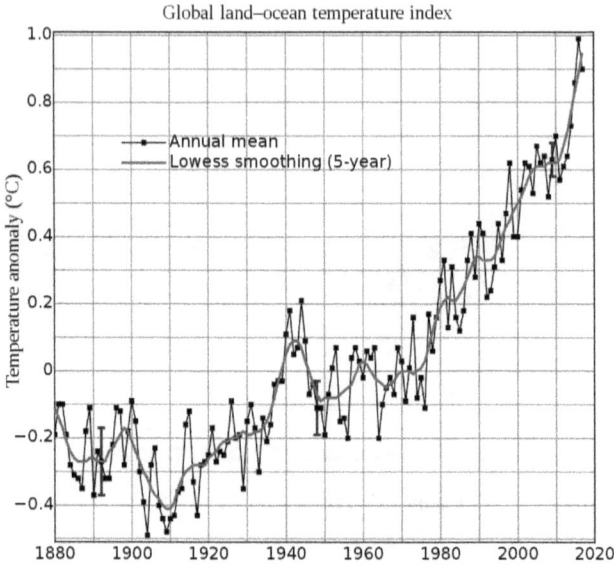

Figure 133: *Global mean surface tempera-*
ture change since 1880. Source: NASA GISS[658]

Studies suggest that precipitation effectiveness (PE) within the Thornthwaite moisture index is overestimated in the summer and underestimated in the winter. This index can be effectively used to determine the number of herbivore and mammal species numbers within a given area. The index is also used in studies of climate change.[657]

Thermal classifications within the Thornthwaite scheme include microthermal, mesothermal, and megathermal regimes. A microthermal climate is one of low annual mean temperatures, generally between 0 °C (32 °F) and 14 °C (57 °F) which experiences short summers and has a potential evaporation between 14 centimetres (5.5 in) and 43 centimetres (17 in). A mesothermal climate lacks persistent heat or persistent cold, with potential evaporation between 57 centimetres (22 in) and 114 centimetres (45 in). A megathermal climate is one with persistent high temperatures and abundant rainfall, with potential annual evaporation in excess of 114 centimetres (45 in).

Record

Modern

Details of the modern climate record are known through the taking of measurements from such weather instruments as thermometers, barometers, and

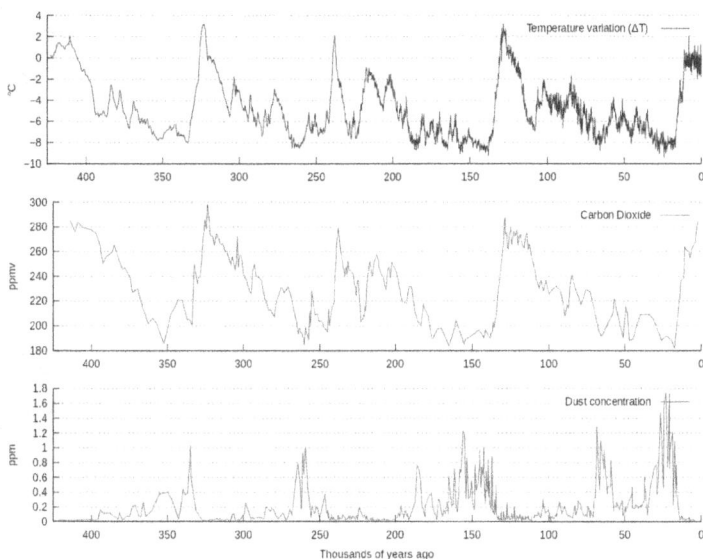

Figure 134: *Variations in CO_2, temperature and dust from the Vostok ice core over the past 450,000 years*

anemometers during the past few centuries. The instruments used to study weather over the modern time scale, their known error, their immediate environment, and their exposure have changed over the years, which must be considered when studying the climate of centuries past.[659]

Paleoclimatology

Paleoclimatology is the study of past climate over a great period of the Earth's history. It uses evidence from ice sheets, tree rings, sediments, coral, and rocks to determine the past state of the climate. It demonstrates periods of stability and periods of change and can indicate whether changes follow patterns such as regular cycles.[660]

Climate change

Climate change is the variation in global or regional climates over time. It reflects changes in the variability or average state of the atmosphere over time scales ranging from decades to millions of years. These changes can be caused by processes internal to the Earth, external forces (e.g. variations in sunlight intensity) or, more recently, human activities.[661]

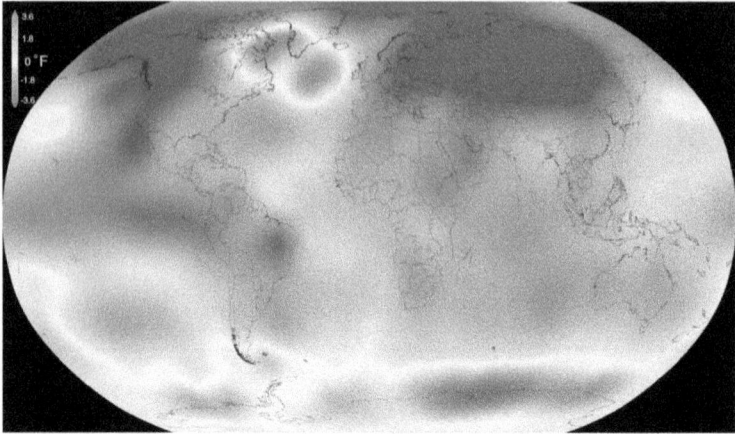

Figure 135:
*2015 – Warmest Global Year on Record (since 1880) – Colors in-
dicate temperature anomalies (NASA/NOAA; 20 January 2016).*

In recent usage, especially in the context of environmental policy, the term
"climate change" often refers only to changes in modern climate, including
the rise in average surface temperature known as global warming. In some
cases, the term is also used with a presumption of human causation, as in the
United Nations Framework Convention on Climate Change (UNFCCC). The
UNFCCC uses "climate variability" for non-human caused variations.

Earth has undergone periodic climate shifts in the past, including four major ice ages. These consisting of glacial periods where conditions are colder
than normal, separated by interglacial periods. The accumulation of snow
and ice during a glacial period increases the surface albedo, reflecting more
of the Sun's energy into space and maintaining a lower atmospheric temperature. Increases in greenhouse gases, such as by volcanic activity, can increase
the global temperature and produce an interglacial period. Suggested causes of
ice age periods include the positions of the continents, variations in the Earth's
orbit, changes in the solar output, and volcanism.[662]

Climate models

Climate models use quantitative methods to simulate the interactions of the
atmosphere,[663] oceans, land surface and ice. They are used for a variety of
purposes; from the study of the dynamics of the weather and climate system,
to projections of future climate. All climate models balance, or very nearly
balance, incoming energy as short wave (including visible) electromagnetic

radiation to the earth with outgoing energy as long wave (infrared) electro-magnetic radiation from the earth. Any imbalance results in a change in the average temperature of the earth.

The most talked-about applications of these models in recent years have been their use to infer the consequences of increasing greenhouse gases in the atmosphere, primarily carbon dioxide (see greenhouse gas). These models predict an upward trend in the global mean surface temperature, with the most rapid increase in temperature being projected for the higher latitudes of the Northern Hemisphere.

Models can range from relatively simple to quite complex:

- Simple radiant heat transfer model that treats the earth as a single point and averages outgoing energy
- this can be expanded vertically (radiative-convective models), or horizontally
- finally, (coupled) atmosphere–ocean–sea ice global climate models discretise and solve the full equations for mass and energy transfer and radiant exchange.[664]

Climate forecasting is a way by some scientists are using to predict climate change. In 1997 the prediction division of the International Research Institute for Climate and Society at Columbia University began generating seasonal climate forecasts on a real-time basis. To produce these forecasts an extensive suite of forecasting tools was developed, including a multimodel ensemble approach that required thorough validation of each model's accuracy level in simulating interannual climate variability.[665]

Further reading

- The Study of Climate on Alien Worlds; Characterizing atmospheres beyond our Solar System is now within our reach[666] Kevin Heng July–August 2012 American Scientist
- Reumert, Johannes: "Vahls climatic divisions. An explanation" ([[*Danish Journal of Geography|Geografisk Tidsskrift*[667]], *Band 48; 1946)*]

External links

> Wikimedia Commons has media related to *Climate*.

> Wikisourcehas the text of the 1905 *New International Encyclopedia*article *Climate*.

- NOAA Climate Services Portal[668]
- NOAA State of the Climate[669]
- NASA's Climate change and global warming portal[670]
- Climate Models and modeling groups[671]
- Climate Prediction Project[672]
- ESPERE Climate Encyclopaedia[673]
- Climate index and mode information[674] – Arctic
- A current view of the Bering Sea Ecosystem and Climate[675]
- Climate: Data and charts for world and US locations[676]
- MIL-HDBK-310, Global Climate Data[677] U.S. Department of Defense – Aid to derive natural environmental design criteria
- IPCC Data Distribution Centre[678] – Climate data and guidance on use.
- HistoricalClimatology.com[679] – Past, present and future climates – 2013.
- Globalclimatemonitor[680] – Contains climatic information from 1901.
- ClimateCharts[681] – Webapplication to generate climate charts for recent and historical data.
- International Disaster Database[682]
- Paris Climate Conference[683]

<indicator name="good-star"> ⊕ </indicator>

Gravitational field

Gravity of Earth

The **gravity of Earth**, which is denoted by g, refers to the acceleration that is imparted to objects due to the distribution of mass within Earth. In SI units this acceleration is measured in metres per second squared (in symbols, m/s^2 or $m \cdot s^{-2}$) or equivalently in newtons per kilogram (N/kg or $N \cdot kg^{-1}$). Near Earth's surface, gravitational acceleration is approximately $9.8 \, m/s^2$, which means that, ignoring the effects of air resistance, the speed of an object falling freely will increase by about 9.8 metres per second every second. This quantity is sometimes referred to informally as *little g* (in contrast, the gravitational constant G is referred to as *big G*).

The precise strength of Earth's gravity varies depending on location. The nominal "average" value at Earth's surface, known as standard gravity is, by definition, $9.80665 \, m/s^2$. This quantity is denoted variously as g_n, g_e (though this sometimes means the normal equatorial value on Earth, $9.78033 \, m/s^2$), g_0, gee, or simply g (which is also used for the variable local value). The weight of an object on Earth's surface is the downwards force on that object, given by Newton's second law of motion, or $F = ma$ (*force = mass × acceleration*). Gravitational acceleration contributes to the total acceleration, but other factors, such as the rotation of Earth, also contribute, and, therefore, affect the weight of the object.

Variation in gravity and apparent gravity

A perfect sphere of uniform mass density, or whose density varies solely with distance from the centre (spherical symmetry), would produce a gravitational field of uniform magnitude at all points on its surface, always pointing directly towards the sphere's centre. The Earth is not spherically symmetric, but is slightly flatter at the poles while bulging at the Equator: an oblate spheroid. There are consequently slight deviations in both the magnitude and direction

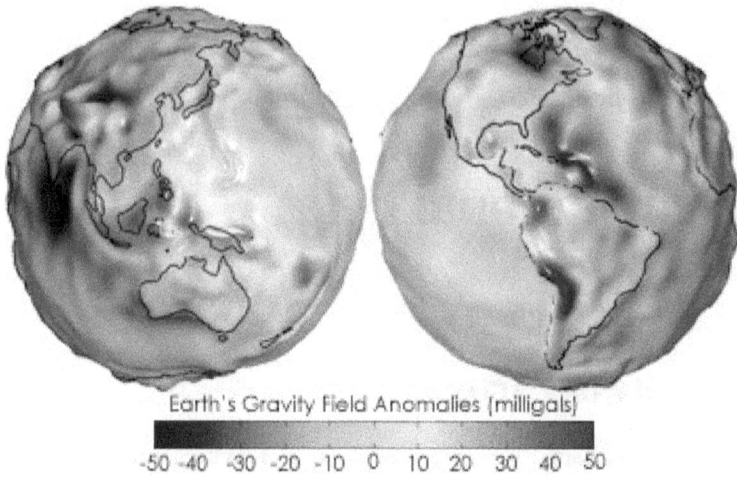

Figure 136: *Earth's gravity measured by NASA GRACE mission, showing deviations from the theoretical gravity of an idealized smooth Earth, the so-called Earth ellipsoid. Red shows the areas where gravity is stronger than the smooth, standard value, and blue reveals areas where gravity is weaker. (**Animated version.**)*

of gravity across its surface. The net force (or corresponding net acceleration) as measured by a scale and plumb bob is called "effective gravity" or "apparent gravity". Effective gravity includes other factors that affect the net force. These factors vary and include things such as centrifugal force at the surface from the Earth's rotation and the gravitational pull of the Moon and Sun.

Effective gravity on the Earth's surface varies by around 0.7%, from 9.7639 m/s^2 on the Nevado Huascarán mountain in Peru to 9.8337 m/s^2 at the surface of the Arctic Ocean. In large cities, it ranges from 9.7760[684] in Kuala Lumpur, Mexico City, and Singapore to 9.825 in Oslo and Helsinki.

Legal definition

In 1901 the third General Conference on Weights and Measures defined a standard gravitational acceleration for the surface of the Earth: g_n = 9.80665 m/s^2. It was based on measurements done at the Pavillon de Breteuil near Paris in 1888, with a theoretical correction applied in order to convert to a latitude of 45° at sea level. This definition is thus not a value of any particular place or carefully worked out average, but an agreement for a value to use if a better actual local value is not known or not important.[685] It is also used to define the units kilogram force and pound force.

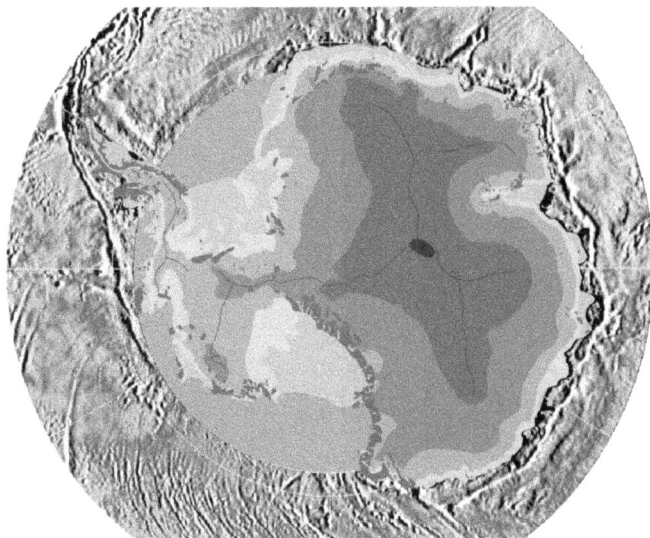

Figure 137: *The differences of Earth's gravity around the Antarctic continent.*

Latitude

The surface of the Earth is rotating, so it is not an inertial frame of reference. At latitudes nearer the Equator, the outward centrifugal force produced by Earth's rotation is larger than at polar latitudes. This counteracts the Earth's gravity to a small degree – up to a maximum of 0.3% at the Equator – and reduces the apparent downward acceleration of falling objects.

The second major reason for the difference in gravity at different latitudes is that the Earth's equatorial bulge (itself also caused by centrifugal force from rotation) causes objects at the Equator to be farther from the planet's centre than objects at the poles. Because the force due to gravitational attraction between two bodies (the Earth and the object being weighed) varies inversely with the square of the distance between them, an object at the Equator experiences a weaker gravitational pull than an object at the poles.

In combination, the equatorial bulge and the effects of the surface centrifugal force due to rotation mean that sea-level effective gravity increases from about 9.780 m/s^2 at the Equator to about 9.832 m/s^2 at the poles, so an object will weigh about 0.5% more at the poles than at the Equator.[686]

The same two factors influence the direction of the effective gravity (as determined by a plumb line or as the perpendicular to the surface of water in a container). Anywhere on Earth away from the Equator or poles, effective

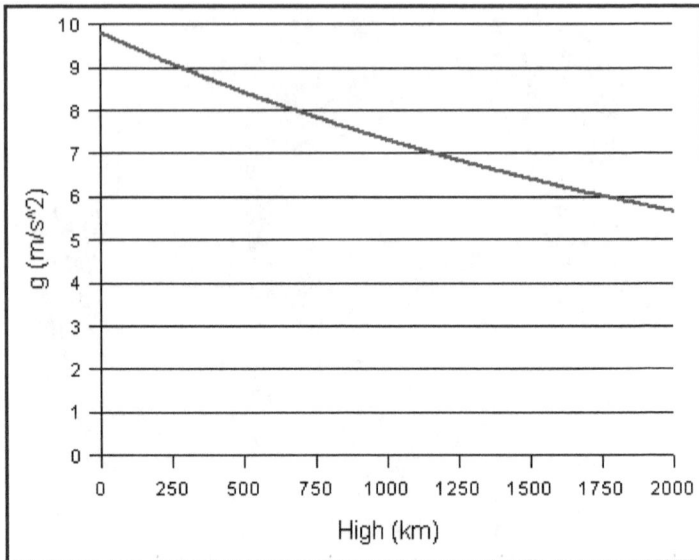

Figure 138: *The graph shows the variation in gravity relative to the height of an object above the surface*

gravity points not exactly toward the centre of the Earth, but rather perpendicular to the surface of the geoid, which, due to the flattened shape of the Earth,
is somewhat toward the opposite pole. About half of the deflection is due to
centrifugal force, and half because the extra mass around the Equator causes a
change in the direction of the true gravitational force relative to what it would
be on a spherical Earth.

Altitude

Gravity decreases with altitude as one rises above the Earth's surface because
greater altitude means greater distance from the Earth's centre. All other things
being equal, an increase in altitude from sea level to 9,000 metres (30,000 ft)
causes a weight decrease of about 0.29%. (An additional factor affecting apparent weight is the decrease in air density at altitude, which lessens an object's
buoyancy.[687] This would increase a person's apparent weight at an altitude of
9,000 metres by about 0.08%)

It is a common misconception that astronauts in orbit are weightless because
they have flown high enough to escape the Earth's gravity. In fact, at an altitude
of 400 kilometres (250 mi), equivalent to a typical orbit of the Space Shuttle,

gravity is still nearly 90% as strong as at the Earth's surface. Weightlessness actually occurs because orbiting objects are in free-fall.[688]

The effect of ground elevation depends on the density of the ground (see Slab correction section). A person flying at 30 000 ft above sea level over mountains will feel more gravity than someone at the same elevation but over the sea. However, a person standing on the earth's surface feels less gravity when the elevation is higher.

The following formula approximates the Earth's gravity variation with altitude:

$$g_h = g_0 \left(\frac{r_e}{r_e + h} \right)^2$$

Where

- g_h is the gravitational acceleration at height h above sea level.
- r_e is the Earth's mean radius.
- g_0 is the standard gravitational acceleration.

The formula treats the Earth as a perfect sphere with a radially symmetric distribution of mass; a more accurate mathematical treatment is discussed below.

Depth

An approximate value for gravity at a distance r from the center of the Earth can be obtained by assuming that the Earth's density is spherically symmetric. The gravity depends only on the mass inside the sphere of radius r. All the contributions from outside cancel out as a consequence of the inverse-square law of gravitation. Another consequence is that the gravity is the same as if all the mass were concentrated at the center. Thus, the gravitational acceleration at this radius is

$$g(r) = -\frac{GM(r)}{r^2}.$$

where G is the gravitational constant and $M(r)$ is the total mass enclosed within radius r. If the Earth had a constant density ϱ, the mass would be $M(r) = (4/3)\pi\varrho r^3$ and the dependence of gravity on depth would be

$$g(r) = \frac{4\pi}{3} G\rho r.$$

g at depth d is given by $g'=g(1-d/R)$ where g is acceleration due to gravity on surface of the earth, d is depth and R is radius of Earth. If the density decreased linearly with increasing radius from a density ϱ_0 at the center to ϱ_1 at the surface, then $\varrho(r) = \varrho_0 - (\varrho_0 - \varrho_1)\, r\, /\, r_e$, and the dependence would be

$$g(r) = \frac{4\pi}{3} G\rho_0 r - \frac{4\pi}{3} G (\rho_0 - \rho_1) \frac{r^2}{r_e}.$$

The actual depth dependence of density and gravity, inferred from seismic travel times (see Adams–Williamson equation), is shown in the graphs below.

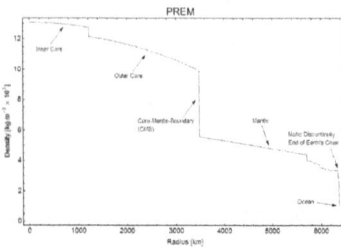

Figure 139: *Earth's radial density distribution according to the Preliminary Reference Earth Model (PREM).*

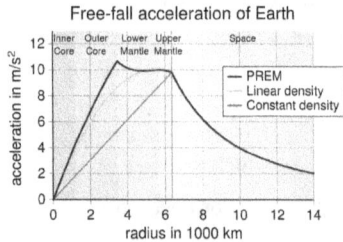

Figure 140: *Earth's gravity according to the Preliminary Reference Earth Model (PREM). Two models for a spherically symmetric Earth are included for comparison. The dark green straight line is for a constant density equal to the Earth's average density. The light green curved line is for a density that decreases linearly from center to surface. The density at the center is the same as in the PREM, but the surface density is chosen so that the mass of the sphere equals the mass of the real Earth.*

Local topography and geology

Local differences in topography (such as the presence of mountains), geology (such as the density of rocks in the vicinity), and deeper tectonic structure cause local and regional differences in the Earth's gravitational field, known as gravitational anomalies. Some of these anomalies can be very extensive, resulting in bulges in sea level, and throwing pendulum clocks out of synchronisation.

The study of these anomalies forms the basis of gravitational geophysics. The fluctuations are measured with highly sensitive gravimeters, the effect of topography and other known factors is subtracted, and from the resulting data conclusions are drawn. Such techniques are now used by prospectors to find oil and mineral deposits. Denser rocks (often containing mineral ores) cause higher than normal local gravitational fields on the Earth's surface. Less dense sedimentary rocks cause the opposite.

Other factors

In air, objects experience a supporting buoyancy force which reduces the apparent strength of gravity (as measured by an object's weight). The magnitude of the effect depends on air density (and hence air pressure); see Apparent weight for details.

The gravitational effects of the Moon and the Sun (also the cause of the tides) have a very small effect on the apparent strength of Earth's gravity, depending on their relative positions; typical variations are 2 μm/s² (0.2 mGal) over the course of a day.

Comparative gravities in various cities around the world

Tools exist for calculating the strength of gravity at various cities around the world.[689]   The effect of latitude can be clearly seen with gravity in high-latitude cities: Anchorage (9.826 m/s²), Helsinki (9.825 m/s²), being about 0.5% greater than that in cities near the equator: Kuala Lumpur (9.776 m/s²), Manila (9.780 m/s²).   The effect of altitude can be seen in Mexico City (9.776 m/s²; altitude 2,240 metres (7,350 ft)), and by comparing Denver (9.798 m/s²; 1,616 metres (5,302 ft)) with Washington, D.C. (9.801 m/s²; 30 metres (98 ft)), both of which are near 39° N. Measured values can be obtained from Physical and Mathematical Tables by T.M. Yarwood and F. Castle, Macmillan, revised edition 1970.[690]

Mathematical models

Latitude model

If the terrain is at sea level, we can estimate $g\{\phi\}$, the acceleration at latitude ϕ:

$$g\{\phi\} = 9.780327 \text{ m} \cdot \text{s}^{-2} \left(1 + 0.0053024 \sin^2 \phi - 0.0000058 \sin^2 2\phi\right),$$
$$= 9.780327 \text{ m} \cdot \text{s}^{-2} \left(1 + 0.0052792 \sin^2 \phi + 0.0000232 \sin^4 \phi\right),$$
$$= 9.780327 \text{ m} \cdot \text{s}^{-2} \left(1.0053024 - 0.0053256 \cos^2 \phi + 0.0000232 \cos^4 \phi\right),$$
$$= 9.780327 \text{ m} \cdot \text{s}^{-2} \left(1.0026454 - 0.0026512 \cos 2\phi + 0.0000058 \cos^2 2\phi\right)$$

This is the International Gravity Formula 1967, the 1967 Geodetic Reference System Formula, Helmert's equation or Clairaut's formula.[691]

An alternative formula for g as a function of latitude is the WGS (World Geodetic System) 84 Ellipsoidal Gravity Formula:[692]

$$g\{\phi\} = \mathbb{G}_e \left[\frac{1 + k \sin^2 \phi}{\sqrt{1 - e^2 \sin^2 \phi}} \right],$$

where,

- a, b are the equatorial and polar semi-axes, respectively;
- $e^2 = 1 - (b/a)^2$ is the spheroid's eccentricity, squared;
- $\mathbb{G}_e$, $\mathbb{G}_p$ is the defined gravity at the equator and poles, respectively;
- $k = \frac{b\, \mathbb{G}_p - a\, \mathbb{G}_e}{a\, \mathbb{G}_e}$ (formula constant);

then, where $\mathbb{G}_p = 9.8321849378$ m $\cdot$ s^{-2} ,

$$g\{\phi\} = 9.7803253359 \text{ m} \cdot \text{s}^{-2} \left[\frac{1+0.00193185265241 \ \sin^2 \phi}{\sqrt{1-0.00669437999013 \ \sin^2 \phi}} \right] \quad .$$

The difference between the WGS-84 formula and Helmert's equation is less than 0.68 μm·s^{-2}.

Free air correction

The first correction to be applied to the model is the free air correction (FAC) that accounts for heights above sea level. Near the surface of the Earth (sea level), gravity decreases with height such that linear extrapolation would give zero gravity at a height of one half of the earth's radius - (9.8 m·s^{-2} per 3,200 km.)[693]

Using the mass and radius of the Earth:

$r_{\text{Earth}} = 6.371 \cdot 10^6$ m

$m_{\text{Earth}} = 5.9722 \cdot 10^{24}$ kg

The FAC correction factor (Δg) can be derived from the definition of the acceleration due to gravity in terms of G, the Gravitational Constant (see Estimating g from the law of universal gravitation, below):

$$g_0 = G \, m_{\text{Earth}}/r_{\text{Earth}}^2 = 9.8196 \, \frac{\text{m}}{\text{s}^2}$$

where:

$$G = 6.67384 \cdot 10^{-11} \, \frac{\text{m}^3}{\text{kg} \cdot \text{s}^2} .$$

At a height h above the nominal surface of the earth g_h is given by:

$$g_h = G \, m_{\text{Earth}}/ \left(r_{\text{Earth}} + h\right)^2$$

So the FAC for a height h above the nominal earth radius can be expressed:

$$\Delta g_h = \left[G \, m_{\text{Earth}}/ \left(r_{\text{Earth}} + h\right)^2 \right] - \left[G \, m_{\text{Earth}}/r_{\text{Earth}}^2 \right]$$

This expression can be readily used for programming or inclusion in a spreadsheet. Collecting terms, simplifying and neglecting small terms ($h \ll r_{\text{Earth}}$), however yields the good approximation:

$$\Delta g_h \approx - \frac{G \, m_{\text{Earth}}}{r_{\text{Earth}}^2} \cdot \frac{2 \, h}{r_{\text{Earth}}}$$

Using the numerical values above and for a height h in metres:

$$\Delta g_h \approx -3.086 \cdot 10^{-6} \, h$$

Grouping the latitude and FAC altitude factors the expression most commonly found in the literature is:

$$g\{\phi, h\} = g\{\phi\} - 3.086 \cdot 10^{-6} h$$

where $g\{\phi, h\}$ = acceleration in m·s^{-2} at latitude ϕ and altitude h in metres.

Slab correction

Note: The section uses the galileo (symbol: "Gal"), which is a cgs unit for acceleration of 1 centimetre/second^2.

For flat terrain above sea level a second term is added for the gravity due to the extra mass; for this purpose the extra mass can be approximated by an infinite horizontal slab, and we get $2\pi G$ times the mass per unit area, i.e. 4.2×10^{-10} m$^3 \cdot$s$^{-2} \cdot$kg^{-1} (0.042 μGal$\cdot$kg$^{-1} \cdot$m^2) (the Bouguer correction). For a mean rock density of 2.67 g$\cdot$cm^{-3} this gives 1.1×10^{-6} s^{-2} (0.11 mGal$\cdot$m^{-1}). Combined with the free-air correction this means a reduction of gravity at the surface of ca. 2 μm$\cdot$s^{-2} (0.20 mGal) for every metre of elevation of the terrain. (The two effects would cancel at a surface rock density of 4/3 times the average density of the whole earth. The density of the whole earth is 5.515 g$\cdot$cm^{-3}, so standing on a slab of something like iron whose density is over 7.35 g$\cdot$cm^{-3} would increase one's weight.)

For the gravity below the surface we have to apply the free-air correction as well as a double Bouguer correction. With the infinite slab model this is because moving the point of observation below the slab changes the gravity due to it to its opposite. Alternatively, we can consider a spherically symmetrical Earth and subtract from the mass of the Earth that of the shell outside the point of observation, because that does not cause gravity inside. This gives the same result.

Estimating g from the law of universal gravitation

From the law of universal gravitation, the force on a body acted upon by Earth's gravity is given by

$$F = G\frac{m_1 m_2}{r^2} = \left(G\frac{m_1}{r^2}\right) m_2$$

where r is the distance between the centre of the Earth and the body (see below), and here we take m_1 to be the mass of the Earth and m_2 to be the mass of the body.

Additionally, Newton's second law, $F = ma$, where m is mass and a is acceleration, here tells us that

$$F = m_2\, g$$

Comparing the two formulas it is seen that:

$$g = G\frac{m_1}{r^2}$$

So, to find the acceleration due to gravity at sea level, substitute the values of the gravitational constant, G, the Earth's mass (in kilograms), m_1, and the Earth's radius (in metres), r, to obtain the value of g:

$$g = G\,\frac{m_1}{r^2} = 6.67384 \cdot 10^{-11}\,\text{m}^3 \cdot \text{kg}^{-1} \cdot \text{s}^{-2}\,\frac{5.9722 \cdot 10^{24}\,\text{kg}}{(6.371 \cdot 10^6\,\text{m})^2} = 9.8196\ \text{m} \cdot \text{s}^{-2}$$

Note that this formula only works because of the mathematical fact that the gravity of a uniform spherical body, as measured on or above its surface, is the same as if all its mass were concentrated at a point at its centre. This is what allows us to use the Earth's radius for r.

The value obtained agrees approximately with the measured value of g. The difference may be attributed to several factors, mentioned above under "Variations":

- The Earth is not homogeneous
- The Earth is not a perfect sphere, and an average value must be used for its radius
- This calculated value of g only includes true gravity. It does not include the reduction of constraint force that we perceive as a reduction of gravity due to the rotation of Earth, and some of gravity being counteracted by centrifugal force.

There are significant uncertainties in the values of r and m_1 as used in this calculation, and the value of G is also rather difficult to measure precisely.

If G, g and r are known then a reverse calculation will give an estimate of the mass of the Earth. This method was used by Henry Cavendish.

External links

- Altitude gravity calculator[694]
- GRACE – Gravity Recovery and Climate Experiment[695]
- GGMplus high resolution data (2013)[696]
- Geoid 2011 model[697] Potsdam Gravity Potato

Earth's magnetic field

<indicator name="good-star"> ⊕ </indicator>

Earth's magnetic field, also known as the **geomagnetic field**, is the magnetic field that extends from the Earth's interior out into space, where it meets the solar wind, a stream of charged particles emanating from the Sun. Its magnitude at the Earth's surface ranges from 25 to 65 microteslas (0.25 to 0.65 gauss). Approximately, it is the field of a magnetic dipole currently tilted at an angle of about 11 degrees with respect to Earth's rotational axis, as if there were a bar magnet placed at that angle at the center of the Earth. The North geomagnetic pole, located near Greenland in the northern hemisphere, is actually the south pole of the Earth's magnetic field, and the South geomagnetic pole is the north pole. The magnetic field is generated by electric currents due to the motion of convection currents of molten iron in the Earth's outer core driven by heat escaping from the core, a natural process called a geodynamo.

While the North and South magnetic poles are usually located near the geographic poles, they can wander widely over geological time scales, but sufficiently slowly for ordinary compasses to remain useful for navigation. However, at irregular intervals averaging several hundred thousand years, the Earth's field reverses and the North and South Magnetic Poles relatively abruptly switch places. These reversals of the geomagnetic poles leave a record in rocks that are of value to paleomagnetists in calculating geomagnetic fields in the past. Such information in turn is helpful in studying the motions of continents and ocean floors in the process of plate tectonics.

The magnetosphere is the region above the ionosphere that is defined by the extent of the Earth's magnetic field in space. It extends several tens of thousands of kilometers into space, protecting the Earth from the charged particles of the solar wind and cosmic rays that would otherwise strip away the upper atmosphere, including the ozone layer that protects the Earth from harmful ultraviolet radiation.

Importance

The Earth's magnetic field serves to deflect most of the solar wind, whose charged particles would otherwise strip away the ozone layer that protects the Earth from harmful ultraviolet radiation. One stripping mechanism is for gas to be caught in bubbles of magnetic field, which are ripped off by solar winds. Calculations of the loss of carbon dioxide from the atmosphere of Mars, resulting from scavenging of ions by the solar wind, indicate that the dissipation of the magnetic field of Mars caused a near total loss of its atmosphere.[698]

Figure 141: *Computer simulation of the Earth's field in a period of normal polarity between reversals. The lines represent magnetic field lines, blue when the field points towards the center and yellow when away. The rotation axis of the Earth is centered and vertical. The dense clusters of lines are within the Earth's core.*

The study of past magnetic field of the Earth is known as paleomagnetism. The polarity of the Earth's magnetic field is recorded in igneous rocks, and reversals of the field are thus detectable as "stripes" centered on mid-ocean ridges where the sea floor is spreading, while the stability of the geomagnetic poles between reversals has allowed paleomagnetists to track the past motion of continents. Reversals also provide the basis for magnetostratigraphy, a way of dating rocks and sediments. The field also magnetizes the crust, and magnetic anomalies can be used to search for deposits of metal ores.

Humans have used compasses for direction finding since the 11th century A.D. and for navigation since the 12th century. Although the magnetic declination does shift with time, this wandering is slow enough that a simple compass remains useful for navigation. Using magnetoreception various other organisms, ranging from some types of bacteria to pigeons, use the Earth's magnetic field for orientation and navigation.

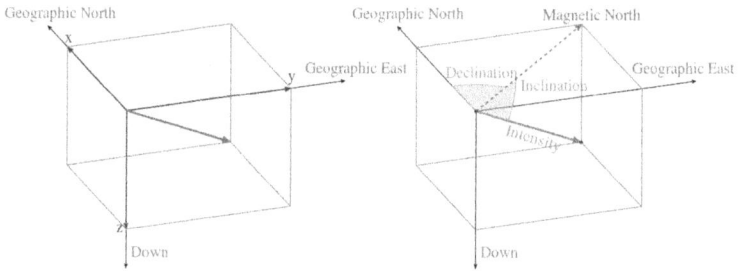

Figure 142: *Common coordinate systems used for representing the Earth's magnetic field.*

Main characteristics

Description

At any location, the Earth's magnetic field can be represented by a three-dimensional vector. A typical procedure for measuring its direction is to use a compass to determine the direction of magnetic North. Its angle relative to true North is the *declination* (D) or *variation*. Facing magnetic North, the angle the field makes with the horizontal is the *inclination* (I) or *magnetic dip*. The *intensity* (F) of the field is proportional to the force it exerts on a magnet. Another common representation is in X (North), Y (East) and Z (Down) coordinates.

Intensity

The intensity of the field is often measured in gauss (G), but is generally reported in nanoteslas (nT), with 1 G = 100,000 nT. A nanotesla is also referred to as a gamma (γ). The tesla is the SI unit of the magnetic field, **B**. The Earth's field ranges between approximately 25,000 and 65,000 nT (0.25–0.65 G). By comparison, a strong refrigerator magnet has a field of about 10,000,000 nanoteslas (100 G).

A map of intensity contours is called an *isodynamic chart*. As the World Magnetic Model shows, the intensity tends to decrease from the poles to the equator. A minimum intensity occurs in the South Atlantic Anomaly over South America while there are maxima over northern Canada, Siberia, and the coast of Antarctica south of Australia.

Inclination

The inclination is given by an angle that can assume values between -90° (up) to 90° (down). In the northern hemisphere, the field points downwards. It is straight down at the North Magnetic Pole and rotates upwards as the latitude decreases until it is horizontal (0°) at the magnetic equator. It continues to rotate upwards until it is straight up at the South Magnetic Pole. Inclination can be measured with a dip circle.

An *isoclinic chart* (map of inclination contours) for the Earth's magnetic field is shown below.

Declination

Declination is positive for an eastward deviation of the field relative to true north. It can be estimated by comparing the magnetic north/south heading on a compass with the direction of a celestial pole. Maps typically include information on the declination as an angle or a small diagram showing the relationship between magnetic north and true north. Information on declination for a region can be represented by a chart with isogonic lines (contour lines with each line representing a fixed declination).

Geographical variation

Components of the Earth's magnetic field at the surface from the World Magnetic Model for 2015.

Dipolar approximation

Near the surface of the Earth, its magnetic field can be closely approximated by the field of a magnetic dipole positioned at the center of the Earth and tilted at an angle of about 11° with respect to the rotational axis of the Earth. The dipole is roughly equivalent to a powerful bar magnet, with its south pole pointing towards the geomagnetic North Pole. This may seem surprising, but the north pole of a magnet is so defined because, if allowed to rotate freely, it points roughly northward (in the geographic sense). Since the north pole of a magnet attracts the south poles of other magnets and repels the north poles, it must be attracted to the south pole of Earth's magnet. The dipolar field accounts for 80–90% of the field in most locations.

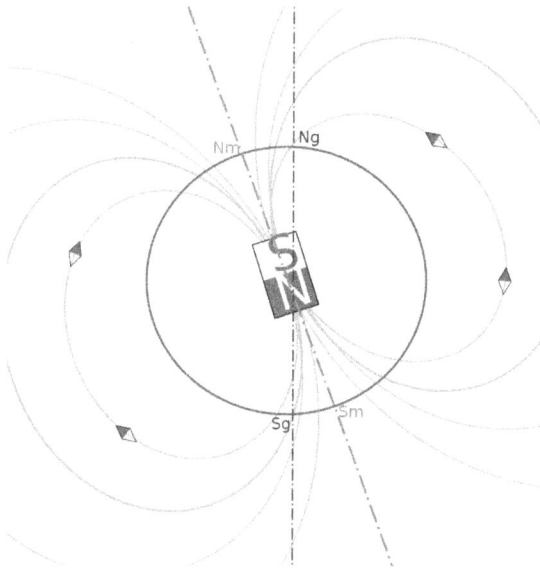

Figure 143: *The variation between magnetic north (N_m) and "true" north (N_g).*

Magnetic poles

Historically, the north and south poles of a magnet were first defined by the Earth's magnetic field, not vice versa, since one of the first uses for a magnet was as a compass needle. Its North pole was defined as the pole that would be attracted by the Earth's North Magnetic Pole when the magnet was suspended so it could turn freely. Since opposite poles attract, the North Magnetic Pole of the Earth is really the south pole of its magnetic field (the place where the field is directed downward into the Earth).

The positions of the magnetic poles can be defined in at least two ways: locally or globally. The local definition is the point where the magnetic field is vertical. This can be determined by measuring the inclination. The inclination of the Earth's field is 90° (downwards) at the North Magnetic Pole and -90° (upwards) at the South Magnetic Pole. The two poles wander independently of each other and are not directly opposite each other on the globe. They can migrate rapidly: movements of up to 40 kilometres (25 mi) per year have been observed for the North Magnetic Pole. Over the last 180 years, the North Magnetic Pole has been migrating northwestward, from Cape Adelaide in the Boothia Peninsula in 1831 to 600 kilometres (370 mi) from Resolute Bay in 2001. The *magnetic equator* is the line where the inclination is zero (the magnetic field is horizontal).

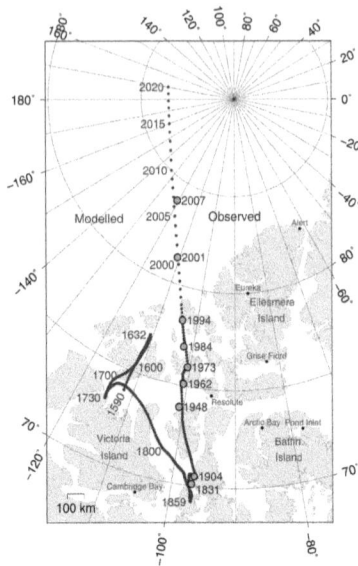

Figure 144: *The movement of Earth's North Magnetic Pole across the Canadian arctic.*

The global definition of the Earth's field is based on a mathematical model. If a line is drawn through the center of the Earth, parallel to the moment of the best-fitting magnetic dipole, the two positions where it intersects the Earth's surface are called the North and South geomagnetic poles. If the Earth's magnetic field were perfectly dipolar, the geomagnetic poles and magnetic dip poles would coincide and compasses would point towards them. However, the Earth's field has a significant non-dipolar contribution, so the poles do not coincide and compasses do not generally point at either.

Magnetosphere

Earth's magnetic field, predominantly dipolar at its surface, is distorted further out by the solar wind. This is a stream of charged particles leaving the Sun's corona and accelerating to a speed of 200 to 1000 kilometres per second. They carry with them a magnetic field, the interplanetary magnetic field (IMF).[699]

The solar wind exerts a pressure, and if it could reach Earth's atmosphere it would erode it. However, it is kept away by the pressure of the Earth's magnetic field. The magnetopause, the area where the pressures balance, is the boundary of the magnetosphere. Despite its name, the magnetosphere is asymmetric,

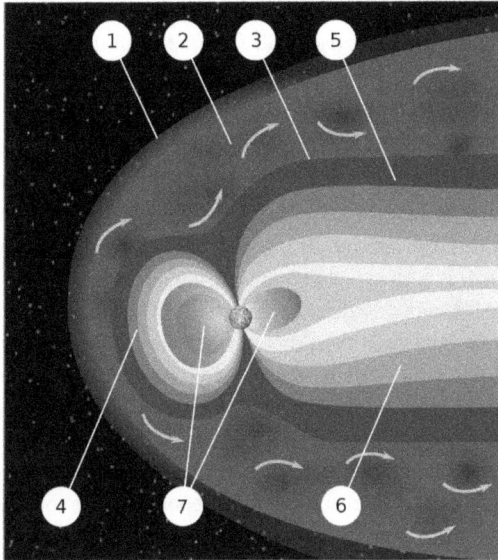

Figure 145: *An artist's rendering of the structure of a magnetosphere.*
1) Bow shock. 2) Magnetosheath. 3) Magnetopause. 4) Magneto-
sphere. 5) Northern tail lobe. 6) Southern tail lobe. 7) Plasmasphere.

with the sunward side being about 10 Earth radii out but the other side stretching out in a magnetotail that extends beyond 200 Earth radii. Sunward of the magnetopause is the bow shock, the area where the solar wind slows abruptly.

Inside the magnetosphere is the plasmasphere, a donut-shaped region containing low-energy charged particles, or plasma. This region begins at a height of 60 km, extends up to 3 or 4 Earth radii, and includes the ionosphere. This region rotates with the Earth. There are also two concentric tire-shaped regions, called the Van Allen radiation belts, with high-energy ions (energies from 0.1 to 10 million electron volts (MeV)). The inner belt is 1–2 Earth radii out while the outer belt is at 4–7 Earth radii. The plasmasphere and Van Allen belts have partial overlap, with the extent of overlap varying greatly with solar activity.

As well as deflecting the solar wind, the Earth's magnetic field deflects cosmic rays, high-energy charged particles that are mostly from outside the Solar system. (Many cosmic rays are kept out of the Solar system by the Sun's magnetosphere, or heliosphere.) By contrast, astronauts on the Moon risk exposure to radiation. Anyone who had been on the Moon's surface during a particularly violent solar eruption in 2005 would have received a lethal dose.

Some of the charged particles do get into the magnetosphere. These spiral around field lines, bouncing back and forth between the poles several times per second. In addition, positive ions slowly drift westward and negative ions drift eastward, giving rise to a ring current. This current reduces the magnetic field at the Earth's surface. Particles that penetrate the ionosphere and collide with the atoms there give rise to the lights of the aurorae and also emit X-rays.

The varying conditions in the magnetosphere, known as space weather, are largely driven by solar activity. If the solar wind is weak, the magnetosphere expands; while if it is strong, it compresses the magnetosphere and more of it gets in. Periods of particularly intense activity, called geomagnetic storms, can occur when a coronal mass ejection erupts above the Sun and sends a shock wave through the Solar System. Such a wave can take just two days to reach the Earth. Geomagnetic storms can cause a lot of disruption; the "Halloween" storm of 2003 damaged more than a third of NASA's satellites. The largest documented storm occurred in 1859. It induced currents strong enough to short out telegraph lines, and aurorae were reported as far south as Hawaii.

Time dependence

Short-term variations

The geomagnetic field changes on time scales from milliseconds to millions of years. Shorter time scales mostly arise from currents in the ionosphere (ionospheric dynamo region) and magnetosphere, and some changes can be traced to geomagnetic storms or daily variations in currents. Changes over time scales of a year or more mostly reflect changes in the Earth's interior, particularly the iron-rich core.

Frequently, the Earth's magnetosphere is hit by solar flares causing geomagnetic storms, provoking displays of aurorae. The short-term instability of the magnetic field is measured with the K-index.

Data from THEMIS show that the magnetic field, which interacts with the solar wind, is reduced when the magnetic orientation is aligned between Sun and Earth – opposite to the previous hypothesis. During forthcoming solar storms, this could result in blackouts and disruptions in artificial satellites.

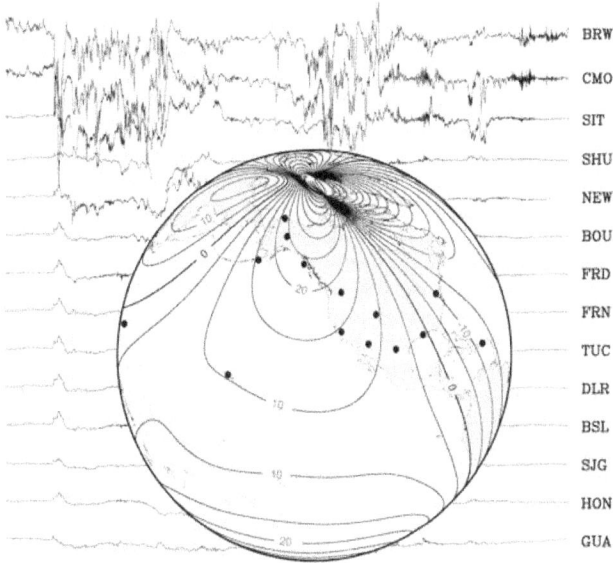

Figure 146: *Background: a set of traces from magnetic observatories showing a magnetic storm in 2000.* *Globe: map showing locations of observatories and contour lines giving horizontal magnetic intensity in μ T.*

Secular variation

Changes in Earth's magnetic field on a time scale of a year or more are referred to as *secular variation*. Over hundreds of years, magnetic declination is observed to vary over tens of degrees. A movie on the right shows how global declinations have changed over the last few centuries.

The direction and intensity of the dipole change over time. Over the last two centuries the dipole strength has been decreasing at a rate of about 6.3% per century. At this rate of decrease, the field would be negligible in about 1600 years. However, this strength is about average for the last 7 thousand years, and the current rate of change is not unusual.

A prominent feature in the non-dipolar part of the secular variation is a *westward drift* at a rate of about 0.2 degrees per year. This drift is not the same everywhere and has varied over time. The globally averaged drift has been westward since about 1400 AD but eastward between about 1000 AD and 1400 AD.

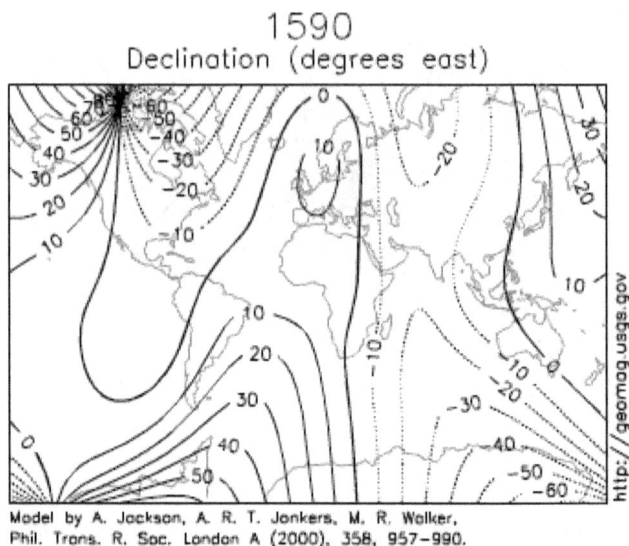

Model by A. Jackson, A. R. T. Jonkers, M. R. Walker,
Phil. Trans. R. Soc. London A (2000), 358, 957–990.

Figure 147: *Estimated declination contours by year, 1590 to 1990 (click to see variation).*

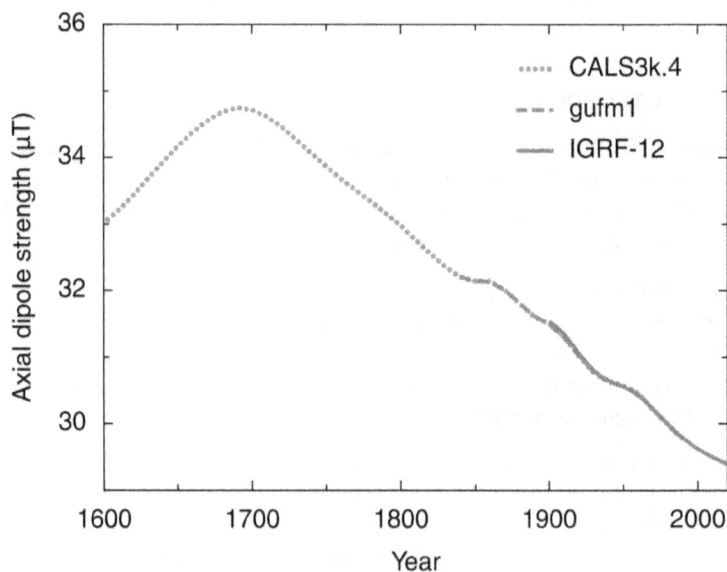

Figure 148: *Strength of the axial dipole component of Earth's magnetic field from 1600 to 2020.*

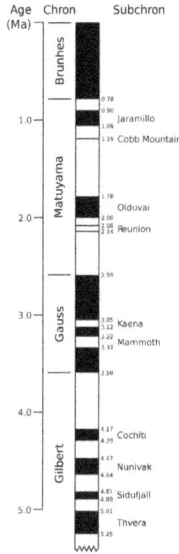

Figure 149: *Geomagnetic polarity during the late Cenozoic Era. Dark areas denote periods where the polarity matches today's polarity, light areas denote periods where that polarity is reversed.*

Changes that predate magnetic observatories are recorded in archaeological and geological materials. Such changes are referred to as *paleomagnetic secular variation* or *paleosecular variation (PSV)*. The records typically include long periods of small change with occasional large changes reflecting geomagnetic excursions and reversals.

Magnetic field reversals

Although generally Earth's field is approximately dipolar, with an axis that is nearly aligned with the rotational axis, occasionally the North and South geomagnetic poles trade places. Evidence for these *geomagnetic reversals* can be found in basalts, sediment cores taken from the ocean floors, and seafloor magnetic anomalies. Reversals occur nearly randomly in time, with intervals between reversals ranging from less than 0.1 million years to as much as 50 million years. The most recent geomagnetic reversal, called the Brunhes–Matuyama reversal, occurred about 780,000 years ago. A related phenomenon, a geomagnetic excursion, amounts to an incomplete reversal, with no change in polarity. The Laschamp event is an example of an excursion, it having occurred during the last ice age (41,000 years ago).

The past magnetic field is recorded mostly by strongly magnetic minerals, particularly iron oxides such as magnetite, that can carry a permanent magnetic moment. This remanent magnetization, or *remanence*, can be acquired in more than one way. In lava flows, the direction of the field is "frozen" in small minerals as they cool, giving rise to a thermoremanent magnetization. In sediments, the orientation of magnetic particles acquires a slight bias towards the magnetic field as they are deposited on an ocean floor or lake bottom. This is called *detrital remanent magnetization*.

Thermoremanent magnetization is the main source of the magnetic anomalies around mid-ocean ridges. As the seafloor spreads, magma wells up from the mantle, cools to form new basaltic crust on both sides of the ridge, and is carried away from it by seafloor spreading. As it cools, it records the direction of the Earth's field. When the Earth's field reverses, new basalt records the reversed direction. The result is a series of stripes that are symmetric about the ridge. A ship towing a magnetometer on the surface of the ocean can detect these stripes and infer the age of the ocean floor below. This provides information on the rate at which seafloor has spread in the past.

Radiometric dating of lava flows has been used to establish a *geomagnetic polarity time scale*, part of which is shown in the image. This forms the basis of magnetostratigraphy, a geophysical correlation technique that can be used to date both sedimentary and volcanic sequences as well as the seafloor magnetic anomalies.

Studies of lava flows on Steens Mountain, Oregon, indicate that the magnetic field could have shifted at a rate of up to 6 degrees per day at some time in Earth's history, which significantly challenges the popular understanding of how the Earth's magnetic field works. This finding was later attributed to unusual rock magnetic properties of the lava flow under study, not rapid field change, by one of the original authors of the 1995 study.

Temporary dipole tilt variations that take the dipole axis across the equator and then back to the original polarity are known as *excursions*.

Earliest appearance

Paleomagnetic studies of Paleoarchean lava in Australia and conglomerate in South Africa have concluded that the magnetic field has been present since at least about 3,450[700] million years ago.

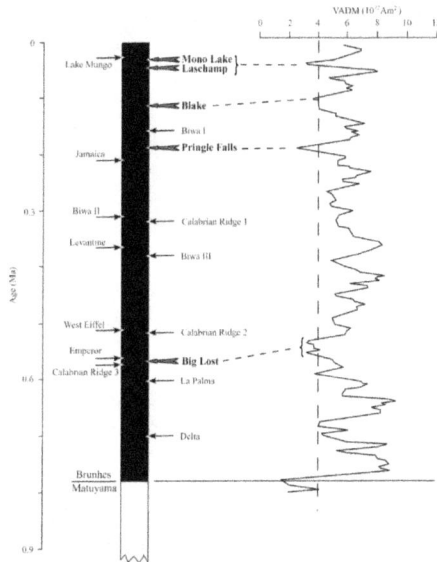

Figure 150: *Variations in virtual axial dipole moment since the last reversal.*

Future

At present, the overall geomagnetic field is becoming weaker; the present strong deterioration corresponds to a 10–15% decline over the last 150 years and has accelerated in the past several years; geomagnetic intensity has declined almost continuously from a maximum 35% above the modern value achieved approximately 2,000 years ago. The rate of decrease and the current strength are within the normal range of variation, as shown by the record of past magnetic fields recorded in rocks.

The nature of Earth's magnetic field is one of heteroscedastic fluctuation. An instantaneous measurement of it, or several measurements of it across the span of decades or centuries, are not sufficient to extrapolate an overall trend in the field strength. It has gone up and down in the past for unknown reasons. Also, noting the local intensity of the dipole field (or its fluctuation) is insufficient to characterize Earth's magnetic field as a whole, as it is not strictly a dipole field. The dipole component of Earth's field can diminish even while the total magnetic field remains the same or increases.

The Earth's magnetic north pole is drifting from northern Canada towards Siberia with a presently accelerating rate—10 kilometres (6.2 mi) per year at the beginning of the 20th century, up to 40 kilometres (25 mi) per year in 2003, and since then has only accelerated.

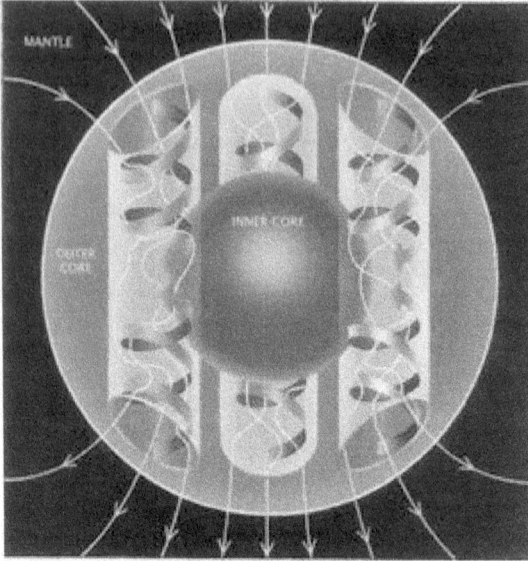

Figure 151: *A schematic illustrating the relationship between motion of conducting fluid, organized into rolls by the Coriolis force, and the magnetic field the motion generates.*

Physical origin

The Earth's magnetic field is believed to be generated by electric currents in the conductive material of its core, created by convection currents due to heat escaping from the core. However the process is complex, and computer models that reproduce some of its features have only been developed in the last few decades.

Earth's core and the geodynamo

The Earth and most of the planets in the Solar System, as well as the Sun and other stars, all generate magnetic fields through the motion of electrically conducting fluids. The Earth's field originates in its core. This is a region of iron alloys extending to about 3400 km (the radius of the Earth is 6370 km). It is divided into a solid inner core, with a radius of 1220 km, and a liquid outer core. The motion of the liquid in the outer core is driven by heat flow from the inner core, which is about 6,000 K (5,730 °C; 10,340 °F), to the core-mantle boundary, which is about 3,800 K (3,530 °C; 6,380 °F). The heat is generated by potential energy released by heavier materials sinking toward the core (planetary differentiation, the iron catastrophe) as well as decay of radioactive elements in the interior. The pattern of flow is organized by the rotation of the Earth and the presence of the solid inner core.

The mechanism by which the Earth generates a magnetic field is known as a dynamo. The magnetic field is generated by a feedback loop: current loops generate magnetic fields (Ampère's circuital law); a changing magnetic field generates an electric field (Faraday's law); and the electric and magnetic fields exert a force on the charges that are flowing in currents (the Lorentz force). These effects can be combined in a partial differential equation for the magnetic field called the *magnetic induction equation*,

$$\frac{\partial \mathbf{B}}{\partial t} = \eta \nabla^2 \mathbf{B} + \nabla \times (\mathbf{u} \times \mathbf{B}),$$

where $\mathbf{u}$ is the velocity of the fluid; $\mathbf{B}$ is the magnetic B-field; and $\eta = 1/\sigma\mu$ is the magnetic diffusivity, which is inversely proportional to the product of the electrical conductivity σ and the permeability μ. The term $\partial \mathbf{B}/\partial t$ is the time derivative of the field; ∇^2 is the Laplace operator and $\nabla \times$ is the curl operator.

The first term on the right hand side of the induction equation is a diffusion term. In a stationary fluid, the magnetic field declines and any concentrations of field spread out. If the Earth's dynamo shut off, the dipole part would disappear in a few tens of thousands of years.

In a perfect conductor ($\sigma = \infty$), there would be no diffusion. By Lenz's law, any change in the magnetic field would be immediately opposed by currents, so the flux through a given volume of fluid could not change. As the fluid moved, the magnetic field would go with it. The theorem describing this effect is called the *frozen-in-field theorem*. Even in a fluid with a finite conductivity, new field is generated by stretching field lines as the fluid moves in ways that deform it. This process could go on generating new field indefinitely, were it not that as the magnetic field increases in strength, it resists fluid motion.

The motion of the fluid is sustained by convection, motion driven by buoyancy. The temperature increases towards the center of the Earth, and the higher temperature of the fluid lower down makes it buoyant. This buoyancy is enhanced by chemical separation: As the core cools, some of the molten iron solidifies and is plated to the inner core. In the process, lighter elements are left behind in the fluid, making it lighter. This is called *compositional convection*. A Coriolis effect, caused by the overall planetary rotation, tends to organize the flow into rolls aligned along the north-south polar axis.

A dynamo can amplify a magnetic field, but it needs a "seed" field to get it started. For the Earth, this could have been an external magnetic field. Early in its history the Sun went through a T-Tauri phase in which the solar wind would have had a magnetic field orders of magnitude larger than the present solar wind. However, much of the field may have been screened out by the Earth's mantle. An alternative source is currents in the core-mantle boundary driven by chemical reactions or variations in thermal or electric conductivity. Such effects may still provide a small bias that are part of the boundary conditions for the geodynamo.

The average magnetic field in the Earth's outer core was calculated to be 25 gausses, 50 times stronger than the field at the surface.

Numerical models

Simulating the geodynamo requires numerically solving a set of nonlinear partial differential equations for the magnetohydrodynamics (MHD) of the Earth's interior. Simulation of the MHD equations is performed on a 3D grid of points and the fineness of the grid, which in part determines the realism of the solutions, is limited mainly by computer power. For decades, theorists were confined to creating *kinematic dynamo* computer models in which the fluid motion is chosen in advance and the effect on the magnetic field calculated. Kinematic dynamo theory was mainly a matter of trying different flow geometries and testing whether such geometries could sustain a dynamo.

The first *self-consistent* dynamo models, ones that determine both the fluid motions and the magnetic field, were developed by two groups in 1995, one in Japan and one in the United States. The latter received attention because it successfully reproduced some of the characteristics of the Earth's field, including geomagnetic reversals.

Currents in the ionosphere and magnetosphere

Electric currents induced in the ionosphere generate magnetic fields (ionospheric dynamo region). Such a field is always generated near where the atmosphere is closest to the Sun, causing daily alterations that can deflect surface magnetic fields by as much as one degree. Typical daily variations of field strength are about 25 nanoteslas (nT) (one part in 2000), with variations over a few seconds of typically around 1 nT (one part in 50,000).

Measurement and analysis

Detection

The Earth's magnetic field strength was measured by Carl Friedrich Gauss in 1832 and has been repeatedly measured since then, showing a relative decay of about 10% over the last 150 years. The Magsat satellite and later satellites have used 3-axis vector magnetometers to probe the 3-D structure of the Earth's magnetic field. The later Ørsted satellite allowed a comparison indicating a

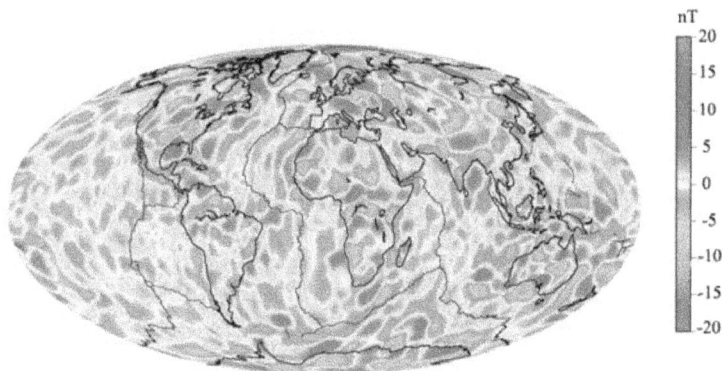

Figure 152: *A model of short-wavelength features of
Earth's magnetic field, attributed to lithospheric anomalies*

dynamic geodynamo in action that appears to be giving rise to an alternate
pole under the Atlantic Ocean west of South Africa.

Governments sometimes operate units that specialize in measurement of the
Earth's magnetic field. These are geomagnetic observatories, typically part
of a national Geological survey, for example the British Geological Survey's
Eskdalemuir Observatory. Such observatories can measure and forecast magnetic conditions such as magnetic storms that sometimes affect communica-
tions, electric power, and other human activities.

The International Real-time Magnetic Observatory Network, with over 100
interlinked geomagnetic observatories around the world, has been recording
the Earth's magnetic field since 1991.

The military determines local geomagnetic field characteristics, in order to detect *anomalies* in the natural background that might be caused by a significant
metallic object such as a submerged submarine. Typically, these magnetic
anomaly detectors are flown in aircraft like the UK's Nimrod or towed as an
instrument or an array of instruments from surface ships.

Commercially, geophysical prospecting companies also use magnetic detectors
to identify naturally occurring anomalies from ore bodies, such as the Kursk
Magnetic Anomaly.

Crustal magnetic anomalies

Magnetometers detect minute deviations in the Earth's magnetic field caused
by iron artifacts, kilns, some types of stone structures, and even ditches and

middens in archaeological geophysics. Using magnetic instruments adapted from airborne magnetic anomaly detectors developed during World War II to detect submarines, the magnetic variations across the ocean floor have been mapped. Basalt — the iron-rich, volcanic rock making up the ocean floor — contains a strongly magnetic mineral (magnetite) and can locally distort compass readings. The distortion was recognized by Icelandic mariners as early as the late 18th century. More important, because the presence of magnetite gives the basalt measurable magnetic properties, these magnetic variations have provided another means to study the deep ocean floor. When newly formed rock cools, such magnetic materials record the Earth's magnetic field.

Statistical models

Each measurement of the magnetic field is at a particular place and time. If an accurate estimate of the field at some other place and time is needed, the measurements must be converted to a model and the model used to make predictions.

Spherical harmonics

The most common way of analyzing the global variations in the Earth's magnetic field is to fit the measurements to a set of spherical harmonics. This was first done by Carl Friedrich Gauss.[701] Spherical harmonics are functions that oscillate over the surface of a sphere. They are the product of two functions, one that depends on latitude and one on longitude. The function of longitude is zero along zero or more great circles passing through the North and South Poles; the number of such *nodal lines* is the absolute value of the *order* m. The function of latitude is zero along zero or more latitude circles; this plus the order is equal to the *degree* ℓ. Each harmonic is equivalent to a particular arrangement of magnetic charges at the center of the Earth. A *monopole* is an isolated magnetic charge, which has never been observed. A *dipole* is equivalent to two opposing charges brought close together and a *quadrupole* to two dipoles brought together. A quadrupole field is shown in the lower figure on the right.

Spherical harmonics can represent any scalar field (function of position) that satisfies certain properties. A magnetic field is a vector field, but if it is expressed in Cartesian components X, Y, Z, each component is the derivative of the same scalar function called the *magnetic potential*. Analyses of the Earth's magnetic field use a modified version of the usual spherical harmonics that differ by a multiplicative factor. A least-squares fit to the magnetic field measurements gives the Earth's field as the sum of spherical harmonics, each multiplied by the best-fitting *Gauss coefficient* $g_m{}^\ell$ or $h_m{}^\ell$.

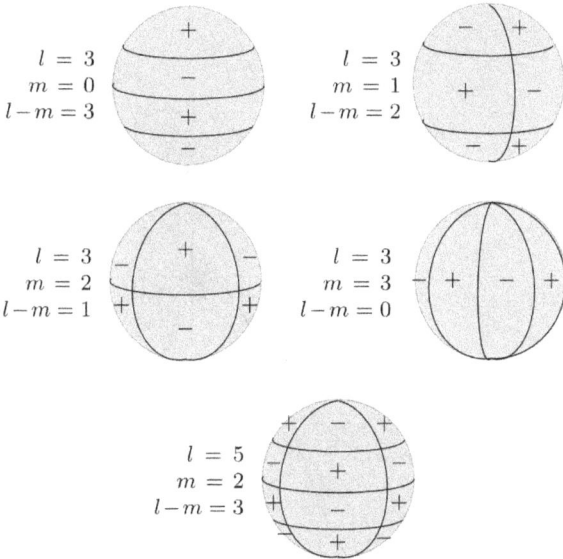

Figure 153: *Schematic representation of spherical harmonics on a sphere and their nodal lines. $P_{l\,m}$ is equal to 0 along m great circles passing through the poles, and along l-m circles of equal latitude. The function changes sign each ltime it crosses one of these lines.*

The lowest-degree Gauss coefficient, $g_0^{\,0}$, gives the contribution of an isolated magnetic charge, so it is zero. The next three coefficients – $g_1^{\,0}$, $g_1^{\,1}$, and $h_1^{\,1}$ – determine the direction and magnitude of the dipole contribution. The best fitting dipole is tilted at an angle of about $10°$ with respect to the rotational axis, as described earlier.

Radial dependence

Spherical harmonic analysis can be used to distinguish internal from external sources if measurements are available at more than one height (for example, ground observatories and satellites). In that case, each term with coefficient $g_m^{\,l}$ or $h_m^{\,l}$ can be split into two terms: one that decreases with radius as $1/r^{l+1}$ and one that *increases* with radius as r^l. The increasing terms fit the external sources (currents in the ionosphere and magnetosphere). However, averaged over a few years the external contributions average to zero.

The remaining terms predict that the potential of a dipole source ($\ell{=}1$) drops off as $1/r^2$. The magnetic field, being a derivative of the potential, drops off as $1/r^3$. Quadrupole terms drop off as $1/r^4$, and higher order terms drop off increasingly rapidly with the radius. The radius of the outer core is about half of

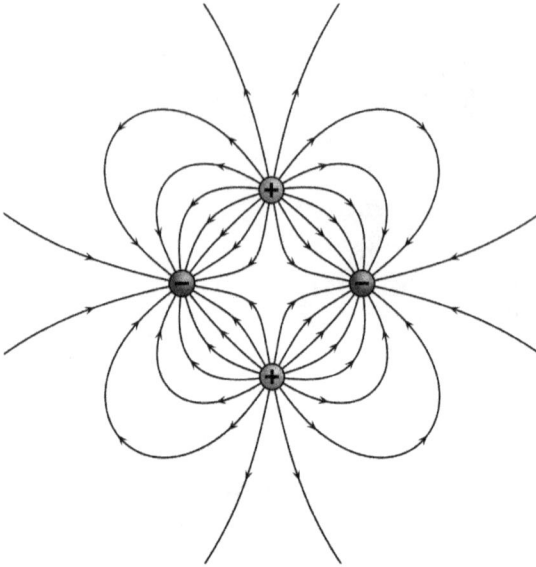

Figure 154: *Example of a quadrupole field. This can also be constructed by moving two dipoles together.*

the radius of the Earth. If the field at the core-mantle boundary is fit to spherical harmonics, the dipole part is smaller by a factor of about 8 at the surface, the quadrupole part by a factor of 16, and so on. Thus, only the components with large wavelengths can be noticeable at the surface. From a variety of arguments, it is usually assumed that only terms up to degree 14 or less have their origin in the core. These have wavelengths of about 2,000 kilometres (1,200 mi) or less. Smaller features are attributed to crustal anomalies.

Global models

The International Association of Geomagnetism and Aeronomy maintains a standard global field model called the International Geomagnetic Reference Field. It is updated every five years. The 11th-generation model, IGRF11, was developed using data from satellites (Ørsted, CHAMP and SAC-C) and a world network of geomagnetic observatories. The spherical harmonic expansion was truncated at degree 10, with 120 coefficients, until 2000. Subsequent models are truncated at degree 13 (195 coefficients).

Another global field model, called the World Magnetic Model, is produced jointly by the United States National Centers for Environmental Information (formerly the National Geophysical Data Center) and the British Geological

Survey. This model truncates at degree 12 (168 coefficients) with an approximate spatial resolution of 3,000 kilometers. It is the model used by the United States Department of Defense, the Ministry of Defence (United Kingdom), the United States Federal Aviation Administration (FAA), the North Atlantic Treaty Organization (NATO), and the International Hydrographic Office as well as in many civilian navigation systems.

A third model, produced by the Goddard Space Flight Center (NASA and GSFC) and the Danish Space Research Institute, uses a "comprehensive modeling" approach that attempts to reconcile data with greatly varying temporal and spatial resolution from ground and satellite sources.

For users with higher accuracy needs, the United States National Centers for Environmental Information developed the Enhanced Magnetic Model (EMM), which extends to degree and order 790 and resolves magnetic anomalies down to a wavelength of 56 kilometers. It was compiled from satellite, marine, aeromagnetic and ground magnetic surveys. As of 2018[702], the latest version, EMM2017, includes data from The European Space Agency's Swarm satellite mission.

Biomagnetism

Animals including birds and turtles can detect the Earth's magnetic field, and use the field to navigate during migration. Some researchers have found that cows and wild deer tend to align their bodies north-south while relaxing, but not when the animals are under high-voltage power lines, suggesting that magnetism is responsible. Other researchers reported in 2011 that they could not replicate those findings using different Google Earth images.

Researchers found out that very weak electromagnetic fields disrupt the magnetic compass used by European robins and other songbirds to navigate using the Earth's magnetic field. Neither power lines nor cellphone signals are to blame for the electromagnetic field effect on the birds; instead, the culprits have frequencies between 2 kHz and 5 MHz. These include AM radio signals and ordinary electronic equipment that might be found in businesses or private homes.

Further reading

<templatestyles src="Template:Refbegin/styles.css" />

- Campbell, Wallace H. (2003). *Introduction to geomagnetic fields* (2nd ed.). New York: Cambridge University Press. ISBN 978-0-521-52953-2.

- Comins, Neil F. (2008). *Discovering the Essential Universe* (Fourth ed.). W. H. Freeman. ISBN 978-1-4292-1797-2.
- Herndon, J. M. (1996-01-23). "Substructure of the inner core of the Earth"[703]. *PNAS*. **93** (2): 646–648. Bibcode: 1996PNAS...93..646H[704]. doi: 10.1073/pnas.93.2.646[705]. PMC 40105[703] ∂. PMID 11607625[706].
- Hollenbach, D. F.; Herndon, J. M. (2001-09-25). "Deep-Earth reactor: Nuclear fission, helium, and the geomagnetic field"[707]. *PNAS*. **98** (20): 11085–90. Bibcode: 2001PNAS...9811085H[708]. doi: 10.1073/pnas.201393998[709]. PMC 58687[707] ∂. PMID 11562483[710].
- Love, Jeffrey J. (2008). "Magnetic monitoring of Earth and space"[711] (PDF). *Physics Today*. **61** (2): 31–37. Bibcode: 2008PhT....61b..31H[712]. doi: 10.1063/1.2883907[713].
- Luhmann, J. G.; Johnson, R. E.; Zhang, M. H. G. (1992). "Evolutionary impact of sputtering of the Martian atmosphere by O+ pickup ions". *Geophysical Research Letters*. **19** (21): 2151–2154. Bibcode: 1992GeoRL..19.2151L[714]. doi: 10.1029/92GL02485[715].
- Merrill, Ronald T. (2010). *Our Magnetic Earth: The Science of Geomagnetism*. University of Chicago Press. ISBN 0-226-52050-1.
- Merrill, Ronald T.; McElhinny, Michael W.; McFadden, Phillip L. (1996). *The magnetic field of the earth: paleomagnetism, the core, and the deep mantle*. Academic Press. ISBN 978-0-12-491246-5.
- "Temperature of the Earth's core"[716]. *NEWTON Ask a Scientist*. 1999.
- Tauxe, Lisa (1998). *Paleomagnetic Principles and Practice*. Kluwer. ISBN 0-7923-5258-0.
- Towle, J. N. (1984). "The Anomalous Geomagnetic Variation Field and Geoelectric Structure Associated with the Mesa Butte Fault System, Arizona". *Geological Society of America Bulletin*. **9** (2): 221–225. Bibcode: 1984GSAB...95..221T[717]. doi: 10.1130/0016-7606(1984)95<221:TAGVFA>2.0.CO;2[718].
- Wait, James R. (1954). "On the relation between telluric currents and the earth's magnetic field". *Geophysics*. **19** (2): 281–289. Bibcode: 1954Geop...19..281W[719]. doi: 10.1190/1.1437994[720].
- Walt, Martin (1994). *Introduction to Geomagnetically Trapped Radiation*. Cambridge University Press. ISBN 978-0-521-61611-9.

External links

> Wikimedia Commons has media related to *Geomagnetism*.

- *Geomagnetism & Paleomagnetism background material*[721]. American Geophysical Union Geomagnetism and Paleomagnetism Section.
- *National Geomagnetism Program*[722]. United States Geological Survey, March 8, 2011.
- *BGS Geomagnetism*[723]. Information on monitoring and modeling the geomagnetic field. British Geological Survey, August 2005.
- William J. Broad, *Will Compasses Point South?*[724]. New York Times, July 13, 2004.
- John Roach, *Why Does Earth's Magnetic Field Flip?*[725]. National Geographic, September 27, 2004.
- *Magnetic Storm*[726]. PBS NOVA, 2003. (*ed*. about pole reversals)
- *When North Goes South*[727]. Projects in Scientific Computing, 1996.
- *The Great Magnet, the Earth*[728], History of the discovery of Earth's magnetic field by David P. Stern.
- *Exploration of the Earth's Magnetosphere*[729], Educational web site by David P. Stern and Mauricio Peredo
- *Dr. Dan Lathrop: The study of the Earth's magnetic field*[730]. Interview with Dr. Dan Lathrop, Geophysicist at the University of Maryland, about his experiments with the Earth's core and magnetic field. July 3, 2008
- International Geomagnetic Reference Field 2011[731]
- Global evolution/anomaly of the Earth's magnetic field[732] Sweeps are in 10 degree steps at 10 years intervals. Based on data from: The Institute of Geophysics, ETH Zurich[733]
- *Patterns in Earth's magnetic field that evolve on the order of 1,000 years*[734]. July 19, 2017

Orbit and rotation

Earth's rotation

Earth's rotation is the rotation of Planet Earth around its own axis. Earth rotates eastward, in prograde motion. As viewed from the north pole star Polaris, Earth turns counter clockwise.

The North Pole, also known as the Geographic North Pole or Terrestrial North Pole, is the point in the Northern Hemisphere where Earth's axis of rotation meets its surface. This point is distinct from Earth's North Magnetic Pole. The South Pole is the other point where Earth's axis of rotation intersects its surface, in Antarctica.

Earth rotates once in about 24 hours with respect to the Sun, but once every 23 hours, 56 minutes, and 4 seconds with respect to the stars (see below). Earth's rotation is slowing slightly with time; thus, a day was shorter in the past. This is due to the tidal effects the Moon has on Earth's rotation. Atomic clocks show that a modern day is longer by about 1.7 milliseconds than a century ago, slowly increasing the rate at which UTC is adjusted by leap seconds. Analysis of historical astronomical records shows a slowing trend of about 2.3 milliseconds per century since the 8th century BCE.

History

Among the ancient Greeks, several of the Pythagorean school believed in the rotation of the earth rather than the apparent diurnal rotation of the heavens. Perhaps the first was Philolaus (470–385 BCE), though his system was complicated, including a counter-earth rotating daily about a central fire.

A more conventional picture was that supported by Hicetas, Heraclides and Ecphantus in the fourth century BCE who assumed that the earth rotated but did not suggest that the earth revolved about the sun. In the third century BCE, Aristarchus of Samos suggested the sun's central place.

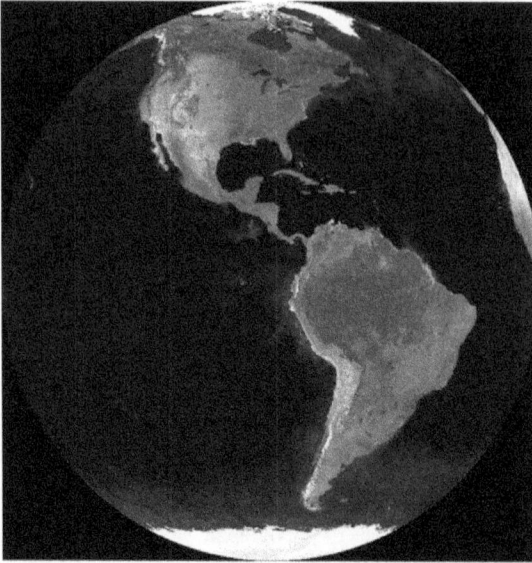

Figure 155: *An animation of Earth's rotation around the planet's axis*

Figure 156: *This long-exposure photo of the northern night sky over the Nepali Himalayas shows the apparent paths of the stars as Earth rotates.*

However, Aristotle in the fourth century BCE criticized the ideas of Philolaus as being based on theory rather than observation. He established the idea of a sphere of fixed stars that rotated about the earth. This was accepted by most of those who came after, in particular Claudius Ptolemy (2nd century CE), who thought the earth would be devastated by gales if it rotated.

In 499 CE, the Indian astronomer Aryabhata wrote that the spherical earth rotates about its axis daily, and that the apparent movement of the stars is a relative motion caused by the rotation of the Earth. He provided the following analogy: "Just as a man in a boat going in one direction sees the stationary things on the bank as moving in the opposite direction, in the same way to a man at Lanka the fixed stars appear to be going westward."[735]

In the 10th century, some Muslim astronomers accepted that the Earth rotates around its axis. According to al-Biruni, Abu Sa'id al-Sijzi (d. circa 1020) invented an astrolabe called *al-zūraqī* based on the idea believed by some of his contemporaries "that the motion we see is due to the Earth's movement and not to that of the sky." The prevalence of this view is further confirmed by a reference from the 13th century which states: "According to the geometers [or engineers] (*muhandisīn*), the earth is in constant circular motion, and what appears to be the motion of the heavens is actually due to the motion of the earth and not the stars." Treatises were written to discuss its possibility, either as refutations or expressing doubts about Ptolemy's arguments against it.[736] At the Maragha and Samarkand observatories, the Earth's rotation was discussed by Tusi (b. 1201) and Qushji (b. 1403); the arguments and evidence they used resemble those used by Copernicus.

In medieval Europe, Thomas Aquinas accepted Aristotle's view[737] and so, reluctantly, did John Buridan[738] and Nicole Oresme[739] in the fourteenth century. Not until Nicolaus Copernicus in 1543 adopted a heliocentric world system did the contemporary understanding of earth's rotation begin to be established. Copernicus pointed out that if the movement of the earth is violent, then the movement of the stars must be very much more so. He acknowledged the contribution of the Pythagoreans and pointed to examples of relative motion. For Copernicus this was the first step in establishing the simpler pattern of planets circling a central sun.

Tycho Brahe, who produced accurate observations on which Kepler based his laws, used Copernicus's work as the basis of a system assuming a stationary earth. In 1600, William Gilbert strongly supported the earth's rotation in his treatise on the earth's magnetism and thereby influenced many of his contemporaries. Those like Gilbert who did not openly support or reject the motion of the earth about the sun are often called "semi-Copernicans". A century after Copernicus, Riccioli disputed the model of a rotating earth due to the lack of then-observable eastward deflections in falling bodies;[740] such deflections would later be called the Coriolis effect. However, the contributions of Kepler, Galileo and Newton gathered support for the theory of the rotation of the Earth.

Empirical tests

The earth's rotation implies that the equator bulges and the poles are flattened. In his *Principia*, Newton predicted this flattening would occur in the ratio of 1:230, and pointed to the 1673 pendulum measurements by Richer as corroboration of the change in gravity, but initial measurements of meridian lengths by Picard and Cassini at the end of the 17th century suggested the opposite. However measurements by Maupertuis and the French Geodesic Mission in the 1730s established the flattening, thus confirming both Newton and the Copernican position.

In the Earth's rotating frame of reference, a freely moving body follows an apparent path that deviates from the one it would follow in a fixed frame of reference. Because of the Coriolis effect, falling bodies veer slightly eastward from the vertical plumb line below their point of release, and projectiles veer right in the northern hemisphere (and left in the southern) from the direction in which they are shot. The Coriolis effect is mainly observable at a meteorological scale, where it is responsible for the differing rotation direction of cyclones in the northern and southern hemispheres.

Hooke, following a 1679 suggestion from Newton, tried unsuccessfully to verify the predicted eastward deviation of a body dropped from a height of 8.2 meters, but definitive results were only obtained later, in the late 18th and early 19th century, by Giovanni Battista Guglielmini in Bologna, Johann Friedrich Benzenberg in Hamburg and Ferdinand Reich in Freiberg, using taller towers and carefully released weights.[741] A ball dropped from a height of 158.5 m (520 ft) departed by 27.4 mm (1.08 in) from the vertical compared with a calculated value of 28.1 mm (1.11 in).

The most celebrated test of Earth's rotation is the Foucault pendulum first built by physicist Léon Foucault in 1851, which consisted of a lead-filled brass sphere suspended 67 m from the top of the Panthéon in Paris. Because of the Earth's rotation under the swinging pendulum, the pendulum's plane of oscillation appears to rotate at a rate depending on latitude. At the latitude of Paris the predicted and observed shift was about 11 degrees clockwise per hour. Foucault pendulums now swing in museums around the world.

Figure 157: *Starry circles arc around the south celestial pole, seen overhead at ESO's La Silla Observatory.*

Periods

True solar day

Earth's rotation period relative to the Sun (solar noon to solar noon) is its *true solar day* or *apparent solar day*. It depends on the Earth's orbital motion and is thus affected by changes in the eccentricity and inclination of Earth's orbit. Both vary over thousands of years, so the annual variation of the true solar day also varies. Generally, it is longer than the mean solar day during two periods of the year and shorter during another two.[742] The true solar day tends to be longer near perihelion when the Sun apparently moves along the ecliptic through a greater angle than usual, taking about 10 seconds longer to do so. Conversely, it is about 10 seconds shorter near aphelion. It is about 20 seconds longer near a solstice when the projection of the Sun's apparent motion along the ecliptic onto the celestial equator causes the Sun to move through a greater angle than usual. Conversely, near an equinox the projection onto the equator is shorter by about 20 seconds. Currently, the perihelion and solstice effects combine to lengthen the true solar day near 22 December by 30 mean solar seconds, but the solstice effect is partially cancelled by the aphelion effect near 19 June when it is only 13 seconds longer. The effects of the equinoxes shorten it near 26 March and 16 September by 18 seconds and 21 seconds, respectively.[743,744]

Mean solar day

The average of the true solar day during the course of an entire year is the *mean solar day*, which contains 86,400 mean solar seconds. Currently, each of these seconds is slightly longer than an SI second because Earth's mean solar day is now slightly longer than it was during the 19th century due to tidal friction. The average length of the mean solar day since the introduction of the leap second in 1972 has been about 0 to 2 ms longer than 86,400 SI seconds.[745,746,747] Random fluctuations due to core-mantle coupling have an amplitude of about 5 ms.[748,749] The mean solar second between 1750 and 1892 was chosen in 1895 by Simon Newcomb as the independent unit of time in his Tables of the Sun. These tables were used to calculate the world's ephemerides between 1900 and 1983, so this second became known as the ephemeris second. In 1967 the SI second was made equal to the ephemeris second.[750]

The apparent solar time is a measure of the Earth's rotation and the difference between it and the mean solar time is known as the equation of time.

Stellar and sidereal day

Earth's rotation period relative to the fixed stars, called its *stellar day* by the International Earth Rotation and Reference Systems Service (IERS), is 86,164.098 903 691 seconds of mean solar time (UT1) (23h 56m 4.098 903 691s, 0.997 269 663 237 16 mean solar days).[751,752]</ref> Earth's rotation period relative to the precessing or moving mean vernal equinox, named *sidereal day*, is 86,164.090 530 832 88 seconds of mean solar time (UT1) (23h 56m 4.090 530 832 88s, 0.997 269 566 329 08 mean solar days). Thus the sidereal day is shorter than the stellar day by about 8.4 ms.[753]

Both the stellar day and the sidereal day are shorter than the mean solar day by about 3 minutes 56 seconds. The mean solar day in SI seconds is available from the IERS for the periods 1623–2005[754] and 1962–2005.[755]

Recently (1999–2010) the average annual length of the mean solar day in excess of 86,400 SI seconds has varied between 0.25 ms and 1 ms, which must be added to both the stellar and sidereal days given in mean solar time above to obtain their lengths in SI seconds (see Fluctuations in the length of day).

Angular speed

The angular speed of Earth's rotation in inertial space is $(7.2921150 \pm 0.0000001) \times 10^{-5}$ radians per SI second (mean solar second). Multiplying by $(180°/\pi \text{ radians}) \times (86,400 \text{ seconds/mean solar day})$ yields 360.9856°/mean solar day, indicating that Earth rotates more than 360° relative to the fixed stars in one solar day. Earth's movement along its nearly circular orbit while it is

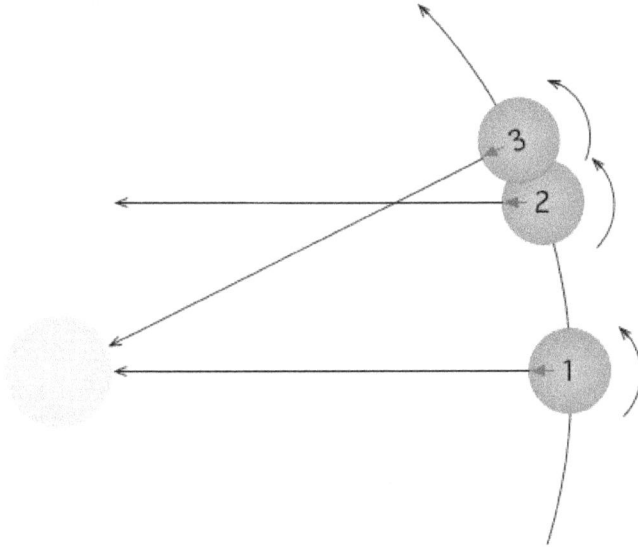

Figure 158: *On a prograde planet like the Earth, the stellar day is shorter than the solar day. At time 1, the Sun and a certain distant star are both overhead. At time 2, the planet has rotated 360° and the distant star is overhead again but the Sun is not (1→2 = one stellar day). It is not until a little later, at time 3, that the Sun is overhead again (1→3 = one solar day).*

rotating once around its axis requires that Earth rotate slightly more than once relative to the fixed stars before the mean Sun can pass overhead again, even though it rotates only once (360°) relative to the mean Sun.[756] Multiplying the value in rad/s by Earth's equatorial radius of 6,378,137 m (WGS84 ellipsoid) (factors of 2π radians needed by both cancel) yields an **equatorial speed of 465.1 m/s (1,526 ft/s)**, or 1,674.4 km/h (1,040.4 mph).[757] Some sources state that Earth's equatorial speed is slightly less, or 1,669.8 km/h.[758] This is obtained by dividing Earth's equatorial circumference by 24 hours. However, the use of only one circumference unwittingly implies only one rotation in inertial space, so the corresponding time unit must be a sidereal hour. This is confirmed by multiplying by the number of sidereal days in one mean solar day, 1.002 737 909 350 795, which yields the equatorial speed in mean solar hours given above of 1,674.4 km/h.

The tangential speed of Earth's rotation at a point on Earth can be approximated by multiplying the speed at the equator by the cosine of the latitude. For example, the Kennedy Space Center is located at latitude 28.59° N, which

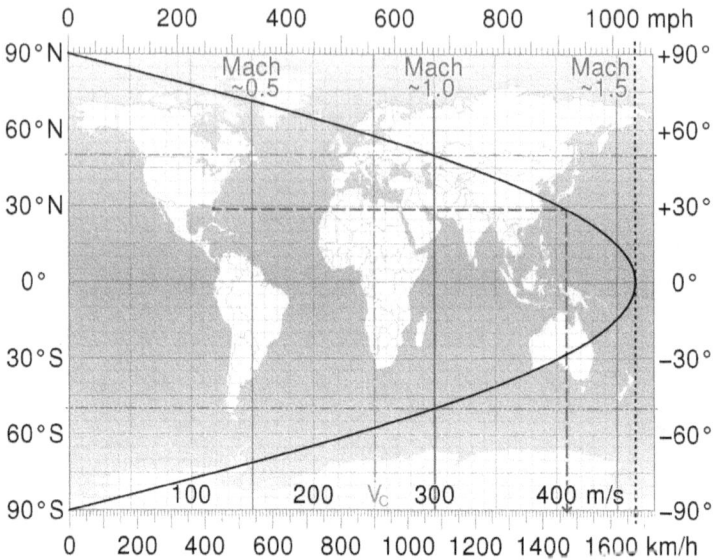

Figure 159: *Plot of latitude vs tangential speed. The dashed line shows that the Kennedy Space Center example. The dot-dash line denotes typical airliner cruise speed.*

yields a speed of: cos 28.59° × 1,674.4 km/h (1,040.4 mph; 465.1 m/s) = 1,470.23 km/h (913.56 mph; 408.40 m/s)

Changes

In rotational axis

The Earth's rotation axis moves with respect to the fixed stars (inertial space); the components of this motion are precession and nutation. It also moves with respect to the Earth's crust; this is called polar motion.

Precession is a rotation of the Earth's rotation axis, caused primarily by external torques from the gravity of the Sun, Moon and other bodies. The polar motion is primarily due to free core nutation and the Chandler wobble.

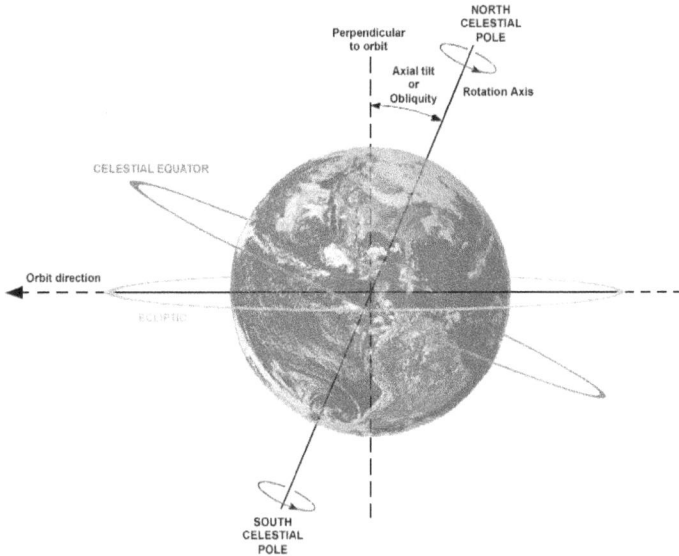

Figure 160: *Earth's axial tilt is about 23.4°. It oscillates between 22.1° and 24.5° on a 41,000-year cycle and is currently decreasing.*

In rotational velocity

Tidal interactions

Over millions of years, the Earth's rotation slowed significantly by tidal acceleration through gravitational interactions with the Moon. In this process, angular momentum is slowly transferred to the Moon at a rate proportional to r^{-6}, where r is the orbital radius of the Moon. This process gradually increased the length of day to its current value and resulted in the Moon being tidally locked with the Earth.

This gradual rotational deceleration is empirically documented with estimates of day lengths obtained from observations of tidal rhythmites and stromatolites; a compilation of these measurements found the length of day to increase steadily from about 21 hours at 600Myr ago to the current 24 hour value. By counting the microscopic lamina that form at higher tides, tidal frequencies (and thus day lengths) can be estimated, much like counting tree rings, though these estimates can be increasingly unreliable at older ages.

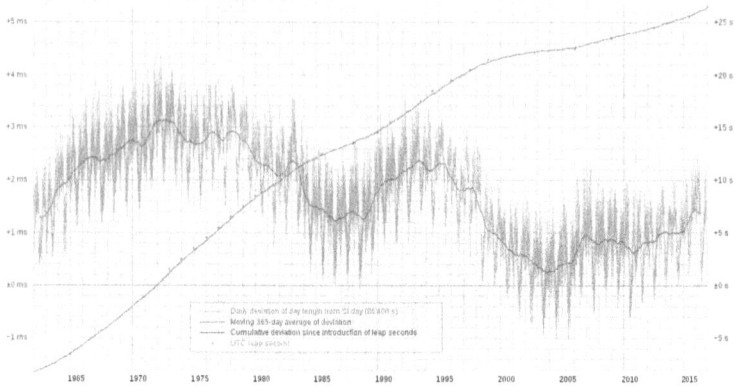

Figure 161: *Deviation of day length from SI based day*

Resonant stabilization

The current rate of tidal deceleration is anomalously high, implying the Earth's rotational velocity must have decreased more slowly in the past. Empirical data tentatively shows a sharp increase in rotational deceleration about 600Myr ago. Some models suggest that the Earth maintained a constant day length of 21 hours throughout much of the Precambrian. This day length corresponds to the semidiurnal resonant period of the thermally-driven atmospheric tide; at this day length, the decelerative lunar torque could have been canceled by an accelerative torque from the atmospheric tide, resulting in no net torque and a constant rotational period. This stabilizing effect could have been broken by a sudden change in global temperature. Recent computational simulations support this hypothesis and suggest the Marinoan or Sturtian glaciations broke this stable configuration about 600Myr ago, citing the resemblance of simulated results and existing paleorotational data.

Global events

Additionally, some large-scale events, such as the 2004 Indian Ocean earthquake, have caused the length of a day to shorten by 3 microseconds by affecting the Earth's moment of inertia.[759] Post-glacial rebound, ongoing since the last Ice age, is also changing the distribution of the Earth's mass thus affecting the moment of inertia of the Earth and, by the conservation of angular momentum, the Earth's rotation period.

Figure 162: *An artist's rendering of the protoplanetary disk.*

Measurement

The primary monitoring of the Earth's rotation is performed with very-long-baseline interferometry coordinated with the Global Positioning System, satellite laser ranging, and other satellite techniques. This provides an absolute reference for the determination of universal time, precession, and nutation.[760]

Ancient observations

There are recorded observations of solar and lunar eclipses by Babylonian and Chinese astronomers beginning in the 8th century BCE, as well as from the medieval Islamic world and elsewhere. These observations can be used to determine changes in the Earth's rotation over the last 27 centuries, since the length of the day is a critical parameter in the calculation of the place and time of eclipses. A change in day length of milliseconds per century shows up as a change of hours and thousands of kilometers in eclipse observations. The ancient data is consistent with a shorter day, meaning the Earth was turning faster throughout the past.

Origin

The Earth's original rotation was a vestige of the original angular momentum of the cloud of dust, rocks, and gas that coalesced to form the Solar System. This primordial cloud was composed of hydrogen and helium produced in the Big Bang, as well as heavier elements ejected by supernovas. As this interstellar dust is heterogeneous, any asymmetry during gravitational accretion resulted in the angular momentum of the eventual planet.

However, if the giant-impact hypothesis for the origin of the Moon is correct, this primordial rotation rate would have been reset by the Theia impact 4.5 billion years ago. Regardless of the speed and tilt of the Earth's rotation before the impact, it would have experienced a day some five hours long after the impact. Tidal effects would then have slowed this rate to its modern value.

External links

- USNO Earth Orientation[761] new site, being populated
- USNO IERS[762] old site, to be abandoned
- IERS Earth Orientation Center: Earth rotation data and interactive analysis[763]
- International Earth Rotation and Reference Systems Service (IERS)[764]
- If the Earth's rotation period is less than 24 hours, why don't our clocks fall out of sync with the Sun?[765]

Earth's orbit

All Celestial bodies in the Solar System, including planets such as our own, orbit around the Solar System's centre of mass. The sun makes up 99.76% of this mass which is why the centre of mass is extremely close to the sun.

Earth's orbit is the trajectory along which Earth travels around the Sun. The average distance between the Earth and the Sun is 149.60 million km (92.96 million mi), and one complete orbit takes 365.256 days (1 sidereal year), during which time Earth has traveled 940 million km (584 million mi).[766] Earth's orbit has an eccentricity of 0.0167.

As seen from Earth, the planet's orbital prograde motion makes the Sun appear to move with respect to other stars at a rate of about 1° (or a Sun or Moon diameter every 12 hours) eastward per solar day.[767] Earth's orbital speed averages about 30 km/s (108,000 km/h; 67,000 mph), which is fast enough to cover the planet's diameter in 7 minutes and the distance to the Moon in 4 hours.

From a vantage point above the north pole of either the Sun or Earth, Earth would appear to revolve in a counterclockwise direction around the Sun. From the same vantage point, both the Earth and the Sun would appear to rotate also in a counterclockwise direction about their respective axes.

Figure 163: *The Earth at different points in its orbit*

History of study

Heliocentrism is the scientific model that first placed the Sun at the center of the Solar System and put the planets, including Earth, in its orbit. Historically, heliocentrism is opposed to geocentrism, which placed the Earth at the center. Aristarchus of Samos already proposed a heliocentric model in the 3rd century BC. In the 16th century, Nicolaus Copernicus' *De revolutionibus* presented a full discussion of a heliocentric model of the universe in much the same way as Ptolemy had presented his geocentric model in the 2nd century. This "Copernican revolution" resolved the issue of planetary retrograde motion by arguing that such motion was only perceived and apparent. "Although Copernicus's groundbreaking book...had been [printed] over a century earlier, [the Dutch mapmaker] Joan Blaeu was the first mapmaker to incorporate his revolutionary heliocentric theory into a map of the world."[768]

Influence on Earth

Because of Earth's axial tilt (often known as the obliquity of the ecliptic), the inclination of the Sun's trajectory in the sky (as seen by an observer on Earth's surface) varies over the course of the year. For an observer at a northern latitude, when the north pole is tilted toward the Sun the day lasts longer and the Sun appears higher in the sky. This results in warmer average temperatures, as additional solar radiation reaches the surface. When the north pole is tilted away from the Sun, the reverse is true and the weather is generally cooler. Above the Arctic Circle and below the Antarctic Circle, an extreme case is reached in which there is no sunlight at all for part of the year. This is called

Figure 164: *Heliocentric Solar System*

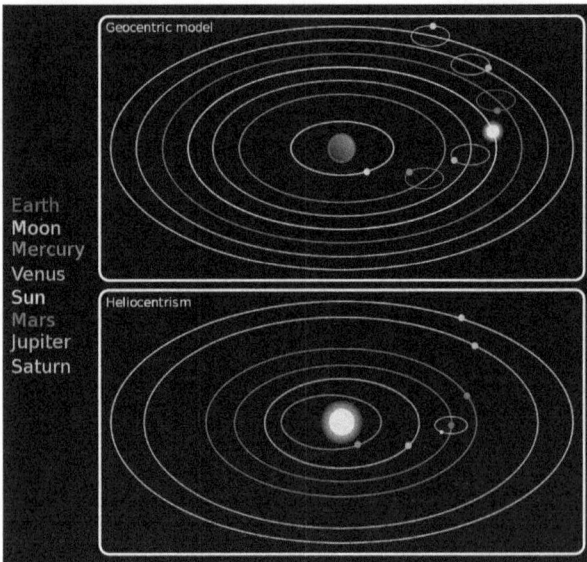

Figure 165: *Heliocentrism (lower panel) in comparison to the geocentric model (upper panel)*

a polar night. This variation in the weather (because of the direction of the Earth's axial tilt) results in the seasons.

Events in the orbit

By astronomical convention, the four seasons are determined by the solstices (the two points in the Earth's orbit of the maximum tilt of the Earth's axis, towards the Sun or away from the Sun) and the equinoxes (the two points in the Earth's orbit where the Earth's tilted axis and an imaginary line drawn from the Earth to the Sun are exactly perpendicular to one another). The solstices and equinoxes divide the year up into four approximately equal parts. In the northern hemisphere winter solstice occurs on or about December 21; summer solstice is near June 21; spring equinox is around March 20; and autumnal equinox is about September 23. The effect of the Earth's axial tilt in the southern hemisphere is the opposite of that in the northern hemisphere, thus the seasons of the solstices and equinoxes in the southern hemisphere are the reverse of those in the northern hemisphere (e.g. the northern summer solstice is at the same time as the southern winter solstice).

In modern times, Earth's perihelion occurs around January 3, and the aphelion around July 4 (for other eras, see precession and Milankovitch cycles). The changing Earth–Sun distance results in an increase of about 6.9%[769] in total solar energy reaching the Earth at perihelion relative to aphelion. Since the southern hemisphere is tilted toward the Sun at about the same time that the Earth reaches the closest approach to the Sun, the southern hemisphere receives slightly more energy from the Sun than does the northern over the course of a year. However, this effect is much less significant than the total energy change due to the axial tilt, and most of the excess energy is absorbed by the higher proportion of water in the southern hemisphere.

The Hill sphere (gravitational sphere of influence) of the Earth is about 1,500,000 kilometers (0.01 AU) in radius, or approximately 4 times the average distance to the moon.[770] This is the maximal distance at which the Earth's gravitational influence is stronger than the more distant Sun and planets. Objects orbiting the Earth must be within this radius, otherwise they can become unbound by the gravitational perturbation of the Sun.

Orbital characteristics

epoch	J2000.0
aphelion	152.10×10^6 km (94.51×10^6 mi) 1.0167 AU
perihelion	147.10×10^6 km (91.40×10^6 mi) 0.98329 AU
semimajor axis	149.60×10^6 km (92.96×10^6 mi) 1.000001018 AU
eccentricity	0.0167086
inclination	7.155° to Sun's equator 1.578690° to invariable plane
longitude of the ascending node	174.9°
longitude of perihelion	102.9°
argument of periapsis	288.1°
period	365.256363004 days[771]
average speed	29.78 km/s (18.50 mi/s) 107,200 km/h (66,600 mph)

The following diagram shows the relation between the line of solstice and the line of apsides of Earth's elliptical orbit. The orbital ellipse goes through each of the six Earth images, which are sequentially the perihelion (periapsis — nearest point to the Sun) on anywhere from January 2 to January 5, the point of March equinox on March 19, 20, or 21, the point of June solstice on June 20, 21, or 22, the aphelion (apoapsis — farthest point from the Sun) on anywhere from July 3 to July 5, the September equinox on September 22, 23, or 24, and the December solstice on December 21, 22, or 23. The diagram shows an exaggerated shape of Earth's orbit; the actual orbit is less eccentric than pictured.

Because of the axial tilt of the Earth in its orbit, the maximal intensity of Sun rays hits the Earth 23.4 degrees north of equator at the June Solstice (at the **Tropic of Cancer**), and 23.4 degrees south of equator at the December Solstice (at the **Tropic of Capricorn**).

Future

Mathematicians and astronomers (such as Laplace, Lagrange, Gauss, Poincaré, Kolmogorov, Vladimir Arnold, and Jürgen Moser) have searched for evidence for the stability of the planetary motions, and this quest led to many mathematical developments and several successive "proofs" of stability for the Solar System. By most predictions, Earth's orbit will be relatively stable over long periods.

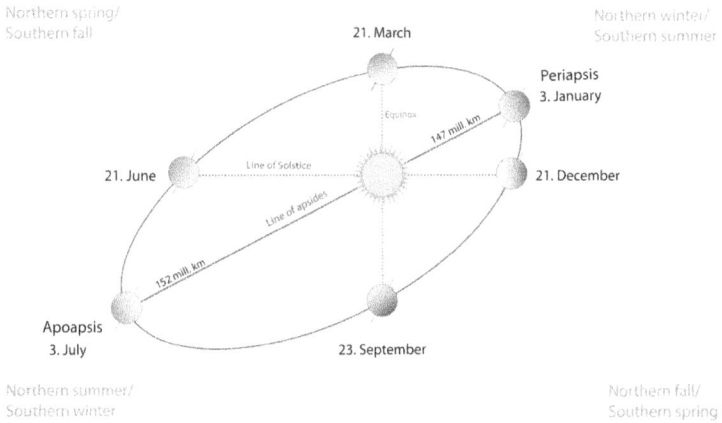

In 1989, Jacques Laskar's work indicated that the Earth's orbit (as well as the orbits of all the inner planets) can become chaotic and that an error as small as 15 meters in measuring the initial position of the Earth today would make it impossible to predict where the Earth would be in its orbit in just over 100 million years' time. Modeling the Solar System is a subject covered by the n-body problem.

External links

 Media related to Earth's orbit at Wikimedia Commons

Habitability

Biosphere

The **biosphere** (from Greek βίος *bíos* "life" and σφαῖρα *sphaira* "sphere")
also known as the **ecosphere** (from Greek οἶκος *oîkos* "environment" and
σφαῖρα), is the worldwide sum of all ecosystems. It can also be termed the
zone of life on Earth, a closed system (apart from solar and cosmic radiation
and heat from the interior of the Earth), and largely self-regulating. By the
most general biophysiological definition, the biosphere is the global ecological system integrating all living beings and their relationships, including their
interaction with the elements of the lithosphere, geosphere, hydrosphere, and
atmosphere. The biosphere is postulated to have evolved, beginning with a
process of biopoiesis (life created naturally from non-living matter, such as
simple organic compounds) or biogenesis (life created from living matter), at
least some 3.5 billion years ago.

In a general sense, biospheres are any closed, self-regulating systems containing ecosystems. This includes artificial biospheres such as Biosphere 2 and
BIOS-3, and potentially ones on other planets or moons.

Origin and use of the term

The term "biosphere" was coined by geologist Eduard Suess in 1875, which
he defined as the place on Earth's surface where life dwells.[772]

While the concept has a geological origin, it is an indication of the effect of
both Charles Darwin and Matthew F. Maury on the Earth sciences. The biosphere's ecological context comes from the 1920s (see Vladimir I. Vernadsky),
preceding the 1935 introduction of the term "ecosystem" by Sir Arthur Tans-
ley (see ecology history). Vernadsky defined ecology as the science of the
biosphere. It is an interdisciplinary concept for integrating astronomy, geophysics, meteorology, biogeography, evolution, geology, geochemistry, hy-
drology and, generally speaking, all life and Earth sciences.

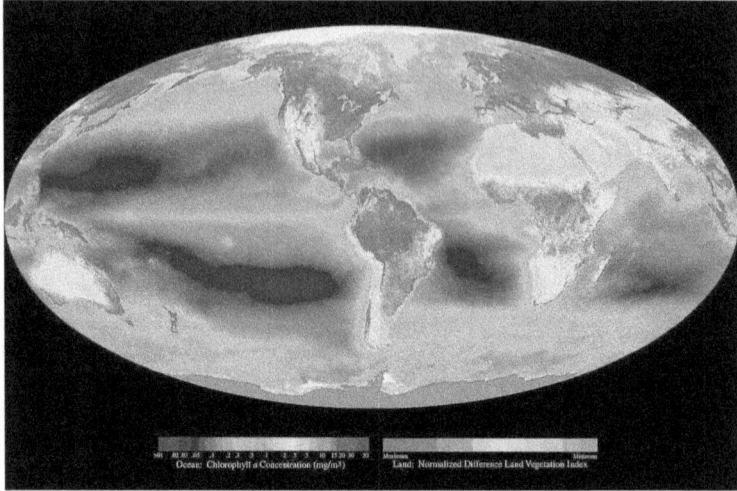

Figure 166: *A false-color composite of global oceanic and terrestrial photoautotroph abundance, from September 2001 to August 2017. Provided by the SeaWiFS Project, NASA/Goddard Space Flight Center and ORBIMAGE.Wikipedia:Citation needed*

Figure 167: *A beach scene on Earth, simultaneously showing the lithosphere (ground), hydrosphere (ocean) and atmosphere (air)*

Figure 168: *Stromatolite fossil estimated at 3.2–3.6 billion years old*

Narrow definition

Geochemists define the biosphere as being the total sum of living organisms (the "biomass" or "biota" as referred to by biologists and ecologists). In this sense, the biosphere is but one of four separate components of the geochemical model, the other three being *geosphere*, *hydrosphere*, and *atmosphere*. When these four component spheres are combined into one system, it is known as the Ecosphere. This term was coined during the 1960s and encompasses both biological and physical components of the planet.

The Second International Conference on Closed Life Systems defined *biospherics* as the science and technology of analogs and models of Earth's biosphere; i.e., artificial Earth-like biospheres. Others may include the creation of artificial non-Earth biospheres—for example, human-centered biospheres or a native Martian biosphere—as part of the topic of biospherics.Wikipedia:Citation needed

Earth's biosphere

Age

The earliest evidence for life on Earth includes biogenic graphite found in 3.7 billion-year-old metasedimentary rocks from Western Greenland and micro-

Figure 169: *Rüppell's vulture*

bial mat fossils found in 3.48 billion-year-old sandstone from Western Australia. More recently, in 2015, "remains of biotic life" were found in 4.1 billion-year-old rocks in Western Australia.[773] In 2017, putative fossilized microorganisms (or microfossils) were announced to have been discovered in hydrothermal vent precipitates in the Nuvvuagittuq Belt of Quebec, Canada that were as old as 4.28 billion years, the oldest record of life on earth, suggesting "an almost instantaneous emergence of life" after ocean formation 4.4 billion years ago, and not long after the formation of the Earth 4.54 billion years ago. According to biologist Stephen Blair Hedges, "If life arose relatively quickly on Earth ... then it could be common in the universe."

Extent

Every part of the planet, from the polar ice caps to the equator, features life of some kind. Recent advances in microbiology have demonstrated that microbes live deep beneath the Earth's terrestrial surface, and that the total mass of microbial life in so-called "uninhabitable zones" may, in biomass, exceed all animal and plant life on the surface. The actual thickness of the biosphere on earth is difficult to measure. Birds typically fly at altitudes as high as 1,800 m (5,900 ft; 1.1 mi) and fish live as much as 8,372 m (27,467 ft; 5.202 mi) underwater in the Puerto Rico Trench.

Figure 170: *Xenophyophore, a barophilic organism, from the Galapagos Rift.*

There are more extreme examples for life on the planet: Rüppell's vulture has been found at altitudes of 11,300 m (37,100 ft; 7.0 mi); bar-headed geese migrate at altitudes of at least 8,300 m (27,200 ft; 5.2 mi); yaks live at elevations as high as 5,400 m (17,700 ft; 3.4 mi) above sea level; mountain goats live up to 3,050 m (10,010 ft; 1.90 mi). Herbivorous animals at these elevations depend on lichens, grasses, and herbs.

Life forms live in every part of the Earth's biosphere, including soil, hot springs, inside rocks at least 19 km (12 mi) deep underground, the deepest parts of the ocean, and at least 64 km (40 mi) high in the atmosphere. Microorganisms, under certain test conditions, have been observed to survive the vacuum of outer space. The total amount of soil and subsurface bacterial carbon is estimated as 5×10^{17} g, or the "weight of the United Kingdom". The mass of prokaryote microorganisms—which includes bacteria and archaea, but not the nucleated eukaryote microorganisms—may be as much as 0.8 trillion tons of carbon (of the total biosphere mass, estimated at between 1 and 4 trillion tons). Barophilic marine microbes have been found at more than a depth of 10,000 m (33,000 ft; 6.2 mi) in the Mariana Trench, the deepest spot in the Earth's oceans. In fact, single-celled life forms have been found in the deepest part of the Mariana Trench, by the Challenger Deep, at depths of 11,034 m (36,201 ft; 6.856 mi).[774] Other researchers reported related studies that microorganisms thrive inside rocks up to 580 m (1,900 ft; 0.36 mi) below the sea floor under 2,590 m (8,500 ft; 1.61 mi) of ocean off the coast of the northwestern United States, as well as 2,400 m (7,900 ft; 1.5 mi) beneath the seabed off Japan. Culturable thermophilic microbes have been extracted from cores drilled more than 5,000 m (16,000 ft; 3.1 mi) into the Earth's crust in Sweden, from rocks between 65–75 °C (149–167 °F). Temperature increases with increasing depth into the Earth's crust. The rate at which the temperature increases depends on many factors, including type of crust (continental vs. oceanic), rock type, geographic location, etc. The greatest known temperature at which microbial life can exist is 122 °C (252 °F) (*Methanopyrus kandleri* Strain 116), and it is likely that the limit of life in the "deep biosphere" is defined by temperature rather than absolute depth.Wikipedia:Citation needed On 20 August 2014, scientists confirmed the existence of microorganisms living 800 m (2,600 ft; 0.50 mi) below the ice of Antarctica. According to one researcher, "You can find microbes everywhere — they're extremely adaptable to conditions, and survive wherever they are."

Our biosphere is divided into a number of biomes, inhabited by fairly similar flora and fauna. On land, biomes are separated primarily by latitude. Terrestrial biomes lying within the Arctic and Antarctic Circles are relatively barren of plant and animal life, while most of the more populous biomes lie near the equator.

Annual variation

<templatestyles src="Multiple_image/styles.css" />

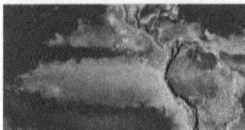

Figure 171: *Biosphere 2 in Arizona.*

On land, vegetation appears on a scale from brown (low vegetation) to dark green (lots of vegetation); at the ocean surface, phytoplankton are indicated on a scale from purple (low) to yellow (high). This visualization was created with data from satellites including SeaWiFS, and instruments including the NASA/NOAA Visible Infrared Imaging Radiometer Suite and the Moderate Resolution Imaging Spectroradiometer.

Artificial biospheres

Experimental biospheres, also called closed ecological systems, have been created to study ecosystems and the potential for supporting life outside the earth. These include spacecraft and the following terrestrial laboratories:

- Biosphere 2 in Arizona, United States, 3.15 acres (13,000 m^2).
- BIOS-1, BIOS-2 and BIOS-3 at the Institute of Biophysics in Krasnoyarsk, Siberia, in what was then the Soviet Union.
- Biosphere J (CEEF, Closed Ecology Experiment Facilities), an experiment in Japan.

- Micro-Ecological Life Support System Alternative (MELiSSA) at Universitat Autònoma de Barcelona

Extraterrestrial biospheres

No biospheres have been detected beyond the Earth; therefore, the existence of extraterrestrial biospheres remains hypothetical. The rare Earth hypothesis suggests they should be very rare, save ones composed of microbial life only. On the other hand, Earth analogs may be quite numerous, at least in the Milky Way galaxy, given the large number of planets. Three of the planets discovered orbiting TRAPPIST-1 could possibly contain biospheres. Given limited understanding of abiogenesis, it is currently unknown what percentage of these planets actually develop biospheres.

Based on observations by the Kepler Space Telescope team, it has been calculated that provided the probability of abiogenesis is higher than 1 to 1000, the closest alien biosphere should be within 100 light-years from the Earth.[775]

It is also possible that artificial biospheres will be created during the future, for example on Mars. The process of creating an uncontained system that mimics the function of Earth's biosphere is called terraforming.

Further reading

- *The Biosphere* (A *Scientific American* Book), San Francisco, W.H. Freeman and Co., 1970, ISBN 0-7167-0945-7. This book, originally the December 1970 *Scientific American* issue, covers virtually every major concern and concept since debated regarding materials and energy resources (including solar energy), population trends, and environmental degradation (including global warming).

External links

ソ ゆ ゆ	Look up *biosphere* in Wiktionary, the free dictionary.
ㄙ ル じ	
ㄆ 維 ㄉ	

- Biosphere Definition[776]
- Article on the Biosphere at Encyclopedia of Earth[777]
- GLOBIO.info[778], an ongoing programme to map the past, current and future impacts of human activities on the biosphere
- Paul Crutzen Interview[779], freeview video of Paul Crutzen Nobel Laureate for his work on decomposition of ozone talking to Harry Kroto Nobel Laureate by the Vega Science Trust.
- Atlas of the Biosphere[780]

Land use

Land use involves the management and modification of natural environment or wilderness into built environment such as settlements and semi-natural habitats such as arable fields, pastures, and managed woods. It also has been defined as "the total of arrangements, activities, and inputs that people undertake in a certain land cover type."[781]

Regulation

Land Use practices vary considerably across the world. The United Nations' Food and Agriculture Organization Water Development Division explains that "Land use concerns the products and/or benefits obtained from use of the land as well as the land management actions (activities) carried out by humans to produce those products and benefits."[782] As of the early 1990s, about 13% of the Earth was considered arable land, with 26% in pasture, 32% forests and woodland, and 1.5% urban areas.

As Albert Guttenberg (1959) wrote many years ago, "'Land use' is a key term in the language of city planning."[783] Commonly, political jurisdictions will undertake land-use planning and regulate the use of land in an attempt to avoid land-use conflicts. Land use plans are implemented through land division and use ordinances and regulations, such as zoning regulations. Management consulting firms and non-governmental organizations will frequently seek to influence these regulations before they are codified.

United States

In colonial America, few regulations existed to control the use of land, due to the seemingly endless amounts of it. As society shifted from rural to urban, public land regulation became important, especially to city governments trying to control industry, commerce, and housing within their boundaries. The first zoning ordinance was passed in New York City in 1916,[784,785] and, by the 1930s, most states had adopted zoning laws.[786] In the 1970s, concerns about the environment and historic preservation led to further regulation.

Today, federal, state, and local governments regulate growth and development through statutory law. The majority of controls on land, however, stem from the actions of private developers and individuals. Three typical situations bringing such private entities into the court system are: suits brought by one neighbor against another; suits brought by a public official against a neighboring landowner on behalf of the public; and suits involving individuals who share ownership of a particular parcel of land. In these situations,

Figure 172: *Habitat fragmented by numerous roads near the Indiana Dunes National Lakeshore*

Figure 173: *A land use map of Europe—major non-natural land uses include arable farmland (yellow) and pasture (light green).*

judicial decisions and enforcement of private land-use arrangements can reinforce public regulation, and achieve forms and levels of control that regulatory zoning cannot.

Two major federal laws have been passed in the last half century that limit the use of land significantly. These are the National Historic Preservation Act of 1966 (today embodied in 16 U.S.C. 461 et seq.) and the National Environmental Policy Act of 1969 (42 U.S.C. 4321 et seq.).

The US Department of Agriculture has identified six major types of land use in the US. Acreage statistics for each type of land use in the contiguous 48 states in 2017 were as follows:

- Pasture/range: 654 M
- Forest: 538.6 M
- Cropland: 391.5 M
- Special use: 168.8 M
- Miscellaneous: 68.9 M
- Urban: 69.4 M

Environment

Land use and land management practices have a major impact on natural resources including water, soil, nutrients, plants and animals. Land use information can be used to develop solutions for natural resource management issues such as salinity and water quality. For instance, water bodies in a region that has been deforested or having erosion will have different water quality than those in areas that are forested. Forest gardening, a plant-based food production system, is believed to be the oldest form of land use in the world.

The major effect of land use on land cover since 1750 has been deforestation of temperate regions.Wikipedia:Link rot[787] More recent significant effects of land use include urban sprawl, soil erosion, soil degradation, salinization, and desertification.[788] Land-use change, together with use of fossil fuels, are the major anthropogenic sources of carbon dioxide, a dominant greenhouse gas.[789]

According to a report by the United Nations' Food and Agriculture Organisation, land degradation has been exacerbated where there has been an absence of any land use planning, or of its orderly execution, or the existence of financial or legal incentives that have led to the wrong land use decisions, or one-sided central planning leading to over-utilization of the land resources - for instance for immediate production at all costs. As a consequence the result has often been misery for large segments of the local population and destruction of valuable ecosystems. Such narrow approaches should be replaced by a technique for the planning and management of land resources that is integrated and holistic and where land users are central. This will ensure the long-term quality of the land for human use, the prevention or resolution of social conflicts related to land use, and the conservation of ecosystems of high biodiversity value.

File:Kastellet_cph.jpg

The citadel of Kastellet, Copenhagen that has been converted into a park, showing multiple examples of suburban land use

Urban growth boundaries

The urban growth boundary is one form of land-use regulation. For example, Portland, Oregon is required to have an urban growth boundary which contains at least 20,000 acres (81 km^2) of vacant land. Additionally, Oregon restricts the development of farmland. The regulations are controversial, but an economic analysis concluded that farmland appreciated similarly to the other land.[790]

Further reading

- Guttenberg, Albert Z. 1959. "A Multiple Land Use Classification System", *Journal of the American Planning Association*, 25:3, 143–150

External links

- Land-use and land-cover change defined at Encyclopedia of Earth[791]
- Land Use Law News Alert[792]
- Land Use Law[793] by Prof. Daniel R. Mandelker (Washington University in St. Louis School of Law)
- Land Use Accountability Project[794] The Center for Public Integrity
- Schindler's Land Use Page[795] (Michigan State University Extension Land Use Team)
- Land Policy Institute at Michigan State University[796]

- Powell, W. Gabe. 2009. Identifying Land Use/Land Cover (LULC) Using National Agriculture Imagery Program (NAIP) Data as a Hydrologic Model Input for Local Flood Plain Management. Applied Research Project. Texas State University–San Marcos. http://ecommons.txstate.edu/arp/296/
- Journal of Transport and Land Use[797]
- Land Use[798], Cornell University Law School

Human geography

Human geography

Human geography is the branch of geography that deals with the study of people and their communities, cultures, economies, and interactions with the environment by studying their relations with and across space and place. Human geography attends to human patterns of social interaction, as well as spatial level interdependencies, and how they influence or affect the earth's environment. As an intellectual discipline, geography is divided into the sub-fields of physical geography and human geography, the latter concentrating upon the study of human activities, by the application of qualitative and quantitative research methods.

History

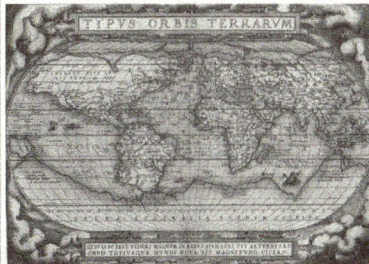

History of geography

- Graeco-Roman
- Chinese
- Islamic
- Age of Discovery

- History of cartography
- Environmental determinism
- Regional geography
- Quantitative revolution
- Critical geography
- $\underline{v}$
- $\underline{t}$
- $\underline{e}^{799}$

Geography was not recognized as a formal academic discipline until the 18th century, although many scholars had undertaken geographical scholarship for much longer, particularly through cartography.

The Royal Geographical Society was founded in England in 1830, although the United Kingdom did not get its first full Chair of geography until 1917. The first real geographical intellect to emerge in United Kingdom's geographical minds was Halford John Mackinder, appointed reader at Oxford University in 1887.

The National Geographic Society was founded in the United States in 1888 and began publication of the *National Geographic* magazine which became, and continues to be, a great popularizer of geographic information. The society has long supported geographic research and education on geographical topics.

The Association of American Geographers was founded in 1904 and was re-named the American Association of Geographers in 2016 to better reflect the increasingly international character of its membership.

One of the first examples of geographic methods being used for purposes other than to describe and theorize the physical properties of the earth is John Snow's map of the 1854 Broad Street cholera outbreak. Though a physician and a pioneer of epidemiology, the map is probably one of the earliest examples of health geography.

The now fairly distinct differences between the subfields of physical and human geography have developed at a later date. This connection between both physical and human properties of geography is most apparent in the theory of environmental determinism, made popular in the 19th century by Carl Ritter and others, and has close links to the field of evolutionary biology of the time. Environmental determinism is the theory, that people's physical, mental and moral habits are directly due to the influence of their natural environment. However, by the mid-19th century, environmental determinism was under attack for lacking methodological rigor associated with modern science, and later as a means to justify racism and imperialism.

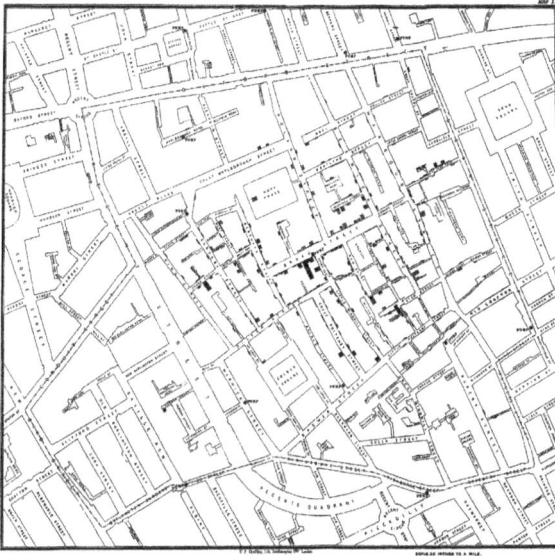

Figure 174: *Original map by John Snow showing the clusters of cholera cases in the London epidemic of 1854, which is a classical case of using human geography*

A similar concern with both human and physical aspects is apparent during the later 19th and first half of the 20th centuries focused on regional geography.The goal of regional geography, through something known as regionalisation, was to delineate space into regions and then understand and describe the unique characteristics of each region through both human and physical aspects. With links to (possibilism) (geography) and cultural ecology some of the same notions of causal effect of the environment on society and culture remain with environmental determinism.

By the 1960s, however, the quantitative revolution led to strong criticism of regional geography. Due to a perceived lack of scientific rigor in an overly descriptive nature of the discipline, and a continued separation of geography from its two subfields of physical and human geography and from geology, geographers in the mid-20th century began to apply statistical and mathematical models in order to solve spatial problems. Much of the development during the quantitative revolution is now apparent in the use of geographic information systems; the use of statistics, spatial modeling, and positivist approaches are still important to many branches of human geography. Well-known geographers from this period are Fred K. Schaefer, Waldo Tobler, William Garrison, Peter Haggett, Richard J. Chorley, William Bunge, and Torsten Hägerstrand.

From the 1970s, a number of critiques of the positivism now associated with geography emerged. Known under the term 'critical geography,' these critiques signaled another turning point in the discipline. Behavioral geography emerged for some time as a means to understand how people made perceived spaces and places, and made locational decisions. The more influential 'radical geography' emerged in the 1970s and 1980s. It draws heavily on Marxist's theory and techniques, and is associated with geographers such as David Harvey and Richard Peet. Radical geographers seek to say meaningful things about problems recognized through quantitative methods, provide explanations rather than descriptions, put forward alternatives and solutions, and be politically engaged, rather than using the detachment associated with positivists. (The detachment and objectivity of the quantitative revolution was itself critiqued by radical geographers as being a tool of capital). Radical geography and the links to Marxism and related theories remain an important part of contemporary human geography (See: *Antipode*). Critical geography also saw the introduction of 'humanistic geography', associated with the work of Yi-Fu Tuan, which pushed for a much more qualitative approach in methodology.

The changes under critical geography have led to contemporary approaches in the discipline such as feminist geography, new cultural geography, "demonic" geographies[800], and the engagement with postmodern and post-structural theories and philosophies.

Fields

The primary fields of study in human geography focus around the core fields of:

Culture

Cultural geography is the study of cultural products and norms - their variation across spaces and places, as well as their relations. It focuses on describing and analyzing the ways language, religion, economy, government, and other cultural phenomena vary or remain constant from one place to another and on explaining how humans function spatially.

- Subfields include: Social geography, Animal geographies, Language geography, Sexuality and space, Children's geographies, and Religion and geography.

Figure 175: *This picture shows terraced rice agriculture in Asia.*

Development

Development geography is the study of the Earth's geography with reference to the standard of living and the quality of life of its human inhabitants, study of the location, distribution and spatial organization of economic activities, across the Earth. The subject matter investigated is strongly influenced by the researcher's methodological approach.

Economic

Economic geography examines relationships between human economic systems, states, and other factors, and the biophysical environment.

• Subfields include: Marketing geography and Transportation geography

Health

Health geography is the application of geographical information, perspectives, and methods to the study of health, disease, and health care. Health geography deals with the spatial relations and patterns between people and the environment. This is a sub-discipline of human geography, researching how and why diseases are spread.

Figure 176: *Economic Geography: Shan Street bazaar, market in Myanmar*

Historical

Historical geography is the study of the human, physical, fictional, theoretical, and "real" geographies of the past. Historical geography studies a wide variety of issues and topics. A common theme is the study of the geographies of the past and how a place or region changes through time. Many historical geographers study geographical patterns through time, including how people have interacted with their environment, and created the cultural landscape.

Political

Political geography is concerned with the study of both the spatially uneven outcomes of political processes and the ways in which political processes are themselves affected by spatial structures.

- Subfields include: Electoral geography, Geopolitics, Strategic geography and Military geography

Population

Population geography is the study of ways in which spatial variations in the distribution, composition, migration, and growth of populations are related to their environment or location.

Settlement

Settlement geography, including urban geography, is the study of urban and rural areas with specific regards to spatial, relational and theoretical aspects of settlement. That is the study of areas which have a concentration of buildings and infrastructure. These are areas where the majority of economic activities are in the secondary sector and tertiary sectors. In case of urban settlement, they probably have a high population density. Wikipedia:Citation needed

Urban

Urban geography is the study of cities, towns, and other areas of relatively dense settlement. Two main interests are site (how a settlement is positioned relative to the physical environment) and situation (how a settlement is positioned relative to other settlements). Another area of interest is the internal organization of urban areas with regard to different demographic groups and the layout of infrastructure. This subdiscipline also draws on ideas from other branches of Human Geography to see their involvement in the processes and patterns evident in an urban area.

- Subfields include: Economic geography, Population geography, and Settlement geography. These are clearly not the only subfields that could be used to assist in the study of Urban geography, but they are some major players.

Philosophical and theoretical approaches

Within each of the subfields, various philosophical approaches can be used in research; therefore, an urban geographer could be a Feminist or Marxist geographer, etc.

Such approaches are:

- Animal geographies
- Behavioral geography
- Cognitive geography
- Critical geography
- Feminist geography
- Marxist geography
- Non-representational theory
- Positivism
- Postcolonialism
- Poststructuralist geography
- Psychoanalytic geography
- Psychogeography

Figure 177: *Carl Ritter - considered to be one of the founders of modern geography*

- Spatial analysis
- Time geography

List of notable human geographers

- Alexander von Humboldt (1769-1859), one of the founders of modern geography, he traveled extensively and pioneered empirical research methods that would later develop primarily into biogeography and physical geography but also anticipated population geography and economic geography. Humboldt University of Berlin is named after Alexander and his brother Wilhelm von Humboldt.
- Carl Ritter (1779–1859), considered to be one of the founders of modern geography and first chair in geography at the Humboldt University of Berlin, also noted for his use of organic analogy in his works.
- Xavier Hommaire de Hell (1812–1848), research in Turkey, southern Russia and Persia
- Élisée Reclus (1830–1905), known for his monumental 19 volume The Earth and Its Inhabitants[801], he coined the term social geography and his thinking anticipated the social ecology and animal rights movements, where he advocated anarchism and veganism as part of an ethical life.

- Peter Kropotkin (1842–1921), one of the first radical geographers, he was a proponent of anarchism and notable for his introduction of the concept of mutual aid.
- Friedrich Ratzel (1844–1904), environmental determinist, invented the term *Lebensraum*
- Paul Vidal de la Blache (1845–1918), founder of the French School of geopolitics and possibilism.
- Sir Halford John Mackinder (1861–1947), author of *The Geographical Pivot of History*, co-founder of the London School of Economics, along with the Geographical Association.
- Jovan Cvijić (1865–1927), a Serbian geographer and a world-renowned scientist. He started his scientific career as a geographer and geologist, and continued his activity as an anthropogeographer and sociologist.
- Carl O. Sauer (1889–1975), critic of environmental determinism and proponent of cultural ecology.
- Walter Christaller (1893–1969), economic geographer and developer of the central place theory.
- Richard Hartshorne (1899–1992), scholar of the history and philosophy of geography.
- Torsten Hägerstrand (1916–2004), key figure in the quantitative revolution and regional science, developer of time geography and indirect contributor to aspects of critical geography.
- Milton Santos (1926–2001) winner of the Vautrin Lud Prize in 1994, one of the most important geographers in South America.
- Waldo R. Tobler (born 1930), developer of the First law of geography.
- Gamal Hamdan (born 1928), an Egyptian thinker, intellect and professor of geography. Best known for *The Character of Egypt, Studies of the Arab World, and The Contemporary Islamic World Geography*, which form a trilogy on Egypt's natural, economic, political and cultural character and its position in the world.
- Yi-Fu Tuan (born 1930) Professor Emeritus at University of Wisconsin–Madison, key figure behind the development of humanist and phenomenological geography and the most prominent Chinese-American geographer. Recipient of the Vautrin Lud Prize in 2012.
- David Harvey (born 1935), world's most cited academic geographer and winner of the Lauréat Prix International de Géographie Vautrin Lud, also noted for his work in critical geography and critique of global capitalism.
- Evelyn Stokes (1936–2005). Professor of geography at the University of Waikato in New Zealand. Known for recognizing inequality with marginalised groups including women and Māori using geography.
- Allen J. Scott (born 1938), winner of Vautrin Lud Prize in 2003 and the Anders Retzius Gold medal 2009; author of numerous books and papers

on economic and urban geography, known for his work on regional development, new industrial spaces, agglomeration theory, global city-regions and the cultural economy.

- Edward Soja (1941-2015), noted for his work on regional development, planning and governance, along with coining the terms synekism and postmetropolis.
- Doreen Massey (1944-2016), key scholar in the space and places of globalization and its pluralities, winner of the Vautrin Lud Prize.
- Denis Cosgrove (1948–2008), Alexander von Humboldt Professor of geography at UCLA in California. Specialized in cultural geography and landscapes.
- Michael Watts, Class of 1963 Professor of Geography and Development Studies, University of California, Berkeley
- Nigel Thrift (born 1949), developer of non-representational theory.
- Derek Gregory (born 1951), famous for writing on the Israeli, U.S. and UK actions in the Middle East after 9/11, influenced by Edward Said and has contributed work on imagined geographies.
- Cindi Katz (born 1954), who writes on social reproduction and the production of space. Writing on children's geographies, place and nature, everyday life and security.
- Gillian Rose (born 1962), most famous for her critique: *Feminism & Geography: The Limits of Geographical Knowledge* (1993), which was one of the first moves towards a development of feminist geography.

Journals

As with all social sciences, human geographers publish research and other written work in a variety of academic journals. Whilst human geography is interdisciplinary, there are a number of journals that focus on human geography.

These include:

- *ACME: An International E-Journal for Critical Geographies*[802]
- *Antipode*
- *Area*
- *Economic geography*
- *Environment and Planning*
- *Geografiska Annaler*
- *GeoHumanities*[803]
- *Global Environmental Change: Human and Policy Dimensions*[804]
- *Migration Letters*
- *Social & Cultural Geography*[805]

- *Transactions of the Institute of British Geographers*
- *Geoforum*
- *Progress in Human Geography*
- *Tijdschrift voor economische en sociale geografie*

Notes

- Urbanization is a major proponent to human and population geography, especially over the past 100 years as population shift has moved to urban areas.[806]

Further reading

- Blij, Harm Jan, De (2008). *Geography: realms, regions, and concepts.* Hoboken, NJ: John Wiley. ISBN 978-0-470-12905-0.
- Clifford, N.J.; Holloway, S.L.; Rice, S.P.; Valentine, G., eds. (2009). *Key Concepts in Geography* (2nd ed.). London: SAGE. ISBN 978-1-4129-3021-5.
- Cloke, Paul J.; Crang, Philip; Goodwin, Mark (2004). *Envisioning human geographies.* London: Arnold. ISBN 978-0-340-72013-4.
- Cloke, Paul J.; Crang, Phil; Crang, Philip; Goodwin, Mark (2005). *Introducing human geographies* (2nd ed.). London: Hodder Arnold. ISBN 978-0-340-88276-4.
- Crang, Mike; Thrift, Nigel J. (2000). *Thinking space.* London: Routledge. ISBN 978-0-415-16016-2.
- Daniels, Peter; Bradshaw, Michael; Shaw, Denis J. B.; Sidaway, James D. (2004). *An Introduction to Human Geography: issues for the 21st century* (2nd ed.). Prentice Hall. ISBN 978-0-13-121766-9.
- Flowerdew, Robin; Martin, David (2005). *Methods in human geography: a guide for students doing a research project* (2nd ed.). Harlow: Prentice Hall. ISBN 978-0-582-47321-8.
- Gregory, Derek; Martin, Ron G.; Smith, Graham (1994). *Human geography: society, space and social science.* Basingstoke: Macmillan. ISBN 978-0-333-45251-6.
- Harvey, David D. (1996). *Justice, Nature and the Geography of Difference.* Blackwell Pub. ISBN 978-1-55786-680-6.
- Johnston, R.J. (1979). *Geography and Geographers. Anglo-American Human Geography since 1945.* Edward Arnold, London.
- Johnston, R.J. (2009). *The Dictionary of Human Geography* (5th ed.). Blackwell Publishers, London.
- Johnston, R.J (2002). *Geographies of Global Change: Remapping the World.* Blackwell Publishers, London.

- Moseley, William W.; Lanegran, David A.; Pandit, Kavita (2007). *The Introductory Reader in Human Geography: Contemporary Debates and Classic Writings*. Malden, MA: Blackwell Publishing Limited. ISBN 978-1-4051-4922-8.
- Peet, Richard, ed. (1998). *Modern Geographical Thought*. Oxford: Wiley-Blackwell. ISBN 978-1-55786-378-2.
- Soja, Edward (1989). *Postmodern Geographies: The Reassertion of Space in Critical Social Theory*. Verso, London.

External links

- Media related to Human geography at Wikimedia Commons
- Worldmapper[807] – Mapping project using social data sets

Moon

Moon

<indicator name="pp-default"> 🔒 </indicator>

Moon

Full moon as seen from North America in Earth's Northern Hemisphere

Designations	
Adjectives	• Lunar • selenic
Orbital characteristics	
Perigee	362600 km (356400–370400 km)
Apogee	405400 km (404000–406700 km)
Semi-major axis	384399 km (0.00257 AU)
Eccentricity	0.0549
Orbital period	27.321661 d (27 d 7 h 43 min 11.5 s)
Synodic period	29.530589 d (29 d 12 h 44 min 2.9 s)

Average orbital speed	1.022 km/s
Inclination	5.145° to the ecliptic
Longitude of ascending node	Regressing by one revolution in 18.61 years
Argument of perigee	Progressing by one revolution in 8.85 years
Satellite of	Earth
Physical characteristics	
Mean radius	1737.1 km (0.273 of Earth's)
Equatorial radius	1738.1 km (0.273 of Earth's)
Polar radius	1736.0 km (0.273 of Earth's)
Flattening	0.0012
Circumference	10921 km (equatorial)
Surface area	3.793×10^7 km^2 (0.074 of Earth's)
Volume	2.1958×10^{10} km^3 (0.020 of Earth's)
Mass	7.342×10^{22} kg (0.012300 of Earth's)[808]
Mean density	3.344 g/cm^3 0.606 × Earth
Surface gravity	1.62 m/s^2 (0.1654 g)
Moment of inertia factor	0.3929±0.0009
Escape velocity	2.38 km/s
Sidereal rotation period	27.321661 d (synchronous)
Equatorial rotation velocity	4.627 m/s
Axial tilt	• 1.5424° to ecliptic • 6.687° to orbit plane
Albedo	0.136
	<table><tr><td>Surface temp.</td><td>min</td><td>mean</td><td>max</td></tr><tr><td>Equator</td><td>100 K</td><td>220 K</td><td>390 K</td></tr><tr><td>85°N</td><td></td><td>150 K</td><td>230 K</td></tr></table>
Apparent magnitude	• −2.5 to −12.9 • −12.74 (mean full moon)
Angular diameter	29.3 to 34.1 arcminutes
Atmosphere	
Surface pressure	• 10^{-7} Pa (1 picobar) (day) • 10^{-10} Pa (1 femtobar) (night)

Composition by volume	• He • Ar • Ne • Na • K • H • Rn

The **Moon** is an astronomical body that orbits planet Earth and is Earth's only permanent natural satellite. It is the fifth-largest natural satellite in the Solar System, and the largest among planetary satellites relative to the size of the planet that it orbits (its primary). The Moon is after Jupiter's satellite Io the second-densest satellite in the Solar System among those whose densities are known.

The Moon is thought to have formed about 4.51 billion years ago, not long after Earth. The most widely accepted explanation is that the Moon formed from the debris left over after a giant impact between Earth and a Mars-sized body called Theia.

The Moon is in synchronous rotation with Earth, and thus always shows the same side to earth, the near side. The near side is marked by dark volcanic maria that fill the spaces between the bright ancient crustal highlands and the prominent impact craters. After the Sun, the Moon is the second-brightest regularly visible celestial object in Earth's sky. Its surface is actually dark, although compared to the night sky it appears very bright, with a reflectance just slightly higher than that of worn asphalt. Its gravitational influence produces the ocean tides, body tides, and the slight lengthening of the day.

The Moon's average orbital distance is 384,402 km (238,856 mi),[809] or 1.28 light-seconds. This is about thirty times the diameter of Earth. The Moon's apparent size in the sky is almost the same as that of the Sun (because it is 400x farther and larger). Therefore, the Moon covers the Sun nearly precisely during a total solar eclipse. This matching of apparent visual size will not continue in the far future, because the Moon's distance from Earth is slowly increasing.

The Moon was first reached in 1959 by an unmanned spacecraft of the Soviet Union's Luna program; the United States' NASA Apollo program achieved the only manned lunar missions to date, beginning with the first manned orbital mission by Apollo 8 in 1968, and six manned landings between 1969 and 1972, with the first being Apollo 11. These missions returned lunar rocks which have been used to develop a geological understanding of the Moon's origin, internal structure, and the Moon's later history. Since the Apollo 17 mission in 1972, the Moon has been visited only by unmanned spacecraft.

Figure 178: *The Moon, tinted reddish, during a lunar eclipse*

Both the Moon's natural prominence in the earthly sky and its regular cycle of phases as seen from Earth have provided cultural references and influences for human societies and cultures since time immemorial. Such cultural influences can be found in language, lunar based calendar systems, art, and mythology.

Name and etymology

The usual English proper name for Earth's natural satellite is "the Moon", which in nonscientific texts is usually not capitalized.[810,811,812] The noun *moon* is derived from Old English *mōna*, which (like all Germanic language cognates) stems from Proto-Germanic **mēnô*, which comes from Proto-Indo-European **méh₁n̥s* "moon", "month", which comes from the Proto-Indo-European root **meh₁-* "to measure", the month being the ancient unit of time measured by the Moon.[813] Occasionally, the name "Luna" is used. In literature, especially science fiction, "Luna" is used to distinguish it from other moons, while in poetry, the name has been used to denote personification of our moon.[814]

The modern English adjective pertaining to the Moon is *lunar*, derived from the Latin word for the Moon, *luna*. The adjective *selenic* (usually only used to refer to the chemical element selenium) is so rarely used to refer to the Moon that this meaning is not recorded in most major dictionaries.[815,816,817] It is derived from the Ancient Greek word for the Moon, σελήνη (selénē), from which is however also derived the prefix "seleno-", as in *selenography*, the study of the physical features of the Moon, as well as the element name *selenium*. Both the Greek goddess Selene and the Roman goddess Diana were alternatively called Cynthia. The names Luna, Cynthia, and Selene are reflected in terminology for lunar orbits in words such as *apolune*, *pericynthion*, and *selenocentric*. The name Diana comes from the Proto-Indo-European **diw-yo*, "heavenly", which comes from the PIE root *dyeu- "to shine," which in many derivatives means "sky, heaven, and god" and is also the origin of Latin *dies*, "day".

<templatestyles src="Multiple_image/styles.css" />

The Moon

Near side of the Moon

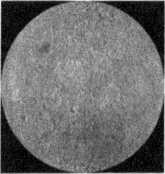

Far side of the Moon

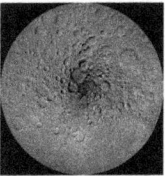

Lunar north pole

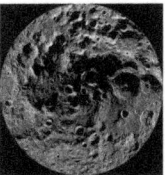

Lunar south pole

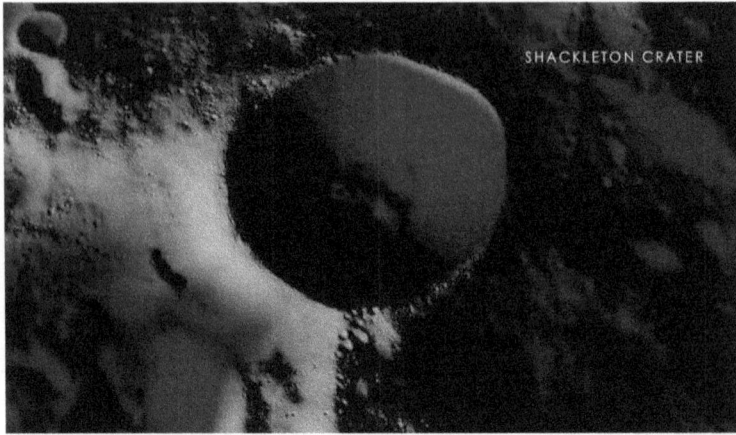

Figure 179: *The evolution of the Moon and a tour of the Moon*

Formation

The Moon formed 4.51 billion years ago, some 60 million years after the origin of the Solar System. Several forming mechanisms have been proposed, including the fission of the Moon from Earth's crust through centrifugal force (which would require too great an initial spin of Earth), the gravitational capture of a pre-formed Moon (which would require an unfeasibly extended atmosphere of Earth to dissipate the energy of the passing Moon), and the co-formation of Earth and the Moon together in the primordial accretion disk (which does not explain the depletion of metals in the Moon). These hypotheses also cannot account for the high angular momentum of the Earth–Moon system.

The prevailing hypothesis is that the Earth–Moon system formed after an impact of a Mars-sized body (named *Theia*) with the proto-Earth (giant impact). The impact blasted material into Earth's orbit and then the material accreted and formed the Moon.

The Moon's far side has a crust that is 30 mi (48 km) thicker than that of the near side. This is thought to be because the Moon fused from two different bodies.

This hypothesis, although not perfect, perhaps best explains the evidence. Eighteen months prior to an October 1984 conference on lunar origins, Bill Hartmann, Roger Phillips, and Jeff Taylor challenged fellow lunar scientists: "You have eighteen months. Go back to your Apollo data, go back to your computer, do whatever you have to, but make up your mind. Don't come to our conference unless you have something to say about the Moon's birth." At the 1984 conference at Kona, Hawaii, the giant impact hypothesis emerged as the most popular.

Before the conference, there were partisans of the three "traditional" theories, plus a few people who were starting to take the giant impact seriously, and there was a huge apathetic middle who didn't think the debate would ever be resolved. Afterward, there were essentially only two groups: the giant impact camp and the agnostics.

Giant impacts are thought to have been common in the early Solar System. Computer simulations of giant impacts have produced results that are consistent with the mass of the lunar core and the angular momentum of the Earth–Moon system. These simulations also show that most of the Moon derived from the impactor, rather than the proto-Earth. However, more recent simulations suggest a larger fraction of the Moon derived from the proto-Earth. Other bodies of the inner Solar System such as Mars and Vesta have, according to meteorites from them, very different oxygen and tungsten isotopic compositions compared to Earth. However, Earth and the Moon have nearly identical isotopic compositions. The isotopic equalization of the Earth-Moon system might be explained by the post-impact mixing of the vaporized material that formed the two, although this is debated.

The impact released a lot of energy and then the released material re-accreted into the Earth–Moon system. This would have melted the outer shell of Earth, and thus formed a magma ocean. Similarly, the newly formed Moon would also have been affected and had its own lunar magma ocean; its depth is estimated from about 500 km (300 miles) to 1,737 km (1,079 miles).

While the giant impact hypothesis might explain many lines of evidence, some questions are still unresolved, most of which involve the Moon's composition. <templatestyles src="Multiple_image/styles.css" />

Oceanus Procellarum ("Ocean of Storms")

Ancient rift valleys – rectangular structure (visible – topography – GRAIL gravity gradients)

Ancient rift valleys – context.

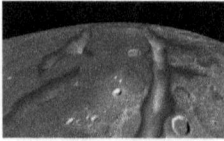

Ancient rift valleys – closeup (artist's concept).

In 2001, a team at the Carnegie Institute of Washington reported the most precise measurement of the isotopic signatures of lunar rocks. To their surprise, the rocks from the Apollo program had the same isotopic signature as rocks from Earth, however they differed from almost all other bodies in the Solar System. Indeed, this observation was unexpected, because most of the material that formed the Moon was thought to come from Theia and it was announced in 2007 that there was less than a 1% chance that Theia and Earth had identical isotopic signatures. Other Apollo lunar samples had in 2012 the same titanium isotopes composition as Earth, which conflicts with what is expected if the Moon formed far from Earth or is derived from Theia. These discrepancies may be explained by variations of the giant impact hypothesis.

Physical characteristics

Internal structure

Chemical composition of the lunar surface regolith (derived from crustal rocks)

Compound	Formula	Composition (wt %)	
		Maria	**Highlands**
silica	SiO_2	45.4%	45.5%
alumina	Al_2O_3	14.9%	24.0%
lime	CaO	11.8%	15.9%
iron(II) oxide	FeO	14.1%	5.9%
magnesia	MgO	9.2%	7.5%
titanium dioxide	TiO_2	3.9%	0.6%
sodium oxide	Na_2O	0.6%	0.6%
Total		**99.9%**	**100.0%**

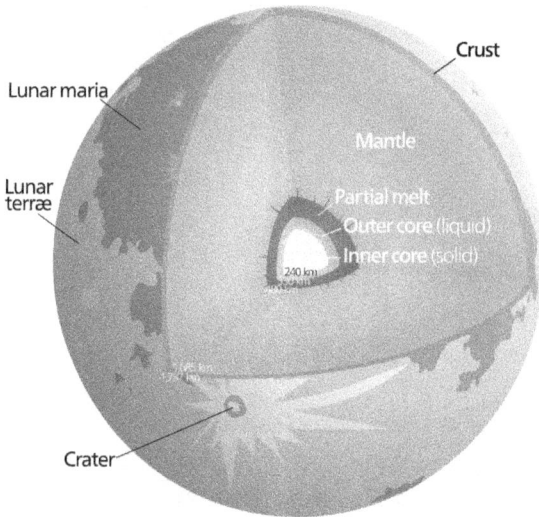

Figure 180: *Structure of the Moon*

The Moon is a differentiated body: it has a geochemically distinct crust, mantle, and core. The Moon has a solid iron-rich inner core with a radius possibly as small as 240 km (150 mi) and a fluid outer core primarily made of liquid iron with a radius of roughly 300 km (190 mi). Around the core is a partially molten boundary layer with a radius of about 500 km (310 mi). This structure is thought to have developed through the fractional crystallization of a global magma ocean shortly after the Moon's formation 4.5 billion years ago. Crystallization of this magma ocean would have created a mafic mantle from the precipitation and sinking of the minerals olivine, clinopyroxene, and orthopyroxene; after about three-quarters of the magma ocean had crystallised, lower-density plagioclase minerals could form and float into a crust atop. The final liquids to crystallise would have been initially sandwiched between the crust and mantle, with a high abundance of incompatible and heat-producing elements. Consistent with this perspective, geochemical mapping made from orbit suggests the crust of mostly anorthosite. The Moon rock samples of the flood lavas that erupted onto the surface from partial melting in the mantle confirm the mafic mantle composition, which is more iron-rich than that of Earth. The crust is on average about 50 km (31 mi) thick.

The Moon is the second-densest satellite in the Solar System, after Io. However, the inner core of the Moon is small, with a radius of about 350 km

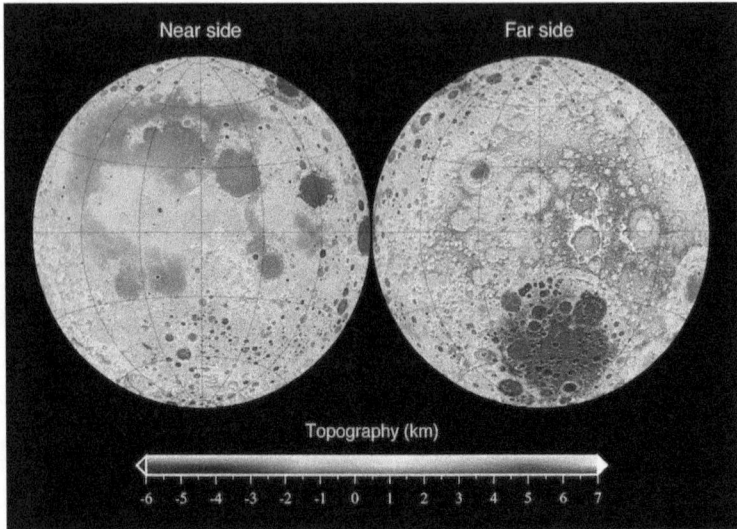

Figure 181: *Topography of the Moon*

(220 mi) or less, around 20% of the radius of the Moon. Its composition is not well defined, but is probably metallic iron alloyed with a small amount of sulfur and nickel; analyses of the Moon's time-variable rotation suggest that it is at least partly molten.

Surface geology

The topography of the Moon has been measured with laser altimetry and stereo image analysis. Its most visible topographic feature is the giant far-side South Pole–Aitken basin, some 2,240 km (1,390 mi) in diameter, the largest crater on the Moon and the second-largest confirmed impact crater in the Solar System. At 13 km (8.1 mi) deep, its floor is the lowest point on the surface of the Moon. The highest elevations of the Moon's surface are located directly to the northeast, and it has been suggested might have been thickened by the oblique formation impact of the South Pole–Aitken basin. Other large impact basins, such as Imbrium, Serenitatis, Crisium, Smythii, and Orientale, also possess regionally low elevations and elevated rims. The far side of the lunar surface is on average about 1.9 km (1.2 mi) higher than that of the near side.

The discovery of fault scarp cliffs by the Lunar Reconnaissance Orbiter suggest that the Moon has shrunk within the past billion years, by about 90 metres (300 ft). Similar shrinkage features exist on Mercury.

Figure 182: *STL 3D model of the Moon with 10× elevation exaggeration rendered with data from the Lunar Orbiter Laser Altimeter of the Lunar Reconnaissance Orbiter*

Volcanic features

The dark and relatively featureless lunar plains, clearly seen with the naked
eye, are called *maria* (Latin for "seas"; singular *mare*), as they were once
believed to be filled with water; they are now known to be vast solidified pools
of ancient basaltic lava. Although similar to terrestrial basalts, lunar basalts
have more iron and no minerals altered by water. The majority of these lavas
erupted or flowed into the depressions associated with impact basins. Several
geologic provinces containing shield volcanoes and volcanic domes are found
within the near side "maria".

Almost all maria are on the near side of the Moon, and cover 31% of the surface of the near side, compared with 2% of the far side. This is thought to
be due to a concentration of heat-producing elements under the crust on the
near side, seen on geochemical maps obtained by *Lunar Prospector*'s gamma-
ray spectrometer, which would have caused the underlying mantle to heat up,
partially melt, rise to the surface and erupt. Most of the Moon's mare basalts
erupted during the Imbrian period, 3.0–3.5 billion years ago, although some
radiometrically dated samples are as old as 4.2 billion years. Until recently,
the youngest eruptions, dated by crater counting, appeared to have been only

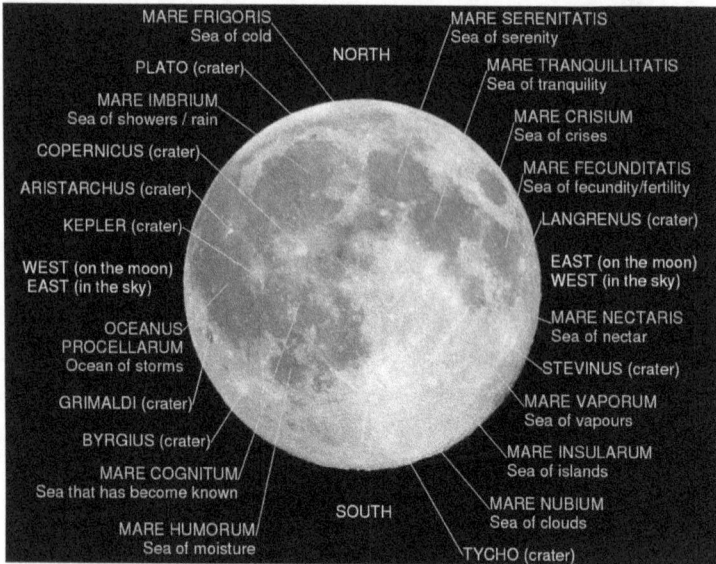

Figure 183: *Lunar nearside with major maria and craters labeled*

Figure 184: *Evidence of young lunar volcanism*

1.2 billion years ago. In 2006, a study of Ina, a tiny depression in Lacus Felicitatis, found jagged, relatively dust-free features that, because of the lack of erosion by infalling debris, appeared to be only 2 million years old. Moonquakes and releases of gas also indicate some continued lunar activity. In 2014 NASA announced "widespread evidence of young lunar volcanism" at 70 irregular mare patches identified by the Lunar Reconnaissance Orbiter, some less than 50 million years old. This raises the possibility of a much warmer lunar mantle than previously believed, at least on the near side where the deep crust is substantially warmer because of the greater concentration of radioactive elements. Just prior to this, evidence has been presented for 2–10 million years younger basaltic volcanism inside Lowell crater, Orientale basin, located in the transition zone between the near and far sides of the Moon. An initially hotter mantle and/or local enrichment of heat-producing elements in the mantle could be responsible for prolonged activities also on the far side in the Orientale basin.

The lighter-coloured regions of the Moon are called *terrae*, or more commonly *highlands*, because they are higher than most maria. They have been radiometrically dated to having formed 4.4 billion years ago, and may represent plagioclase cumulates of the lunar magma ocean. In contrast to Earth, no major lunar mountains are believed to have formed as a result of tectonic events.

The concentration of maria on the Near Side likely reflects the substantially thicker crust of the highlands of the Far Side, which may have formed in a slow-velocity impact of a second moon of Earth a few tens of millions of years after their formation.

Impact craters

The other major geologic process that has affected the Moon's surface is impact cratering, with craters formed when asteroids and comets collide with the lunar surface. There are estimated to be roughly 300,000 craters wider than 1 km (0.6 mi) on the Moon's near side alone. The lunar geologic timescale is based on the most prominent impact events, including Nectaris, Imbrium, and Orientale, structures characterized by multiple rings of uplifted material, between hundreds and thousands of kilometres in diameter and associated with a broad apron of ejecta deposits that form a regional stratigraphic horizon. The lack of an atmosphere, weather and recent geological processes mean that many of these craters are well-preserved. Although only a few multi-ring basins have been definitively dated, they are useful for assigning relative ages. Because impact craters accumulate at a nearly constant rate, counting the number of craters per unit area can be used to estimate the age of the surface. The radiometric ages of impact-melted rocks collected during the Apollo

Figure 185: *Lunar crater Daedalus on the Moon's far side*

missions cluster between 3.8 and 4.1 billion years old: this has been used to propose a Late Heavy Bombardment of impacts.

Blanketed on top of the Moon's crust is a highly comminuted (broken into ever smaller particles) and impact gardened surface layer called regolith, formed by impact processes. The finer regolith, the lunar soil of silicon dioxide glass, has a texture resembling snow and a scent resembling spent gunpowder. The regolith of older surfaces is generally thicker than for younger surfaces: it varies in thickness from 10–20 km (6.2–12.4 mi) in the highlands and 3–5 km (1.9–3.1 mi) in the maria. Beneath the finely comminuted regolith layer is the *megaregolith*, a layer of highly fractured bedrock many kilometres thick.

Comparison of high-resolution images obtained by the Lunar Reconnaissance Orbiter has shown a contemporary crater-production rate significantly higher than previously estimated. A secondary cratering process caused by distal ejecta is thought to churn the top two centimetres of regolith a hundred times more quickly than previous models suggested—on a timescale of 81,000 years.

Lunar swirls

Lunar swirls are enigmatic features found across the Moon's surface. They are characterized by a high albedo, appear optically immature (i.e. the optical characteristics of a relatively young regolith), and have often a sinuous shape.

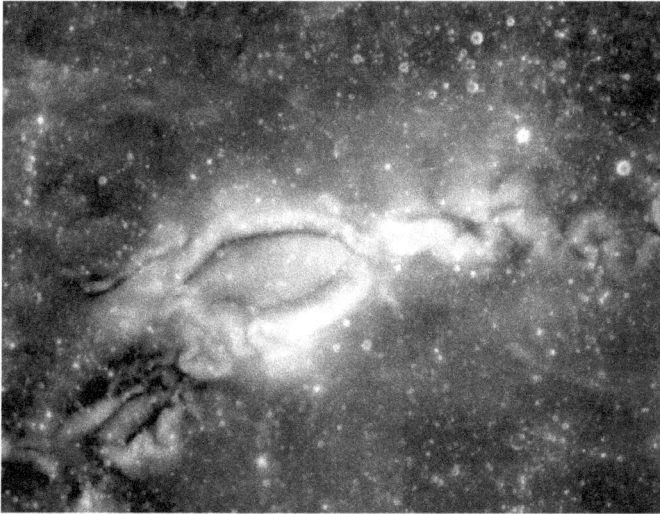

Figure 186: *Lunar swirls at Reiner Gamma*

Their shape is often accentuated by low albedo regions that wind between the bright swirls.

Presence of water

Liquid water cannot persist on the lunar surface. When exposed to solar radiation, water quickly decomposes through a process known as photodissociation and is lost to space. However, since the 1960s, scientists have hypothesized that water ice may be deposited by impacting comets or possibly produced by the reaction of oxygen-rich lunar rocks, and hydrogen from solar wind, leaving traces of water which could possibly persist in cold, permanently shadowed craters at either pole on the Moon. Computer simulations suggest that up to 14,000 km^2 (5,400 sq mi) of the surface may be in permanent shadow. The presence of usable quantities of water on the Moon is an important factor in rendering lunar habitation as a cost-effective plan; the alternative of transporting water from Earth would be prohibitively expensive.

In years since, signatures of water have been found to exist on the lunar surface. In 1994, the bistatic radar experiment located on the *Clementine* spacecraft, indicated the existence of small, frozen pockets of water close to the surface. However, later radar observations by Arecibo, suggest these findings may rather be rocks ejected from young impact craters. In 1998, the neutron spectrometer on the *Lunar Prospector* spacecraft showed that high concentrations of hydrogen are present in the first meter of depth in the regolith near the polar regions. Volcanic lava beads, brought back to Earth aboard Apollo 15, showed small amounts of water in their interior.

The 2008 *Chandrayaan-1* spacecraft has since confirmed the existence of surface water ice, using the on-board Moon Mineralogy Mapper. The spectrometer observed absorption lines common to hydroxyl, in reflected sunlight, providing evidence of large quantities of water ice, on the lunar surface. The spacecraft showed that concentrations may possibly be as high as 1,000 ppm. Using the mapper's reflectance spectra, indirect lighting of areas in shadow confirmed water ice within 20° latitude of both poles in 2018. In 2009, *LCROSS* sent a 2,300 kg (5,100 lb) impactor into a permanently shadowed polar crater, and detected at least 100 kg (220 lb) of water in a plume of ejected material. Another examination of the LCROSS data showed the amount of detected water to be closer to 155 ± 12 kg (342 ± 26 lb).

In May 2011, 615–1410 ppm water in melt inclusions in lunar sample 74220 was reported, the famous high-titanium "orange glass soil" of volcanic origin collected during the Apollo 17 mission in 1972. The inclusions were formed during explosive eruptions on the Moon approximately 3.7 billion years ago. This concentration is comparable with that of magma in Earth's upper mantle. Although of considerable selenological interest, this announcement affords little comfort to would-be lunar colonists—the sample originated many kilometers below the surface, and the inclusions are so difficult to access that it took 39 years to find them with a state-of-the-art ion microprobe instrument.

Analysis of the findings of the Moon Mineralogy Mapper (M3) revealed in August 2018 for the first time "definitive evidence" for water-ice on the lunar surface. The data revealed the distinct reflective signatures of water-ice, as opposed to dust and other reflective substances. The ice deposits were found on the North and South poles, although it is more abundant in the South, where water is trapped in permanently shadowed craters and cravices, allowing it to persist as ice on the surface since they are shielded from the sun.

Gravitational field

The gravitational field of the Moon has been measured through tracking the Doppler shift of radio signals emitted by orbiting spacecraft. The main lunar gravity features are mascons, large positive gravitational anomalies associated with some of the giant impact basins, partly caused by the dense mare basaltic lava flows that fill those basins. The anomalies greatly influence the orbit of spacecraft about the Moon. There are some puzzles: lava flows by themselves cannot explain all of the gravitational signature, and some mascons exist that are not linked to mare volcanism.

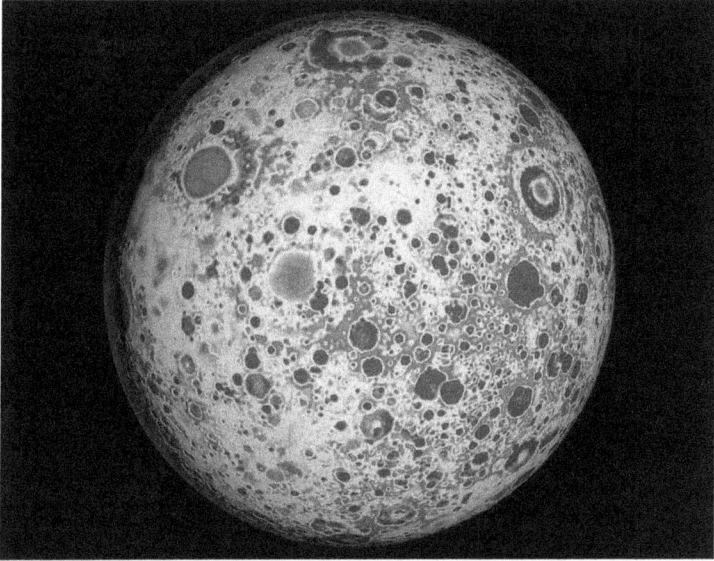

Figure 187: *GRAIL's gravity map of the Moon*

Magnetic field

The Moon has an external magnetic field of about 1–100 nanoteslas, less than one-hundredth that of Earth. The Moon does not currently have a global dipolar magnetic field and only has crustal magnetization, probably acquired early in its history when a dynamo was still operating. Alternatively, some of the remnant magnetization may be from transient magnetic fields generated during large impacts through the expansion of an impact-generated plasma cloud in an ambient magnetic field. This is supported by the apparent location of the largest crustal magnetizations near the antipodes of the giant impact basins.

Atmosphere

The Moon has an atmosphere so tenuous as to be nearly vacuum, with a total mass of less than 10 metric tons (9.8 long tons; 11 short tons). The surface pressure of this small mass is around 3×10^{-15} atm (0.3 nPa); it varies with the lunar day. Its sources include outgassing and sputtering, a product of the bombardment of lunar soil by solar wind ions. Elements that have been detected include sodium and potassium, produced by sputtering (also found in the atmospheres of Mercury and Io); helium-4 and neon from the solar wind; and argon-40, radon-222, and polonium-210, outgassed after their creation by radioactive decay within the crust and mantle. The absence of such neutral

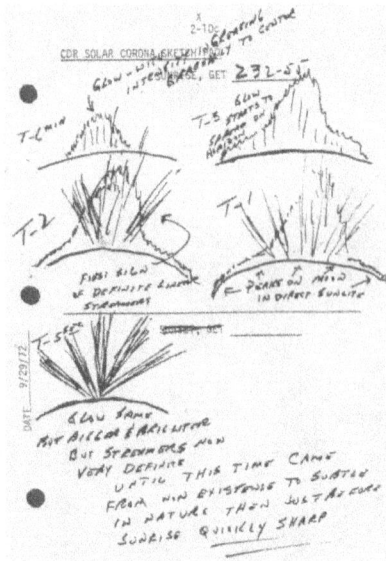

Figure 188: *Sketch by the Apollo 17 astronauts.*
The lunar atmosphere was later studied by LADEE.

species (atoms or molecules) as oxygen, nitrogen, carbon, hydrogen and magnesium, which are present in the regolith, is not understood. Water vapour has been detected by *Chandrayaan-1* and found to vary with latitude, with a maximum at ∼60–70 degrees; it is possibly generated from the sublimation of water ice in the regolith. These gases either return into the regolith because of the Moon's gravity or are lost to space, either through solar radiation pressure or, if they are ionized, by being swept away by the solar wind's magnetic field.

Dust

A permanent asymmetric moon dust cloud exists around the Moon, created by small particles from comets. Estimates are 5 tons of comet particles strike the Moon's surface each 24 hours. The particles strike the Moon's surface ejecting moon dust above the Moon. The dust stays above the Moon approximately 10 minutes, taking 5 minutes to rise, and 5 minutes to fall. On average, 120 kilograms of dust are present above the Moon, rising to 100 kilometers above the surface. The dust measurements were made by LADEE's Lunar Dust EXperiment (LDEX), between 20 and 100 kilometers above the surface, during a six-month period. LDEX detected an average of one 0.3 micrometer moon dust particle each minute. Dust particle counts peaked during the Geminid, Quadrantid, Northern Taurid, and Omicron Centaurid meteor showers, when the Earth, and Moon, pass through comet debris. The cloud is asymmetric, more dense near the boundary between the Moon's dayside and nightside.

Past thicker atmosphere

In October 2017, NASA scientists at the Marshall Space Flight Center and the Lunar and Planetary Institute in Houston announced their finding, based on studies of Moon magma samples retrieved by the Apollo missions, that the Moon had once possessed a relatively thick atmosphere for a period of 70 million years between 3 and 4 billion years ago. This atmosphere, sourced from gases ejected from lunar volcanic eruptions, was twice the thickness of that of present-day Mars. The ancient lunar atmosphere was eventually stripped away by solar winds and dissipated into space.[818]

Seasons

The Moon's axial tilt with respect to the ecliptic is only 1.5424°, much less than the 23.44° of Earth. Because of this, the Moon's solar illumination varies much less with season, and topographical details play a crucial role in seasonal effects. From images taken by *Clementine* in 1994, it appears that four mountainous regions on the rim of Peary Crater at the Moon's north pole may remain illuminated for the entire lunar day, creating peaks of eternal light. No such regions exist at the south pole. Similarly, there are places that remain in permanent shadow at the bottoms of many polar craters, and these "craters of eternal darkness" are extremely cold: *Lunar Reconnaissance Orbiter* measured the lowest summer temperatures in craters at the southern pole at 35 K (–238 °C; –397 °F) and just 26 K (–247 °C; –413 °F) close to the winter solstice in north polar Hermite Crater. This is the coldest temperature in the Solar System ever measured by a spacecraft, colder even than the surface of Pluto. Average temperatures of the Moon's surface are reported, but temperatures of different areas will vary greatly depending upon whether they are in sunlight or shadow.

Relationship to Earth

Orbit

The Moon makes a complete orbit around Earth with respect to the fixed stars about once every 27.3 days (its sidereal period). However, because Earth is moving in its orbit around the Sun at the same time, it takes slightly longer for the Moon to show the same phase to Earth, which is about 29.5 days (its synodic period). Unlike most satellites of other planets, the Moon orbits closer to the ecliptic plane than to the planet's equatorial plane. The Moon's orbit is subtly perturbed by the Sun and Earth in many small, complex and interacting

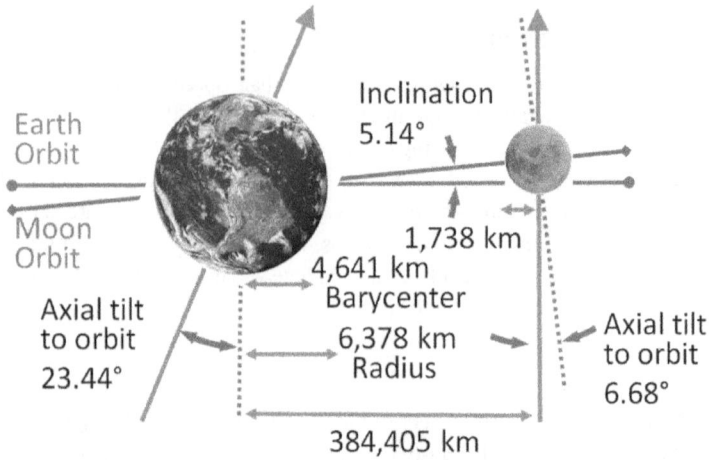

Figure 189: *Earth–Moon system (schematic)*

Figure 190: *DSCOVR satellite sees the Moon passing in front of Earth*

Figure 191: *Moon setting in western sky over the High Desert in California*

ways. For example, the plane of the Moon's orbit gradually rotates once every 18.61[819] years, which affects other aspects of lunar motion. These follow-on effects are mathematically described by Cassini's laws.

Relative size

The Moon is exceptionally large relative to Earth: Its diameter is a quarter and its mass is 1/81 of Earth's. It is the largest moon in the Solar System relative to the size of its planet,[820] though Charon is larger relative to the dwarf planet Pluto, at 1/9 Pluto's mass. The Earth and the Moon's barycentre, their common centre of mass, is located 1,700 km (1,100 mi) (about a quarter of Earth's radius) beneath Earth's surface.

The Earth revolves around the Earth-Moon barycentre once a sidereal month, with 1/81 the speed of the Moon, or about 12.5 metres (41 ft) per second. This motion is superimposed on the much larger revolution of the Earth around the Sun at a speed of about 30 kilometres (19 mi) per second.

Appearance from Earth

The Moon is in synchronous rotation as it orbits Earth; it rotates about its axis in about the same time it takes to orbit Earth. This results in it always keeping nearly the same face turned towards Earth. However, because of the effect of libration, about 59% of the Moon's surface can actually be seen from Earth. The side of the Moon that faces Earth is called the near side, and the opposite the far side. The far side is often inaccurately called the "dark side", but it is in

Figure 192: *The Moon is prominently featured in Vincent van Gogh's 1889 painting, The Starry Night*

fact illuminated as often as the near side: once every 29.5 Earth days. During new moon, the near side is dark.

The Moon had once rotated at a faster rate, but early in its history, its rotation slowed and became tidally locked in this orientation as a result of frictional effects associated with tidal deformations caused by Earth. With time, the energy of rotation of the Moon on its axis was dissipated as heat, until there was no rotation of the Moon relative to Earth. In 2016, planetary scientists, using data collected on the much earlier NASA Lunar Prospector mission, found two hydrogen-rich areas on opposite sides of the Moon, probably in the form of water ice. It is speculated that these patches were the poles of the Moon billions of years ago, before it was tidally locked to Earth.

The Moon has an exceptionally low albedo, giving it a reflectance that is slightly brighter than that of worn asphalt. Despite this, it is the brightest object in the sky after the Sun. This is due partly to the brightness enhancement of the opposition surge; the Moon at quarter phase is only one-tenth as bright, rather than half as bright, as at full moon. Additionally, color constancy in the visual system recalibrates the relations between the colors of an object and its surroundings, and because the surrounding sky is comparatively dark, the sunlit Moon is perceived as a bright object. The edges of the full moon seem

Figure 193: *14 November 2016 supermoon was 356,511 kilometres (221,526 mi) away from the center of Earth, the closest occurrence since 26 January 1948. It will not be closer until 25 November 2034.*

as bright as the centre, without limb darkening, because of the reflective properties of lunar soil, which retroreflects light more towards the Sun than in other directions. The Moon does appear larger when close to the horizon, but this is a purely psychological effect, known as the moon illusion, first described in the 7th century BC. The full Moon's angular diameter is about 0.52° (on average) in the sky, roughly the same apparent size as the Sun (see § Eclipses).

The Moon's highest altitude at culmination varies by its phase and time of year. The full moon is highest in the sky during winter (for each hemisphere). The 18.61-year nodal cycle has an influence on lunar standstill. When the ascending node of the lunar orbit is in the vernal equinox, the lunar declination can reach up to plus or minus 28° each month. This means the Moon can pass overhead if viewed from latitudes up to 28° north or south (of the Equator), instead of only 18°. The orientation of the Moon's crescent also depends on the latitude of the viewing location; an observer in the tropics can see a smile-shaped crescent Moon. The Moon is visible for two weeks every 27.3 days at the North and South Poles. Zooplankton in the Arctic use moonlight when the Sun is below the horizon for months on end.

The distance between the Moon and Earth varies from around 356,400 km (221,500 mi) to 406,700 km (252,700 mi) at perigee (closest) and apogee (farthest), respectively. On 14 November 2016, it was closer to Earth when at full phase than it has been since 1948, 14% closer than its farthest position in apogee. Reported as a "supermoon", this closest point coincided within an hour of a full moon, and it was 30% more luminous than when at its greatest distance because its angular diameter is 14% greater and $1.14^2 \approx 1.30$. At lower levels, the human perception of reduced brightness as a percentage is provided by the following formula:

$$\text{perceived reduction}\% = 100 \times \sqrt{\frac{\text{actual reduction}\%}{100}}$$

When the actual reduction is 1.00 / 1.30, or about 0.770, the perceived reduction is about 0.877, or 1.00 / 1.14. This gives a maximum perceived increase of 14% between apogee and perigee moons of the same phase.

There has been historical controversy over whether features on the Moon's surface change over time. Today, many of these claims are thought to be illusory, resulting from observation under different lighting conditions, poor astronomical seeing, or inadequate drawings. However, outgassing does occasionally occur and could be responsible for a minor percentage of the reported lunar transient phenomena. Recently, it has been suggested that a roughly 3 km (1.9 mi) diameter region of the lunar surface was modified by a gas release event about a million years ago.

The Moon's appearance, like the Sun's, can be affected by Earth's atmosphere. Common optical effects are the 22° halo ring, formed when the Moon's light is refracted through the ice crystals of high cirrostratus clouds, and smaller coronal rings when the Moon is seen through thin clouds.

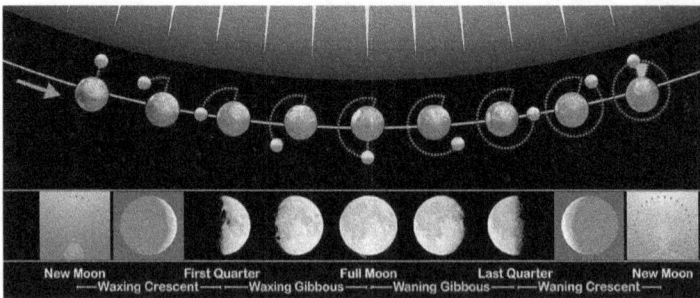

File:Moon_phases_en.jpg

The monthly changes in the angle between the direction of sunlight and view from Earth, and the phases of the Moon that result, as viewed from the Northern Hemisphere. The Earth–Moon distance is not to scale.

Figure 194: *The libration of the Moon over a single lunar month. Also visible is the slight variation in the Moon's visual size from Earth.*

The illuminated area of the visible sphere (degree of illumination) is given by $(1 - \cos e)/2 = \sin^2(e/2)$, where e is the elongation (i.e., the angle between Moon, the observer (on Earth) and the Sun).

Tidal effects

The gravitational attraction that masses have for one another decreases inversely with the square of the distance of those masses from each other. As a result, the slightly greater attraction that the Moon has for the side of Earth closest to the Moon, as compared to the part of the Earth opposite the Moon, results in tidal forces. Tidal forces affect both the Earth's crust and oceans.

The most obvious effect of tidal forces is to cause two bulges in the Earth's oceans, one on the side facing the Moon and the other on the side opposite. This results in elevated sea levels called ocean tides. As the Earth spins on its axis, one of the ocean bulges (high tide) is held in place "under" the Moon, while another such tide is opposite. As a result, there are two high tides, and two low tides in about 24 hours. Since the Moon is orbiting the Earth in the same direction of the Earth's rotation, the high tides occur about every 12 hours and 25 minutes; the 25 minutes is due to the Moon's time to orbit the

Earth. The Sun has the same tidal effect on the Earth, but its forces of attraction are only 40% that of the Moon's; the Sun's and Moon's interplay is responsible for spring and neap tides. If the Earth were a water world (one with no continents) it would produce a tide of only one meter, and that tide would be very predictable, but the ocean tides are greatly modified by other effects: the frictional coupling of water to Earth's rotation through the ocean floors, the inertia of water's movement, ocean basins that grow shallower near land, the sloshing of water between different ocean basins. As a result, the timing of the tides at most points on the Earth is a product of observations that are explained, incidentally, by theory.

While gravitation causes acceleration and movement of the Earth's fluid oceans, gravitational coupling between the Moon and Earth's solid body is mostly elastic and plastic. The result is a further tidal effect of the Moon on the Earth that causes a bulge of the solid portion of the Earth nearest the Moon that acts as a torque in opposition to the Earth's rotation. This "drains" angular momentum and rotational kinetic energy from Earth's spin, slowing the Earth's rotation. That angular momentum, lost from the Earth, is transferred to the Moon in a process (confusingly known as tidal acceleration), which lifts the Moon into a higher orbit and results in its lower orbital speed about the Earth. Thus the distance between Earth and Moon is increasing, and the Earth's spin is slowing in reaction. Measurements from laser reflectors left during the Apollo missions (lunar ranging experiments) have found that the Moon's distance increases by 38 mm (1.5 in) per year (roughly the rate at which human fingernails grow). Atomic clocks also show that Earth's day lengthens by about 15 microseconds every year, slowly increasing the rate at which UTC is adjusted by leap seconds. Left to run its course, this tidal drag would continue until the spin of Earth and the orbital period of the Moon matched, creating mutual tidal locking between the two. As a result, the Moon would be suspended in the sky over one meridian, as is already currently the case with Pluto and its moon Charon. However, the Sun will become a red giant engulfing the Earth-Moon system long before this occurrence.

In a like manner, the lunar surface experiences tides of around 10 cm (4 in) amplitude over 27 days, with two components: a fixed one due to Earth, because they are in synchronous rotation, and a varying component from the Sun. The Earth-induced component arises from libration, a result of the Moon's orbital eccentricity (if the Moon's orbit were perfectly circular, there would only be solar tides). Libration also changes the angle from which the Moon is seen, allowing a total of about 59% of its surface to be seen from Earth over time. The cumulative effects of stress built up by these tidal forces produces moonquakes. Moonquakes are much less common and weaker than are earthquakes, although moonquakes can last for up to an hour—significantly longer than terrestrial quakes—because of the absence of water to damp out the seismic vibrations. The existence of moonquakes was an unexpected discovery from seismometers placed on the Moon by Apollo astronauts from 1969 through 1972.

Eclipses

<templatestyles src="Multiple_image/styles.css" />

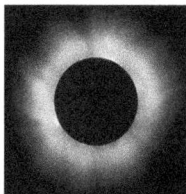

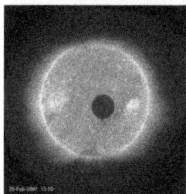

From Earth, the Moon and the Sun appear the same size, as seen in the 1999 solar eclipse (left), whereas from the *STEREO-B* spacecraft in an Earth-trailing orbit, the Moon appears much smaller than the Sun (right).

Eclipses only occur when the Sun, Earth, and Moon are all in a straight line (termed "syzygy"). Solar eclipses occur at new moon, when the Moon is between the Sun and Earth. In contrast, lunar eclipses occur at full moon, when Earth is between the Sun and Moon. The apparent size of the Moon is roughly the same as that of the Sun, with both being viewed at close to one-half a degree wide. The Sun is much larger than the Moon but it is the vastly greater distance that gives it the same apparent size as the much closer and much smaller Moon from the perspective of Earth. The variations in apparent size, due to the non-circular orbits, are nearly the same as well, though occurring in different cycles. This makes possible both total (with the Moon appearing larger than the Sun) and annular (with the Moon appearing smaller than the Sun) solar eclipses. In a total eclipse, the Moon completely covers the disc of the Sun and the solar corona becomes visible to the naked eye. Because the distance between the Moon and Earth is very slowly increasing over time, the angular diameter of the Moon is decreasing. Also, as it evolves toward becoming a red giant, the size of the Sun, and its apparent diameter in the sky, are slowly increasing. The combination of these two changes means that hundreds of millions of years ago, the Moon would always completely cover the Sun on solar eclipses, and no annular eclipses were possible. Likewise, hundreds of millions of years in the future, the Moon will no longer cover the Sun completely, and total solar eclipses will not occur.

Because the Moon's orbit around Earth is inclined by about 5.145° (5° 9') to the orbit of Earth around the Sun, eclipses do not occur at every full and new moon. For an eclipse to occur, the Moon must be near the intersection of the two orbital planes. The periodicity and recurrence of eclipses of the Sun by the Moon, and of the Moon by Earth, is described by the saros, which has a period of approximately 18 years.

Because the Moon is continuously blocking our view of a half-degree-wide circular area of the sky, the related phenomenon of occultation occurs when a bright star or planet passes behind the Moon and is occulted: hidden from view. In this way, a solar eclipse is an occultation of the Sun. Because the Moon is comparatively close to Earth, occultations of individual stars are not visible everywhere on the planet, nor at the same time. Because of the precession of the lunar orbit, each year different stars are occulted.

Observation and exploration

Ancient and medieval studies

Understanding of the Moon's cycles was an early development of astronomy: by the 5th century BC, Babylonian astronomers had recorded the 18-year Saros cycle of lunar eclipses, and Indian astronomers had described the Moon's monthly elongation. The Chinese astronomer Shi Shen (fl. 4th century BC) gave instructions for predicting solar and lunar eclipses.[821] Later, the physical form of the Moon and the cause of moonlight became understood. The ancient Greek philosopher Anaxagoras (d. 428 BC) reasoned that the Sun and Moon were both giant spherical rocks, and that the latter reflected the light of the former.[822] Although the Chinese of the Han Dynasty believed the Moon to be energy equated to *qi*, their 'radiating influence' theory also recognized that the light of the Moon was merely a reflection of the Sun, and Jing Fang (78–37 BC) noted the sphericity of the Moon.[823] In the 2nd century AD, Lucian wrote the novel *A True Story*, in which the heroes travel to the Moon and meet its inhabitants. In 499 AD, the Indian astronomer Aryabhata mentioned in his *Aryabhatiya* that reflected sunlight is the cause of the shining of the Moon. The astronomer and physicist Alhazen (965–1039) found that sunlight was not reflected from the Moon like a mirror, but that light was emitted from every part of the Moon's sunlit surface in all directions. Shen Kuo (1031–1095) of the Song dynasty created an allegory equating the waxing and

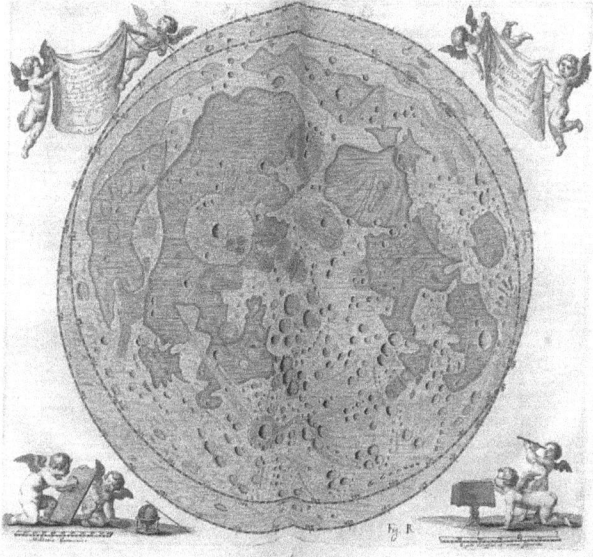

Figure 195: *Map of the Moon by Johannes Hevelius from his Se-
lenographia (1647), the first map to include the libration zones*

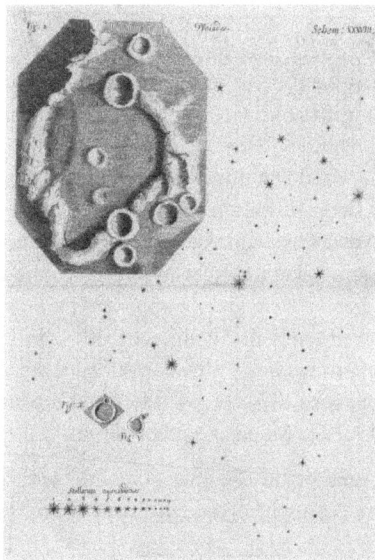

Figure 196: *A study of the Moon in Robert Hooke's Micrographia, 1665*

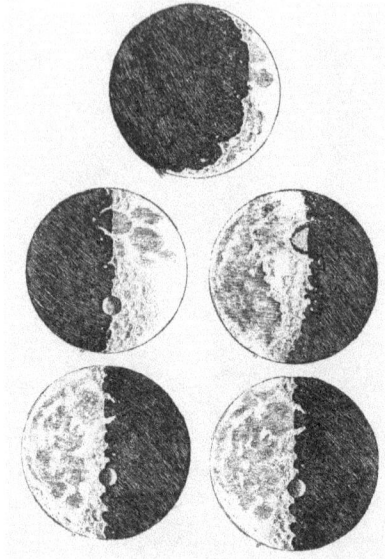

Figure 197: *Galileo's sketches of the Moon from Sidereus Nuncius*

waning of the Moon to a round ball of reflective silver that, when doused with white powder and viewed from the side, would appear to be a crescent.[824]

In Aristotle's (384–322 BC) description of the universe, the Moon marked the boundary between the spheres of the mutable elements (earth, water, air and fire), and the imperishable stars of aether, an influential philosophy that would dominate for centuries. However, in the 2nd century BC, Seleucus of Seleucia correctly theorized that tides were due to the attraction of the Moon, and that their height depends on the Moon's position relative to the Sun. In the same century, Aristarchus computed the size and distance of the Moon from Earth, obtaining a value of about twenty times the radius of Earth for the distance. These figures were greatly improved by Ptolemy (90–168 AD): his values of a mean distance of 59 times Earth's radius and a diameter of 0.292 Earth diameters were close to the correct values of about 60 and 0.273 respectively. Archimedes (287–212 BC) designed a planetarium that could calculate the motions of the Moon and other objects in the Solar System.

During the Middle Ages, before the invention of the telescope, the Moon was increasingly recognised as a sphere, though many believed that it was "perfectly smooth".

In 1609, Galileo Galilei drew one of the first telescopic drawings of the Moon in his book *Sidereus Nuncius* and noted that it was not smooth but had mountains and craters. Telescopic mapping of the Moon followed: later in the 17th century, the efforts of Giovanni Battista Riccioli and Francesco Maria Grimaldi led to the system of naming of lunar features in use today. The more exact 1834–36 *Mappa Selenographica* of Wilhelm Beer and Johann Heinrich Mädler, and their associated 1837 book *Der Mond*, the first trigonometrically accurate study of lunar features, included the heights of more than a thousand mountains, and introduced the study of the Moon at accuracies possible in earthly geography. Lunar craters, first noted by Galileo, were thought to be volcanic until the 1870s proposal of Richard Proctor that they were formed by collisions. This view gained support in 1892 from the experimentation of geologist Grove Karl Gilbert, and from comparative studies from 1920 to the 1940s, leading to the development of lunar stratigraphy, which by the 1950s was becoming a new and growing branch of astrogeology.

By spacecraft

20th century

Soviet missions

<templatestyles src="Multiple_image/styles.css" />

Luna 2, the first human-made object to reach the surface of the Moon (left) and Soviet Moon rover Lunokhod 1

The Cold War-inspired Space Race between the Soviet Union and the U.S. led to an acceleration of interest in exploration of the Moon. Once launchers had the necessary capabilities, these nations sent unmanned probes on both flyby and impact/lander missions. Spacecraft from the Soviet Union's *Luna* program were the first to accomplish a number of goals: following three unnamed, failed missions in 1958, the first human-made object to escape Earth's gravity and pass near the Moon was *Luna 1*; the first human-made object to impact the lunar surface was *Luna 2*, and the first photographs of the normally occluded far side of the Moon were made by *Luna 3*, all in 1959.

The first spacecraft to perform a successful lunar soft landing was *Luna 9* and the first unmanned vehicle to orbit the Moon was *Luna 10*, both in 1966. Rock and soil samples were brought back to Earth by three *Luna* sample return missions (*Luna 16* in 1970, *Luna 20* in 1972, and *Luna 24* in 1976), which returned 0.3 kg total. Two pioneering robotic rovers landed on the Moon in 1970 and 1973 as a part of Soviet Lunokhod programme.

Luna 24 was the last Soviet/Russian mission to the Moon.

United States missions

<templatestyles src="Multiple_image/styles.css" />

Earthrise (Apollo 8, 1968)

Moon rock (Apollo 17, 1972)

During the late 1950s at the height of the Cold War, the United States Army conducted a classified feasibility study that proposed the construction of a manned military outpost on the Moon called Project Horizon with the potential to conduct a wide range of missions from scientific research to nuclear Earth bombardment. The study included the possibility of conducting a lunar-based nuclear test. The Air Force, which at the time was in competition with the Army for a leading role in the space program, developed its own similar plan called Lunex. However, both these proposals were ultimately passed over as the space program was largely transferred from the military to the civilian agency NASA.

Figure 198: *Neil Armstrong working at the lunar module*

Following President John F. Kennedy's 1961 commitment to a manned moon landing before the end of the decade, the United States, under NASA leadership, launched a series of unmanned probes to develop an understanding of the lunar surface in preparation for manned missions: the Jet Propulsion Laboratory's Ranger program produced the first close-up pictures; the Lunar Orbiter program produced maps of the entire Moon; the Surveyor program landed its first spacecraft four months after *Luna 9*. The manned Apollo program was developed in parallel; after a series of unmanned and manned tests of the Apollo spacecraft in Earth orbit, and spurred on by a potential Soviet lunar flight, in 1968 Apollo 8 made the first manned mission to lunar orbit. The subsequent landing of the first humans on the Moon in 1969 is seen by many as the culmination of the Space Race.

	"That's one small step ..."

Problems playing this file? See media help.

Neil Armstrong became the first person to walk on the Moon as the com-
mander of the American mission Apollo 11 by first setting foot on the Moon
at 02:56 UTC on 21 July 1969. An estimated 500 million people worldwide
watched the transmission by the Apollo TV camera, the largest television au-
dience for a live broadcast at that time. The Apollo missions 11 to 17 (except
Apollo 13, which aborted its planned lunar landing) returned 380.05 kilograms
(837.87 lb) of lunar rock and soil in 2,196 separate samples. The American
Moon landing and return was enabled by considerable technological advances
in the early 1960s, in domains such as ablation chemistry, software engineering, and atmospheric re-entry technology, and by highly competent management of the enormous technical undertaking.

Scientific instrument packages were installed on the lunar surface during all the
Apollo landings. Long-lived instrument stations, including heat flow probes,
seismometers, and magnetometers, were installed at the Apollo 12, 14, 15,
16, and 17 landing sites. Direct transmission of data to Earth concluded in
late 1977 because of budgetary considerations, but as the stations' lunar laser
ranging corner-cube retroreflector arrays are passive instruments, they are still
being used. Ranging to the stations is routinely performed from Earth-based
stations with an accuracy of a few centimetres, and data from this experiment
are being used to place constraints on the size of the lunar core.

1980s–2000

After the first Moon race there were years of near quietude but starting in
the 1990s, many more countries have become involved in direct exploration
of the Moon. In 1990, Japan became the third country to place a spacecraft
into lunar orbit with its *Hiten* spacecraft. The spacecraft released a smaller
probe, *Hagoromo*, in lunar orbit, but the transmitter failed, preventing further
scientific use of the mission. In 1994, the U.S. sent the joint Defense De-
partment/NASA spacecraft *Clementine* to lunar orbit. This mission obtained
the first near-global topographic map of the Moon, and the first global multi-
spectral images of the lunar surface. This was followed in 1998 by the *Lunar
Prospector* mission, whose instruments indicated the presence of excess hydrogen at the lunar poles, which is likely to have been caused by the presence of
water ice in the upper few meters of the regolith within permanently shadowed
craters.

India, Japan, China, the United States, and the European Space Agency each
sent lunar orbiters, and especially ISRO's *Chandrayaan-1* has contributed to
confirming the discovery of lunar water ice in permanently shadowed craters at
the poles and bound into the lunar regolith. The post-Apollo era has also seen
two rover missions: the final Soviet Lunokhod mission in 1973, and China's
ongoing Chang'e 3 mission, which deployed its Yutu rover on 14 December

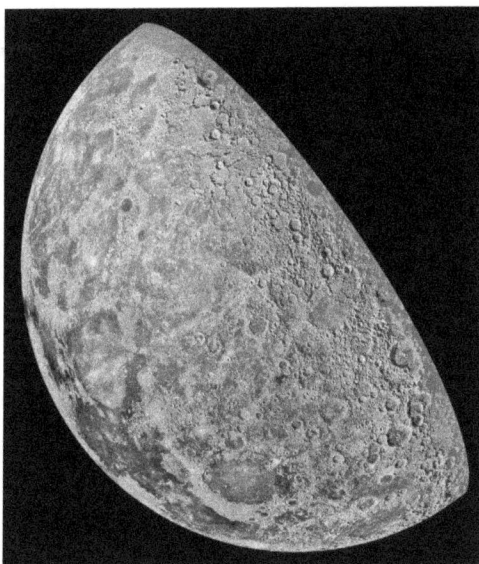

Figure 199: *An artificially coloured mosaic constructed from a series of 53 images taken through three spectral filters by Galileo's imaging system as the spacecraft flew over the northern regions of the Moon on 7 December 1992.*

2013. The Moon remains, under the Outer Space Treaty, free to all nations to explore for peaceful purposes.

21st century

The European spacecraft *SMART-1*, the second ion-propelled spacecraft, was in lunar orbit from 15 November 2004 until its lunar impact on 3 September 2006, and made the first detailed survey of chemical elements on the lunar surface.

The ambitious Chinese Lunar Exploration Program began with *Chang'e 1*, which successfully orbited the Moon from 5 November 2007 until its controlled lunar impact on 1 March 2009. It obtained a full image map of the Moon. *Chang'e 2*, beginning in October 2010, reached the Moon more quickly, mapped the Moon at a higher resolution over an eight-month period, then left lunar orbit for an extended stay at the Earth–Sun L2 Lagrangian point, before finally performing a flyby of asteroid 4179 Toutatis on 13 December 2012, and then heading off into deep space. On 14 December 2013, *Chang'e 3* landed a lunar lander onto the Moon's surface, which in turn deployed a lunar rover, named *Yutu* (Chinese: 玉兔 ; literally "Jade Rabbit"). This was the first lunar soft landing since *Luna 24* in 1976, and the first lunar rover mission since

Figure 200: *Artistic representation of a future Moon colony*

Lunokhod 2 in 1973. China intends to launch another rover mission (*Chang'e 4*) before 2020, followed by a sample return mission (*Chang'e 5*) soon after.

Between 4 October 2007 and 10 June 2009, the Japan Aerospace Exploration Agency's *Kaguya (Selene)* mission, a lunar orbiter fitted with a high-definition video camera, and two small radio-transmitter satellites, obtained lunar geophysics data and took the first high-definition movies from beyond Earth orbit. India's first lunar mission, *Chandrayaan I*, orbited from 8 November 2008 until loss of contact on 27 August 2009, creating a high resolution chemical, mineralogical and photo-geological map of the lunar surface, and confirming the presence of water molecules in lunar soil. The Indian Space Research Organisation planned to launch *Chandrayaan II* in 2013, which would have included a Russian robotic lunar rover. However, the failure of Russia's *Fobos-Grunt* mission has delayed this project.

<templatestyles src="Multiple_image/styles.css" />

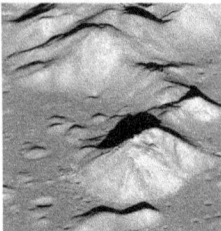

Copernicus's central peaks as observed by the LRO, 2012

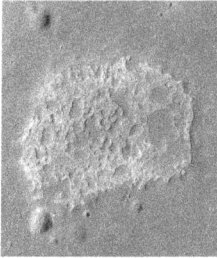

The Ina formation, 2009

The U.S. co-launched the *Lunar Reconnaissance Orbiter* (LRO) and the *LCROSS* impactor and follow-up observation orbiter on 18 June 2009; *LCROSS* completed its mission by making a planned and widely observed impact in the crater Cabeus on 9 October 2009, whereas *LRO* is currently in operation, obtaining precise lunar altimetry and high-resolution imagery. In November 2011, the LRO passed over the large and bright Aristarchus crater. NASA released photos of the crater on 25 December 2011.

Two NASA GRAIL spacecraft began orbiting the Moon around 1 January 2012, on a mission to learn more about the Moon's internal structure. NASA's *LADEE* probe, designed to study the lunar exosphere, achieved orbit on 6 October 2013.

Upcoming lunar missions include Russia's *Luna-Glob*: an unmanned lander with a set of seismometers, and an orbiter based on its failed Martian *Fobos-Grunt* mission. Privately funded lunar exploration has been promoted by the Google Lunar X Prize, announced 13 September 2007, which offers US $20 million to anyone who can land a robotic rover on the Moon and meet other specified criteria. Shackleton Energy Company is building a program to establish operations on the south pole of the Moon to harvest water and supply their Propellant Depots.

NASA began to plan to resume manned missions following the call by U.S. President George W. Bush on 14 January 2004 for a manned mission to the Moon by 2019 and the construction of a lunar base by 2024. The Constellation program was funded and construction and testing begun on a manned spacecraft and launch vehicle, and design studies for a lunar base. However, that program has been cancelled in favor of a manned asteroid landing by 2025 and a manned Mars orbit by 2035. India has also expressed its hope to send a manned mission to the Moon by 2020.

On 28 February 2018, SpaceX, Vodafone, Nokia and Audi announced a collaboration to install a 4G wireless communication network on the Moon, with the aim of streaming live footage on the surface to Earth.[825]

Planned commercial missions

In 2007, the X Prize Foundation together with Google launched the Google Lunar X Prize to encourage commercial endeavors to the Moon. A prize of $20 million will be awarded to the first private venture to get to the Moon with a robotic lander by the end of March 2018, with additional prizes worth $10 million for further milestones. As of August 2016, 16 teams are participating in the competition. In January 2018 the foundation announced that the prize would go unclaimed as none of the finalist teams would be able to make a launch attempt by the deadline.

In August 2016, the US government granted permission to US-based start-up Moon Express to land on the Moon. This marked the first time that a private enterprise was given the right to do so. The decision is regarded as a precedent helping to define regulatory standards for deep-space commercial activity in the future, as thus far companies' operation had been restricted to being on or around Earth.

Astronomy from the Moon

For many years, the Moon has been recognized as an excellent site for telescopes. It is relatively nearby; astronomical seeing is not a concern; certain craters near the poles are permanently dark and cold, and thus especially useful for infrared telescopes; and radio telescopes on the far side would be shielded from the radio chatter of Earth. The lunar soil, although it poses a problem for any moving parts of telescopes, can be mixed with carbon nanotubes and epoxies and employed in the construction of mirrors up to 50 meters in diameter. A lunar zenith telescope can be made cheaply with an ionic liquid.

In April 1972, the Apollo 16 mission recorded various astronomical photos and spectra in ultraviolet with the Far Ultraviolet Camera/Spectrograph.

Legal status

Although *Luna* landers scattered pennants of the Soviet Union on the Moon, and U.S. flags were symbolically planted at their landing sites by the Apollo astronauts, no nation claims ownership of any part of the Moon's surface. Russia and the U.S. are party to the 1967 Outer Space Treaty, which defines the Moon and all outer space as the "province of all mankind". This treaty also restricts the use of the Moon to peaceful purposes, explicitly banning military installations and weapons of mass destruction. The 1979 Moon Agreement was created to restrict the exploitation of the Moon's resources by any single nation, but as of November 2016, it has been signed and ratified by only 18 nations, none of which engages in self-launched human space exploration or

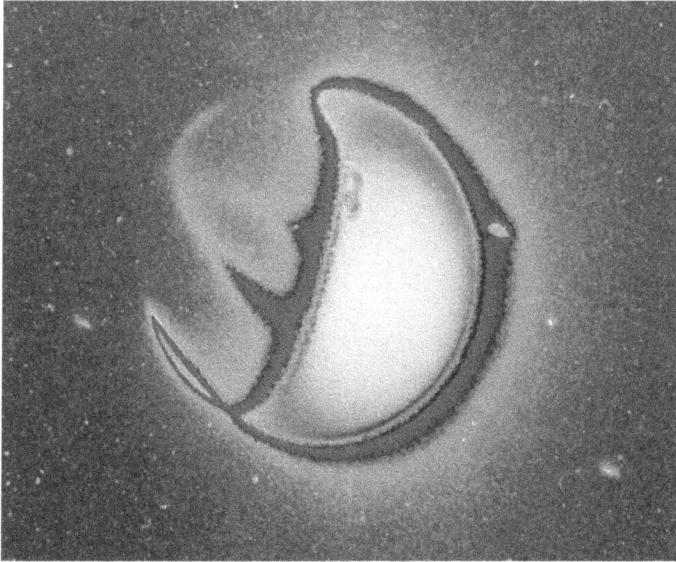

Figure 201: *A false-color image of Earth in ultraviolet light taken from the surface of the Moon on the Apollo 16 mission. The day-side reflects a lot of UV light from the Sun, but the night-side shows faint bands of UV emission from the aurora caused by charged particles.*

has plans to do so. Although several individuals have made claims to the Moon in whole or in part, none of these are considered credible.

In culture

Mythology

A 5,000-year-old rock carving at Knowth, Ireland, may represent the Moon, which would be the earliest depiction discovered. The contrast between the brighter highlands and the darker maria creates the patterns seen by different cultures as the Man in the Moon, the rabbit and the buffalo, among others. In many prehistoric and ancient cultures, the Moon was personified as a deity or other supernatural phenomenon, and astrological views of the Moon continue to be propagated today.

In Proto-Indo-European religion, the moon was personified as the male god *Meh₁not*. The ancient Sumerians believed that the Moon was the god Nanna, who was the father of Inanna, the goddess of the planet Venus, and Utu, the god of the sun. Nanna was later known as Sîn, and was particularly associated

Figure 202: *Luna, the Moon, from a 1550 edition of Guido Bonatti's Liber astronomiae*

Figure 203: *Statue of Chandraprabha (means "as charming as moon")-8th Tirthankara in Jainism with the symbol of crescent moon below it.*

Figure 204: *Sun and Moon with faces (1493 woodcut)*

with magic and sorcery. In Greco-Roman mythology, the Sun and the Moon are represented as male and female, respectively (Helios/Sol and Selene/Luna); this is a development unique to the eastern Mediterranean and traces of an earlier male moon god in the Greek tradition are preserved in the figure of Menelaus.

In Mesopotamian iconography, the crescent was the primary symbol of Nanna-Sîn. In ancient Greek art, the Moon goddess Selene was represented wearing a crescent on her headgear in an arrangement reminiscent of horns. The star and crescent arrangement also goes back to the Bronze Age, representing either the Sun and Moon, or the Moon and planet Venus, in combination. It came to represent the goddess Artemis or Hecate, and via the patronage of Hecate came to be used as a symbol of Byzantium.

An iconographic tradition of representing Sun and Moon with faces developed in the late medieval period.

The splitting of the moon (Arabic: انشقاق القمر) is a miracle attributed to Muhammad.[826]

Figure 205: *Moonrise, 1884, picture by Stanisław Masłowski (National Museum, Kraków, Gallery of Sukiennice Museum)*

Calendar

The Moon's regular phases make it a very convenient timepiece, and the periods of its waxing and waning form the basis of many of the oldest calendars. Tally sticks, notched bones dating as far back as 20–30,000 years ago, are believed by some to mark the phases of the Moon.[827] The ~30-day month is an approximation of the lunar cycle. The English noun *month* and its cognates in other Germanic languages stem from Proto-Germanic *mǣnóth-*, which is connected to the above-mentioned Proto-Germanic *mǣnōn*, indicating the usage of a lunar calendar among the Germanic peoples (Germanic calendar) prior to the adoption of a solar calendar. The PIE root of *moon*, *méh₁nōt*, derives from the PIE verbal root *meh₁-*, "to measure", "indicat[ing] a functional conception of the Moon, i.e. marker of the month" (cf. the English words *measure* and *menstrual*), and echoing the Moon's importance to many ancient cultures in measuring time (see Latin *mensis* and Ancient Greek μείς (*meis*) or μήν (mēn), meaning "month"). Most historical calendars are lunisolar. The 7th-century Islamic calendar is an exceptional example of a purely lunar calendar. Months are traditionally determined by the visual sighting of the hilal, or earliest crescent moon, over the horizon.

Lunacy

The Moon has long been associated with insanity and irrationality; the words *lunacy* and *lunatic* (popular shortening *loony*) are derived from the Latin name for the Moon, *Luna*. Philosophers Aristotle and Pliny the Elder argued that the full moon induced insanity in susceptible individuals, believing that the brain, which is mostly water, must be affected by the Moon and its power over the tides, but the Moon's gravity is too slight to affect any single person. Even today, people who believe in a lunar effect claim that admissions to psychiatric hospitals, traffic accidents, homicides or suicides increase during a full moon, but dozens of studies invalidate these claims.

References

Citations

Bibliography

<templatestyles src="Template:Refbegin/styles.css" />

- Needham, Joseph (1986). *Science and Civilization in China, Volume III: Mathematics and the Sciences of the Heavens and Earth*[828]. Taipei: Caves Books. ISBN 978-0-521-05801-8.
- Terry, paul (2013), *Top 10 Of Everything*, Octopus Publishing Group Ltd 2013, ISBN 9780600628873

Further reading

<templatestyles src="Template:Refbegin/styles.css" />

- "Revisiting the Moon"[829]. *The New York Times*. Retrieved 8 September 2014.
- The Moon[830]. *Discovery 2008*. BBC World Service.
- Bussey, B.; Spudis, P.D. (2004). *The Clementine Atlas of the Moon*. Cambridge University Press. ISBN 0-521-81528-2.
- Cain, Fraser. "Where does the Moon Come From?"[831]. Universe Today. Retrieved 1 April 2008. (podcast and transcript)
- Jolliff, B. (2006). Wieczorek, M.; Shearer, C.; Neal, C., eds. "New views of the Moon"[832]. *Reviews in Mineralogy and Geochemistry*. Chantilly, Virginia: Mineralogy Society of America. **60** (1): 721. Bibcode: 2006RvMG...60D...5J[833]. doi: 10.2138/rmg.2006.60.0[834]. ISBN 0-939950-72-3. Retrieved 12 April 2007.
- Jones, E.M. (2006). "Apollo Lunar Surface Journal"[835]. NASA. Retrieved 12 April 2007.
- "Exploring the Moon"[836]. Lunar and Planetary Institute. Retrieved 12 April 2007.
- Mackenzie, Dana (2003). *The Big Splat, or How Our Moon Came to Be*. Hoboken, New Jersey: John Wiley & Sons. ISBN 0-471-15057-6.

- Moore, P. (2001). *On the Moon*. Tucson, Arizona: Sterling Publishing Co. ISBN 0-304-35469-4.
- "Moon Articles"[837]. *Planetary Science Research Discoveries*. Hawai'i Institute of Geophysics and Planetology.
- Spudis, P. D. (1996). *The Once and Future Moon*. Smithsonian Institution Press. ISBN 1-56098-634-4.
- Taylor, S.R. (1992). *Solar system evolution*. Cambridge University Press. p. 307. ISBN 0-521-37212-7.
- Teague, K. (2006). "The Project Apollo Archive"[838]. Retrieved 12 April 2007.
- Wilhelms, D.E. (1987). "Geologic History of the Moon"[839]. *U.S. Geological Survey Professional paper*. **1348**. Retrieved 12 April 2007.
- Wilhelms, D.E. (1993). *To a Rocky Moon: A Geologist's History of Lunar Exploration*[840]. Tucson, Arizona: University of Arizona Press. ISBN 0-8165-1065-2. Retrieved 10 March 2009.

External links

- NASA images and videos about the Moon[841]
- Albums of images and high-resolution overflight videos by Seán Doran, based on LROC data, on Flickr[842] and YouTube[843]
- Video (04:56) – The Moon in 4K (NASA, April 2018)[844] on YouTube
- Video (04:47) – The Moon in 3D (NASA, July 2018)[845] on YouTube

Cartographic resources

- The Moon on Google Maps[846], a 3-D rendition of the moon akin to Google Earth
- "Consolidated Lunar Atlas"[847]. Lunar and Planetary Institute. Retrieved 26 February 2012.
- Gazetteer of Planetary Nomenclature (USGS)[848] List of feature names.
- "Clementine Lunar Image Browser"[849]. U.S. Navy. 15 October 2003. Retrieved 12 April 2007.
- 3D zoomable globes:
  - "Google Moon"[850]. Google. 2007. Retrieved 12 April 2007.
  - "Moon"[851]. *World Wind Central*. NASA. 2007. Retrieved 12 April 2007.
- Aeschliman, R. "Lunar Maps"[852]. *Planetary Cartography and Graphics*. Retrieved 12 April 2007. Maps and panoramas at Apollo landing sites
- Japan Aerospace Exploration Agency (JAXA)[853] Kaguya (Selene) images
- Large image of the Moon's north pole area[854]

Observation tools

- "NASA's SKYCAL—Sky Events Calendar"[855]. NASA. Archived from the original[856] on 20 August 2007. Retrieved 27 August 2007.
- "Find moonrise, moonset and moonphase for a location"[857]. 2008. Retrieved 18 February 2008.
- "HMNAO's Moon Watch"[858]. 2005. Retrieved 24 May 2009. See when the next new crescent moon is visible for any location.

General

- Lunar shelter[859] (building a lunar base with 3D printing[860])

<indicator name="featured-star"> ⭐ </indicator>

Earth in culture

The cultural perspective on Earth, or the world, varies by society and time period. Religious beliefs often include a creation belief as well as personification in the form of a deity. The exploration of the world has modified many of the perceptions of the planet, resulting in a viewpoint of a globally integrated ecosystem. Unlike the remainder of the planets in the Solar System, mankind didn't perceive the Earth as a planet until the sixteenth century.

Etymology

Unlike the other planets in the Solar System, in English, Earth does not directly share a name with an ancient Roman deity. The name *Earth* derives from the eighth century Anglo-Saxon word *erda*, which means ground or soil. It became *eorthe* later, and then *erthe* in Middle English. These words are all cognates of Jörð, the name of the giantess of Norse myth. Earth was first used as the name of the sphere of the Earth in the early fifteenth century. The planet's name in Latin, used academically and scientifically in the West during the Renaissance, is the same as that of Terra Mater, the Roman goddess, which translates to English as Mother Earth.

Figure 206: *"The Blue Marble" photograph of Earth, taken by the Apollo 17 lunar mission in 1972.*

Figure 207: *Astronomical symbol of Earth*

Figure 208: *The Hindu Earth goddess*

Planetary symbol

The standard astronomical symbol of the Earth consists of a cross circumscribed by a circle. This symbol is known as the wheel cross, sun cross, Odin's cross or Woden's cross. Although it has been used in various cultures for different purposes, it came to represent the compass points, earth and the land. Another version of the symbol is a cross on top of a circle; a stylized globus cruciger that was also used as an early astronomical symbol for the planet Earth.

Religious beliefs

Earth has often been personified as a deity, in particular a goddess. In many cultures the mother goddess is also portrayed as a fertility deity. To the Aztec, Earth was called Tonantzin—"our mother"; to the Incas, Earth was called Pachamama—"mother earth". The Chinese Earth goddess Hou Tu is similar to Gaia, the Greek goddess personifying the Earth. To Hindus it is called Bhuma Devi, the Goddess of Earth. (See also Graha.) The Tuluva people of Tulunadu in Southern India celebrate a Three Day "Earth Day" called Keddaso. This festival comes in usually on 10th,12th,13 February every Calendar year. In Norse mythology, the Earth giantess Jörð was the mother of Thor and the daughter of Annar. Ancient Egyptian mythology is different from that of other cultures because Earth is male, Geb, and sky is female, Nut.

Creation myths in many religions recall a story involving the creation of the world by a supernatural deity or deities. A variety of religious groups, often associated with fundamentalist branches of Protestantism or Islam, assert that their interpretations of the accounts of creation in sacred texts are literal truth and should be considered alongside or replace conventional scientific accounts of the formation of the Earth and the origin and development of life. Such assertions are opposed by the scientific community[861] as well as other religious groups. A prominent example is the creation-evolution controversy.

Physical form

In the ancient past there were varying levels of belief in a flat Earth, with the Mesopotamian culture portraying the world as a flat disk afloat in an ocean. The spherical form of the Earth was suggested by early Greek philosophers; a belief espoused by Pythagoras. By the Middle Ages—as evidenced by thinkers such as Thomas Aquinas—European belief in a spherical Earth was widespread.[862]

Modern perspective

The technological developments of the latter half of the 20th century are widely considered to have altered the public's perception of the Earth. Before space flight, the popular image of Earth was of a green world. Science fiction artist Frank R. Paul provided perhaps the first image of a cloudless *blue* planet (with sharply defined land masses) on the back cover of the July 1940 issue of *Amazing Stories*, a common depiction for several decades thereafter.

Earth was first photographed from space by Explorer 6 in 1959. Yuri Gagarin became the first human to view Earth from space in 1961. The crew of the Apollo 8 was the first to view an Earth-rise from lunar orbit in 1968. In 1972 the crew of the Apollo 17 produced the famous *Blue Marble* photograph of the planet Earth from cislunar space. This became an iconic image of the planet as a marble of cloud-swirled blue ocean broken by green-brown continents. NASA archivist Mike Gentry has speculated that *The Blue Marble* is the most widely distributed image in human history. Inspired by *The Blue Marble* poet-diplomat Abhay K has penned an Earth Anthem describing the planet as a "Cosmic Blue Pearl".[863] A photo taken of a distant Earth by *Voyager 1* in 1990 inspired Carl Sagan to describe the planet as a "Pale Blue Dot."

Figure 209: *The first photograph ever taken of an "Earthrise," on Apollo 8.*

Since the 1960s, Earth has also been described as a massive "Spaceship Earth," with a life support system that requires maintenance, or, in the Gaia hypothesis, as having a biosphere that forms one large organism.

Over the past two centuries a growing environmental movement has emerged that is concerned about humankind's effects on the Earth. The key issues of this socio-political movement are the conservation of natural resources, elimination of pollution, and the usage of land. Although diverse in interests and goals, environmentalists as a group tend to advocate sustainable management of resources and stewardship of the environment through changes in public policy and individual behavior. Of particular concern is the large-scale exploitation of non-renewable resources. Changes sought by the environmental movements are sometimes in conflict with commercial interests due to the additional costs associated with managing the environmental impact of those interests.

Appendix

References

1

[2]Earth's circumference is almost exactly 40,000 km because the metre was calibrated on this measurement—more specifically, 1/10-millionth of the distance between the poles and the equator.

[3]Early edition, published online before print.

[4]Including *eorþe, erþe, erde,* and *erthe.* UNIQ-ref-0-467e96e81ac026b0-QINU

[5]Oxford English Dictionary, "earth, *n.¹*" Oxford University Press (Oxford), 2010.

[6]As in *Beowulf* (1531–33):

Wearp ða wundelmæl wrættum gebunden

yrre oretta, þæt hit on eorðan læg,

stið ond stylecg. UNIQ-ref-1-467e96e81ac026b0-QINU UNIQ-ref-2-467e96e81ac026b0-QINU

"He threw the artfully-wound sword so that it lay upon the **earth**, firm and sharp-edged."<ref name=beo>*Beowulf.* Trans. Chad Matlick in "*Beowulf*: Lines 1399 to 1799" http://www.as.wvu.edu/english/oeoe/english311/1799.html. West Virginia University. Retrieved 5 August 2014. &

[7]As in the Old English glosses of the *Lindisfarne Gospels* (Luke 13:7):

Succidite ergo illam ut quid etiam **terram** *occupat: hrendas* uel *scearfað forðon ðailca* uel *hia to huon uutedlice eorðo gionetað* uel *gemerras.* UNIQ-ref-3-467e96e81ac026b0-QINU

"Remove it. Why should it use up the **soil**?"<ref>*Mounce Reverse-Intralinear New Testament:* " Luke 13:7 https://www.biblegateway.com/passage/?search=Luke%2013:7&version=MOUNCE". Hosted at *Bible Gateway.* 2014. Retrieved 5 August 2014. &

[8]As in Ælfric's *Heptateuch* (Gen. 1:10):

Ond God gecygde ða drignysse **eorðan** *ond ðære wætera gegaderunge he het sæ.*<ref>Ælfric of Eynsham. *Heptateuch.* Reprinted by S.J. Crawford as *The Old English Version of the Heptateuch, Ælfric's Treatise on the Old and New Testament and his Preface to Genesis.* Humphrey Milford (London), 1922. http://wordhord.org/nasb/genesis.html Hosted at *Wordhord.* Retrieved 5 August 2014.

[9]King James Version of the Bible: " Genesis 1:10 https://www.biblegateway.com/passage/?search=Genesis%201:10&version=KJV". Hosted at *Bible Gateway.* 2014. Retrieved 5 August 2014.

[10]As in the Wessex Gospels (Matt. 28:18):

Me is geseald ælc anweald on heofonan & on **eorðan.**

"All authority in heaven and on **earth** has been given to me."<ref>*Mounce Reverse-Intralinear New Testament:* " Matthew 28:18 https://www.biblegateway.com/passage/?search=Matthew+28%3A18&version=MOUNCE". Hosted at *Bible Gateway.* 2014. Retrieved 5 August 2014. &

[11]As in the Codex Junius's *Genesis* (112–16):

her ærest gesceop ece drihten,

helm eallwihta, heofon and **eorðan,**

rodor arærde and þis rume land

gestaþelode strangum mihtum,

frea ælmihtig.<ref>" Genesis A http://www.maldura.unipd.it/dllags/brunetti/OE/TESTI/GenesisA/DATI/testo.html". Hosted at the Dept. of Linguistic Studies at the University of Padua. Retrieved 5 August 2014.

[12]Killings, Douglas. *Codex Junius 11,* I.ii http://www.gutenberg.org/files/618/618-h/618-h.htm. 1996. Hosted at Project Gutenberg. Retrieved 5 August 2014.

[13]As in Ælfric's *On the Seasons of the Year* (Ch. 6, §9):

Seo **eorðe** *stent on gelicnysse anre pinnhnyte, & seo sunne glit onbutan be Godes gesetnysse.*

"The **earth** can be compared to a pine cone, and the Sun glides around it by God's decree.<ref>Ælfric, Abbot of Eynsham. "*De temporibus annis*" Trans. as " On the Seasons of the Year http://faculty.virginia.edu/OldEnglish/aelfric/detemp.html". Hosted at Old English at the University of Virginia, 1998. Retrieved 6 August 2014.

[14]Tacitus. *Germania*, .

[15]Simek, Rudolf. Trans. Angela Hall as *Dictionary of Northern Mythology*, D.S. Brewer, 2007.

[16]*The New Oxford Dictionary of English*, "earth". Oxford University Press (Oxford), 1998.

[17]If Earth were shrunk to the size of a billiard ball, some areas of Earth such as large mountain ranges and oceanic trenches would feel like tiny imperfections, whereas much of the planet, including the Great Plains and the abyssal plains, would feel smoother.<ref>

[18]Locally varies between .

[19]Locally varies between .

[20]   World   http://www.nationalgeographic.com/xpeditions/atlas/index.html?Parent= world&Mode=d&SubMode=w at the Xpeditions Atlas http://www.nationalgeographic.com/ xpeditions/, National Geographic Society, Washington D.C., 2006.

[21]//en.wikipedia.org/w/index.php?title=Earth&action=edit

[22]Lovelock, James. *The Vanishing Face of Gaia*. Basic Books, 2009, p. 255.

[23]https://books.google.com/books?id=xwjlZjFNFlAC

[24]http://adsabs.harvard.edu/abs/2003deu..book.....C

[25]//www.worldcat.org/oclc/52082611

[26]http://education.nationalgeographic.com/education/encyclopedia/earth/?ar_a=1

[27]https://www.webcitation.org/6GVr8TSgV?url=http://solarsystem.nasa.gov/planets/profile. cfm?Object=Earth

[28]http://solarsystem.nasa.gov/

[29]http://www.nasa.gov/centers/goddard/earthandsun/earthshape.html

[30]https://web.archive.org/web/20090430041323/http://eol.jsc.nasa.gov/Coll/weekly.htm

[31]http://earthobservatory.nasa.gov/

[32]http://www.astronomycast.com/stars/episode-51-earth/

[33]https://www.youtube.com/watch?v=74mhQyuyELQ

[34]https://www.youtube.com/watch?v=l6ahFFFQBZY

[35]http://www.usgs.gov/

[36]https://www.google.com/maps/@36.6233227,-44.9959756,5662076m/data=!3m1!1e3

[37]"International Stratigraphic Chart". International Commission on Stratigraphy

[38]Early edition, published online before print.

[39]Charles Frankel, 1996, *Volcanoes of the Solar System*, Cambridge University Press, pp. 7–8,

[40]Pluto's satellite Charon is relatively larger,<ref>

[41]The Sun's evolution http://faculty.wcas.northwestern.edu/~infocom/The%20Website/ evolution.html

[42]http://sp.lyellcollection.org/content/190/1/205.abstract

[43]http://adsabs.harvard.edu/abs/2001GSLSP.190..205D

[44]//doi.org/10.1144/GSL.SP.2001.190.01.14

[45]https://web.archive.org/web/20121028022719/http://www.nysm.nysed.gov/nysgs/resources/ images/geologicaltimescale.pdf

[46]http://www.nysm.nysed.gov/nysgs/resources/images/geologicaltimescale.pdf

[47]http://adsabs.harvard.edu/abs/1991Sci...253..535W

[48]//doi.org/10.1126/science.253.5019.535

[49]//www.ncbi.nlm.nih.gov/pubmed/17745185

[50]https://www.theguardian.com/technology/2005/dec/20/comment.science

[51]http://www.johnkyrk.com/evolution.html

[52]http://www.bbc.com/earth/bespoke/story/20150123-earths-25-biggest-turning-points/

[53]http://historystack.com/30_Major_Events_in_History_of_the_Earth

[54]http://www.bbc.co.uk/programmes/p00547hl

[55]http://www.bbc.co.uk/programmes/p005493g

[56]http://tools.wmflabs.org/timescale/?Ma=750

[57]http://tools.wmflabs.org/timescale/?Ma=600–540

[58]http://tools.wmflabs.org/timescale/?Ma=200

[59]http://tools.wmflabs.org/timescale/?Ma=40

[60]http://tools.wmflabs.org/timescale/?Ma=4,500

[61] http://tools.wmflabs.org/timescale/?Ma=4,400

[62] http://tools.wmflabs.org/timescale/?Ma=4100–3800

[63] http://tools.wmflabs.org/timescale/?Ma=4000–2500

[64] http://tools.wmflabs.org/timescale/?Ma=2500–541

[65] International Stratigraphic Chart 2008, International Commission on Stratigraphy https://www. webcitation.org/5nDdmUWXk?url=http://www.stratigraphy.org/upload/ISChart2008.pdf

[66] http://tools.wmflabs.org/timescale/?Ma=541–252

[67] http://tools.wmflabs.org/timescale/?Ma=447–444

[68] http://tools.wmflabs.org/timescale/?Ma=249

[69] http://tools.wmflabs.org/timescale/?Ma=290

[70] http://tools.wmflabs.org/timescale/?Ma=252–66

[71] Monroe and Wicander, 607.

[72] http://tools.wmflabs.org/timescale/?Ma=145.0

[73] http://tools.wmflabs.org/timescale/?Ma=66

[74] Dougal Dixon et al., *Atlas of Life on Earth*, (New York: Barnes & Noble Books, 2001), p. 215.

[75] http://tools.wmflabs.org/timescale/?Ma=56.0

[76] Hooker, J.J., "Tertiary to Present: Paleocene", pp. 459-465, Vol. 5. of Selley, Richard C., L. Robin McCocks, and Ian R. Plimer, Encyclopedia of Geology, Oxford: Elsevier Limited, 2005.

[77] http://tools.wmflabs.org/timescale/?Ma=33.9

[78] http://tools.wmflabs.org/timescale/?Ma=34

[79] http://tools.wmflabs.org/timescale/?Ma=23

[80] http://tools.wmflabs.org/timescale/?Ma=5.333

[81] http://tools.wmflabs.org/timescale/?Ma=2.588

[82] http://tools.wmflabs.org/timescale/?Ma=2.58

[83] https://www.webcitation.org/5QVjwZCzJ?url=http://www.tufts.edu/as/wright_center/cosmic_ evolution/docs/splash.html

[84] http://www.sciam.com/article.cfm?chanID=sa006&colID=1&articleID=0005FA5D-5F7C- 1333-9F7C83414B7F0000

[85] https://www.theguardian.com/science/story/0,3605,1671164,00.html

[86] http://www.johnkyrk.com/evolution.html

[87] https://web.archive.org/web/20030729055405/http://www.uwmc.uwc.edu/geography/Hutton/ Hutton.htm

[88] http://issuu.com/sergioluisdasilva/docs/paleomaps_mollweide_longitude_0

[89] http://issuu.com/sergioluisdasilva/docs/paleomaps_mollweide_longitude_180

[90] http://www.bbc.co.uk/programmes/p005493g

[91] //en.wikipedia.org/w/index.php?title=Template:Evolutionary_biology&action=edit

[92] Early edition, published online before print.

[93] http://tools.wmflabs.org/timescale/?Ma=525

[94] http://tools.wmflabs.org/timescale/?Ma=252

[95] http://tools.wmflabs.org/timescale/?Ma=66

[96] "This paper was originally presented at a workshop titled *Evolution: A Molecular Point of View*."

[97] • Textbook used for lecture: *Biology Today and Tomorrow With Physiology* (2007), .

[98] Myxozoa were thought to be an exception, but are now thought to be heavily modified members of the Cnidaria.

[99] http://tools.wmflabs.org/timescale/?Ma=580

[100] Paper No. 40-2 presented at the Geological Society of America's 2003 Seattle Annual Meeting (November 2–5, 2003) on November 2, 2003, at the Washington State Convention Center.

[101] http://tools.wmflabs.org/timescale/?Ma=488–444

[102] http://tools.wmflabs.org/timescale/?Ma=385–359

[103]

[104] http://tools.wmflabs.org/timescale/?Ma=423–419

[105] http://tools.wmflabs.org/timescale/?Ma=476

[106] http://tools.wmflabs.org/timescale/?Ma=430

[107] http://tools.wmflabs.org/timescale/?Ma=370

[108] Scheckler 2001, "Afforestation—the First Forests," pp. 67–70

[109] The phrase "Late Devonian wood crisis" is used at
[110] http://tools.wmflabs.org/timescale/?Ma=428
[111] http://tools.wmflabs.org/timescale/?Ma=490
[112] http://tools.wmflabs.org/timescale/?Ma=445
[113] http://tools.wmflabs.org/timescale/?Ma=415
[114]
[115] http://tools.wmflabs.org/timescale/?Ma=370–360
[116] http://tools.wmflabs.org/timescale/?Ma=363
[117] http://tools.wmflabs.org/timescale/?Ma=330–298.9
[118] http://tools.wmflabs.org/timescale/?Ma=313
[119] http://tools.wmflabs.org/timescale/?Ma=298.9–251.902
[120] http://tools.wmflabs.org/timescale/?Ma=201.3–66
[121] http://tools.wmflabs.org/timescale/?Ma=195
[122] http://tools.wmflabs.org/timescale/?Ma=167
[123] http://tools.wmflabs.org/timescale/?Ma=130–90
[124]
[125] http://tools.wmflabs.org/timescale/?Ma=400
[126] http://tools.wmflabs.org/timescale/?Ma=300
[127] http://tools.wmflabs.org/timescale/?Ma=6
[128] http://tools.wmflabs.org/timescale/?Ma=2.5
[129] http://tools.wmflabs.org/timescale/?Ma=542
[130] http://tools.wmflabs.org/timescale/?Ma=542–400
[131] http://tools.wmflabs.org/timescale/?Ma=400–200
[132] http://tools.wmflabs.org/timescale/?Ma=200
[133] http://adsabs.harvard.edu/abs/1980qel..book...32A
[134] //lccn.loc.gov/79057423
[135] //www.worldcat.org/oclc/7121102
[136] https://web.archive.org/web/20170211103042/http://www.cosmonova.org/download/18.
4e32c81078a8d9249800021554/Bengtson2004ESF.pdf
[137] //www.worldcat.org/oclc/57481790
[138] http://www.cosmonova.org/download/18.4e32c81078a8d9249800021554/Bengtson2004ESF.
pdf
[139] //lccn.loc.gov/2007037872
[140] //www.worldcat.org/oclc/172521761
[141] //www.worldcat.org/oclc/37378512
[142] //lccn.loc.gov/2002109744
[143] //www.worldcat.org/oclc/62145244
[144] //lccn.loc.gov/2003028152
[145] //www.worldcat.org/oclc/53970617
[146] https://www.novapublishers.com/catalog/product_info.php?products_id=31918
[147] //lccn.loc.gov/2011038504
[148] //www.worldcat.org/oclc/828424701
[149] //www.worldcat.org/oclc/646754753
[150] //www.ncbi.nlm.nih.gov/pubmed/3324702
[151] //www.worldcat.org/oclc/751583918
[152] //lccn.loc.gov/0632051477
[153] //www.worldcat.org/oclc/43945263
[154] //lccn.loc.gov/71419832
[155] //www.worldcat.org/oclc/230043266
[156] //lccn.loc.gov/2006271630
[157] //www.worldcat.org/oclc/57574490
[158] //lccn.loc.gov/99016542
[159] //www.worldcat.org/oclc/47011068
[160] //lccn.loc.gov/90047051
[161] //www.worldcat.org/oclc/22347190
[162] //lccn.loc.gov/2004029808

[163] //www.worldcat.org/oclc/57311264
[164] //lccn.loc.gov/89001132
[165] //www.worldcat.org/oclc/19129518
[166] //lccn.loc.gov/99049796
[167] //www.worldcat.org/oclc/42476104
[168] //lccn.loc.gov/88037469
[169] //www.worldcat.org/oclc/18983518
[170] //lccn.loc.gov/2004054605
[171] //www.worldcat.org/oclc/56057971
[172] //lccn.loc.gov/2002276846
[173] //www.worldcat.org/oclc/49663129
[174] https://www.ncbi.nlm.nih.gov/books/NBK7908/
[175] //lccn.loc.gov/95050499
[176] //www.worldcat.org/oclc/33838234
[177] //www.ncbi.nlm.nih.gov/pubmed/21413277
[178] https://web.archive.org/web/20070106201614/http://134.106.242.33/krumbein/htdocs/Archive/397/Krumbein_397.pdf
[179] //lccn.loc.gov/2003061870
[180] //www.worldcat.org/oclc/52901566
[181] http://134.106.242.33/krumbein/htdocs/Archive/397/Krumbein_397.pdf
[182] http://www.santafe.edu/media/workingpapers/00-08-044.pdf
[183] //lccn.loc.gov/2001385090
[184] //www.worldcat.org/oclc/44822625
[185] //lccn.loc.gov/94003617
[186] //www.worldcat.org/oclc/30739453
[187] //lccn.loc.gov/80026695
[188] //www.worldcat.org/oclc/6982472
[189] //lccn.loc.gov/96071014
[190] //www.worldcat.org/oclc/36442106
[191] //lccn.loc.gov/2011934330
[192] //www.worldcat.org/oclc/741539226
[193] http://pubs.usgs.gov/gip/geotime/age.html
[194] //www.worldcat.org/oclc/18792528
[195] //www.worldcat.org/oclc/757322661
[196] //lccn.loc.gov/2004049804
[197] //www.worldcat.org/oclc/55000644
[198] //lccn.loc.gov/00062919
[199] //www.worldcat.org/oclc/51667292
[200] //lccn.loc.gov/97151576
[201] //www.worldcat.org/oclc/36463214
[202] //lccn.loc.gov/89016077
[203] //www.worldcat.org/oclc/20012195
[204] //lccn.loc.gov/2004059864
[205] //www.worldcat.org/oclc/56617123
[206] //lccn.loc.gov/2008030270
[207] //www.worldcat.org/oclc/225874308
[208] //lccn.loc.gov/94026965
[209] //www.worldcat.org/oclc/715217397
[210] http://www.fossilmuseum.net/Evolution.htm
[211] http://evolution.berkeley.edu/
[212] http://nationalacademies.org/evolution/
[213] https://web.archive.org/web/20150210112109/http://www.tellapallet.com/tree_of_life.htm
[214] http://tellapallet.com/tree_of_life.htm
[215] https://www.newscientist.com/topic/evolution
[216] http://www.howstuffworks.com/http://science.howstuffworks.com/life/evolution/evolution.htm

[217] http://anthro.palomar.edu/synthetic/

[218] http://darwin-online.org.uk

[219] http://www.rationalrevolution.net/articles/understanding_evolution.htm

[220] Myers 2000, pp. 63–70.

[221] Reaka-Kudla, Wilson & Wilson 1997, pp. 132–33.

[222] Ward & Brownlee 2003, p. 142.

[223] Fishbaugh et al. 2007, p. 114.

[224] Cowie 2007, p. 162.

[225] See also: Life After People, about the decay of structures (if humans disappeared).

[226] Tayler 1993, p. 92.

[227] Hanslmeier 2009, pp. 174–76.

[228] Adams 2008, pp. 33–44.

[229] Hanslmeier 2009, p. 116.

[230] Roberts 1998, p. 60.

[231] Lunine & Lunine 1999, p. 244.

[232] Ward 2006, pp. 231–32.

[233] Ward & Brownlee 2003, pp. 92–96.

[234] Nield 2007, pp. 20–21.

[235] Hoffman 1992, pp. 323–27.

[236] Calkin & Young 1996, pp. 9–75.

[237] Thompson & Perry 1997, pp. 127–28.

[238] Palmer 2003, p. 164.

[239] Gonzalez & Richards 2004, p. 48.

[240] Meadows 2007, p. 34.

[241] Stevenson 2002, p. 605.

[242] van der Maarel 2005, p. 363.

[243] Ward & Brownlee 2003, pp. 117–28.

[244] Brownlee 2010, p. 95.

[245] Brownlee 2010, p. 94.

[246] https://books.google.com/books?id=-Jxc88RuJhgC&pg=PA33

[247] https://books.google.com/books?id=M8NwTYEl0ngC&pg=PA79

[248] https://books.google.com/books?id=KFdu4CyQ1k0C&pg=PA48

[249] https://books.google.com/books?id=PRqVqQKao9QC

[250] http://www.eps.harvard.edu/people/faculty/hoffman/pdfs/supercontinents.pdf

[251] https://books.google.com/books?id=sZgB52BCa0UC&pg=PA244

[252] https://books.google.com/books?id=KTa-jBOBS5UC&pg=PA34

[253] https://books.google.com/books?id=IsKCaK9W0EwC&pg=PA605

[254] https://books.google.com/books?id=mxb1IxSyu7wC&pg=PA92

[255] http://www.scotese.com/

[256] //doi.org/10.1016/S0016-3287%2801%2900050-7

[257] //en.wikipedia.org/w/index.php?title=Template:Geodesy&action=edit

[258] This section is a close paraphrase of Defense Mapping Agency 1983, page 9 of the PDF.

[259] https://web.archive.org/web/20051023083444/http://www.pcigeomatics.com/cgi-bin/pcihlp/
PROJ%7CEARTH+MODELS%7CELLIPSOIDS%7CELLIPSOID+CODES

[260] http://www.google.com/search?q=cache:TjusGxmrm4EJ:www.scanex.ru

[261] http://www.nasa.gov/centers/goddard/earthandsun/earthshape.html

[262] http://www.josleys.com/show_gallery.php?galid=313

[263] Below Jupiter's outer atmosphere, volume fractions are significantly different from mole fractions due to high temperatures (ionization and disproportionation) and high density where the Ideal Gas Law is inapplicable.

[264] What is Dark Energy? http://www.space.com/20929-dark-energy.html , Space.com, 1 May 2013

[265] William F McDonough The composition of the Earth https://web.archive.org/web/20110928074153/http://quake.mit.edu/hilstgroup/CoreMantle/EarthCompo.pdf. quake.mit.edu, archived by the Internet Archive Wayback Machine.

[266] Table data from

[267] http://geopubs.wr.usgs.gov/fact-sheet/fs087-02/

[268] https://web.archive.org/web/20031203202925/http://imagine.gsfc.nasa.gov/docs/dict_ei.html

[269] http://imagine.gsfc.nasa.gov:80/docs/dict_ei.html

[270] https://www.science.co.il/elements/?s=Earth

[271] https://web.archive.org/web/20060901133923/http://www.astro.wesleyan.edu/~bill/courses/astr231/wes_only/element_abundances.pdf

[272] https://web.archive.org/web/20160303223016/http://www.webelements.com/periodicity/

[273] " 2016 Selected Astronomical Constants http://asa.usno.navy.mil/static/files/2016/Astronomical_Constants_2016.pdf" in

[274] Breaking News | Oldest rock shows Earth was a hospitable young planet http://spaceflightnow.com/news/n0101/14earthwater/. Spaceflight Now (2001-01-14). Retrieved on 2012-01-27.

[275] Uwe Walzer, Roland Hendel, John Baumgardner Mantle Viscosity and the Thickness of the Convective Downwellings https://web.archive.org/web/20060826020002/http://www.chemie.uni-jena.de/geowiss/geodyn/poster2.html.

[276] Lawrence Berkeley National Laboratory (Berkeley Lab) is a Department of Energy (DOE) Office of Science lab managed by University of California., What Keeps the Earth Cooking? News Release by Paul Preuss, July 17, 2011 http://newscenter.lbl.gov/2011/07/17/kamland-geoneutrinos/

[277] BBC News, "What is at the centre of the Earth? https://www.bbc.co.uk/news/uk-14678004. Bbc.co.uk (2011-08-31). Retrieved on 2012-01-27.

[278] First Measurement Of Magnetic Field Inside Earth's Core http://www.science20.com/news_articles/first_measurement_magnetic_field_inside_earths_core. Science20.com. Retrieved on 2012-01-27.

[279] Chang, Kenneth (26 August 2005) "Scientists Say Earth's Center Rotates Faster Than Surface" https://www.nytimes.com/2005/08/26/science/26core.html The New York Times Sec. A, Col. 1, p. 13.

[280] http://discovermagazine.com/2007/jun/journey-to-the-center-of-the-earth

[281] https://www.youtube.com/watch?v=BsKyEckDRbo

[282] http://www.bbc.co.uk/programmes/b05s3gyv

[283]

[284] Buffett, B. A. (2007). Taking earth's temperature. Science, 315(5820), 1801–1802.

[285] Lowrie, W. (2007). Fundamentals of geophysics. Cambridge: CUP, 2nd ed.

[286] Thomson, William. (1864). On the secular cooling of the earth http://courses.seas.harvard.edu/climate/eli/Courses/EPS281r/Sources/Earth-age-and-thermal-history/more/Kelvin-1863-excerpts.pdf, read 28 April 1862. Transactions of the Royal Society of Edinburgh, 23, 157–170.

[287] Dye, S. T. (2012). Geoneutrinos and the radioactive power of the Earth. Reviews of Geophysics, 50(3).

[288] Arevalo Jr, R., McDonough, W. F., & Luong, M. (2009). The K/U ratio of the silicate Earth: Insights into mantle composition, structure and thermal evolution. Earth and Planetary Science Letters, 278(3), 361–369.

[289] Jaupart, C., & Mareschal, J. C. (2007). Heat flow and thermal structure of the lithosphere. Treatise on Geophysics, 6, 217–251.

[290] Korenaga, J. (2003). Energetics of mantle convection and the fate of fossil heat. Geophysical Research Letters, 30(8), 1437.

[291] Korenaga, J. (2011). Earth's heat budget: Clairvoyant geoneutrinos. Nature Geoscience, 4(9), 581–582.

[292] Gando, A., Dwyer, D. A., McKeown, R. D., & Zhang, C. (2011). Partial radiogenic heat model for Earth revealed by geoneutrino measurements. Nature Geoscience, 4(9), 647–651.

[293]

[294] Lay, T., Hernlund, J., & Buffett, B. A. (2008). Core–mantle boundary heat flow. Nature Geoscience, 1(1), 25-32.

[295] Pease, V., Percival, J., Smithies, H., Stevens, G., & Van Kranendonk, M. (2008). When did plate tectonics begin? Evidence from the orogenic record. When did plate tectonics begin on planet Earth, 199–208.

[296] Stern, R. J. (2008). Modern-style plate tectonics began in Neoproterozoic time: An alternative interpretation of Earth's tectonic history. When did plate tectonics begin on planet Earth, 265–280.

[297]

[298] Little, Fowler & Coulson 1990.

[299] Read & Watson 1975.

[300] Scalera & Lavecchia 2006.

[301] , .

[302] Turcotte & Schubert 2002, p. 5.

[303] Turcotte & Schubert 2002.

[304] Foulger 2010.

[305] Schmidt & Harbert 1998.

[306] Meissner 2002, p. 100.

[307] http://sideshow.jpl.nasa.gov/mbh/series.html

[308] , .

[309] Tanimoto & Lay 2000.

[310] Meyerhoff et al. 1996.

[311] , .

[312] Conrad & Lithgow-Bertelloni 2002.

[313] , , .

[314] , .

[315] Moore 1973.

[316] Bostrom 1971.

[317] Scoppola et al. 2006.

[318] Torsvik et al. 2010.

[319] ["Kinematics and dynamics of the East Pacific Rise linked to a stable, deep-mantle upwelling", Rowley et al, Science Advances 23 Dec 2016:Vol. 2, no. 12, e1601107,]

[320] Wegener 1929.

[321] Hughes 2001a.

[322] , .

[323] Runcorn 1956.

[324] Carey 1956.

[325] see for example the milestone paper of .

[326] , .

[327] Kious & Tilling 1996.

[328] Frankel 1987.

[329] Joly 1909.

[330] Thomson 1863.

[331] Wegener 1912.

[332] ; see also .

[333] ; see also , .

[334] , .

[335] Heezen 1960.

[336] Dietz 1961.

[337] Hess 1962.

[338] , .

[339] Vine & Matthews 1963.

[340] See summary in

[341] Wilson 1963.

[342] Wilson 1965.

[343] Wilson 1966.

[344] Morgan 1968.

[345] Le Pichon 1967.

[346] McKenzie & Parker 1967.

[347] Moss & Wilson 1998.

[348] Condie 1997.

[349] Lliboutry 2000.

[350] Plate Tectonics May Have Begun a Billion Years After Earth's Birth Pappas, S LiveScience report of PNAS research 21 Sept 2017 https://www.livescience.com/60478-plate-tectonics-gets-new-age.html

[351] Torsvik 2008.

[352] Butler 1992.

[353] http://tools.wmflabs.org/timescale/?Ma=2000–1800

[354] http://tools.wmflabs.org/timescale/?Ma=1500–1300

[355] http://tools.wmflabs.org/timescale/?Ma=600

[356] Valencia, O'Connell & Sasselov 2007.

[357] http://tools.wmflabs.org/timescale/?Ma=500–750

[358] http://tools.wmflabs.org/timescale/?Ma=1,200

[359] Kasting 1988.

[360] Sleep 1994.

[361] Zhong & Zuber 2001.

[362] Andrews-Hanna, Zuber & Banerdt 2008.

[363] Harrison 2000.

[364] Soderblom et al. 2007.

[365] C. O'Neill, A. Lenardic Geological consequences of super-sized Earths http://www.agu.org/pubs/crossref/2007/2007GL030598.shtml *Geophysical Research Letters* 34: L19204

[366] https://web.archive.org/web/20100817084001/http://www.geo.arizona.edu/Paleomag/book/chap10.pdf

[367] http://www.geo.arizona.edu/Paleomag/book/chap10.pdf

[368] https://books.google.com/books?id=QfhGuFwi0DgC&pg=PA6

[369] https://books.google.com/books?id=lp_n-Ng-hhoC&pg=PA203

[370] http://www.mantleplumes.org/WebDocuments/Hess1962.pdf

[371] http://pubs.usgs.gov/gip/dynamic/historical.html

[372] http://pubs.usgs.gov/gip/dynamic/dynamic.html

[373] https://books.google.com/books?id=llos3mwQkxYC&pg=PR9

[374] https://books.google.com/books?id=C7rny3qA6RMC&pg=PA133

[375] https://books.google.com/books?id=IVIBPWyT7BkC

[376] https://web.archive.org/web/20080216095932/http://www.gl.rhul.ac.uk/searg/publications/books/biogeography/books/biogeog_pdfs/Moss_Wilson.pdf

[377] //www.worldcat.org/oclc/317775677

[378] https://web.archive.org/web/20100124060304/http://geoinfo.amu.edu.pl/wpk/pe/a/harbbook/other/contents.html

[379] http://geoinfo.amu.edu.pl/wpk/pe/a/harbbook/other/contents.html

[380] https://web.archive.org/web/20110723122146/http://www.geodynamics.no/indexOld.htm

[381] http://www.geodynamics.no/indexOld.htm

[382] https://web.archive.org/web/20110723121839/http://www.geodynamics.no/guest/Torsvik_SteinbergerGraasteinen12.pdf

[383] http://www.geodynamics.no/guest/Torsvik_SteinbergerGraasteinen12.pdf

[384] http://adsabs.harvard.edu/abs/2008Natur.453.1212A

[385] //doi.org/10.1038/nature07011

[386] //www.ncbi.nlm.nih.gov/pubmed/18580944

[387] http://adsabs.harvard.edu/abs/1971Natur.234..536B

[388] //doi.org/10.1038/234536a0

[389] http://adsabs.harvard.edu/abs/1999Sci...284..794C

[390] //doi.org/10.1126/science.284.5415.794

[391] //www.ncbi.nlm.nih.gov/pubmed/10221909

[392] //www.ncbi.nlm.nih.gov/pmc/articles/PMC1250232

[393] http://adsabs.harvard.edu/abs/2005PNAS..10214970C

[394] //doi.org/10.1073/pnas.0507469102

[395] //www.ncbi.nlm.nih.gov/pubmed/16217034

[396] https://web.archive.org/web/20090920140431/http://www.soest.hawaii.edu/GG/FACULTY/conrad/resproj/forces/forces.html

[397] http://adsabs.harvard.edu/abs/2002Sci...298..207C

[398] //doi.org/10.1126/science.1074161

[399] //www.ncbi.nlm.nih.gov/pubmed/12364804

[400] http://www.soest.hawaii.edu/GG/FACULTY/conrad/resproj/forces/forces.html

[401] http://adsabs.harvard.edu/abs/1961Natur.190..854D

[402] //doi.org/10.1038/190854a0

[403] http://adsabs.harvard.edu/abs/1991Tectp.196...23V

[404] //doi.org/10.1016/0040-1951%2891%2990288-4

[405] https://web.archive.org/web/20130420210655/http://igitur-archive.library.uu.nl/geo/2012-0411-200448/UUindex.html

[406] http://igitur-archive.library.uu.nl/geo/2012-0411-200448/UUindex.html

[407] //doi.org/10.1017/S0007087400016551

[408] //www.jstor.org/stable/4025726

[409] //doi.org/10.1126/science.287.5453.547a

[410] http://adsabs.harvard.edu/abs/1960SciAm.203d..98H

[411] //doi.org/10.1038/scientificamerican1060-98

[412] http://adsabs.harvard.edu/abs/1966DSROA..13..427H

[413] //doi.org/10.1016/0011-7471%2866%2991078-3

[414] http://earthobservatory.nasa.gov/Features/Wegener/wegener_2.php

[415] http://earthobservatory.nasa.gov/Features/Wegener/wegener_4.php

[416] http://adsabs.harvard.edu/abs/1988Icar...74..472K

[417] //doi.org/10.1016/0019-1035%2888%2990116-9

[418] //www.ncbi.nlm.nih.gov/pubmed/11538226

[419] https://web.archive.org/web/20070926055030/http://www.tos.org/oceanography/issues/issue_archive/issue_pdfs/8_1/8.1_korgen.pdf

[420] //doi.org/10.5670/oceanog.1995.29

[421] http://www.tos.org/oceanography/issues/issue_archive/issue_pdfs/8_1/8.1_korgen.pdf

[422] http://www.columbia.edu/cu/alumni/Magazine/Winter2001/ewing.html

[423] http://news.nationalgeographic.com/news/2006/01/0124_060124_moon.html

[424] http://adsabs.harvard.edu/abs/1961GSAB...72.1259M

[425] //doi.org/10.1130/0016-7606%281961%2972%5B1259%3AMSOTWC%5D2.0.CO%3B2

[426] //www.worldcat.org/issn/0016-7606

[427] http://adsabs.harvard.edu/abs/1967Natur.216.1276M

[428] //doi.org/10.1038/2161276a0

[429] http://adsabs.harvard.edu/abs/1973Geo.....1...99M

[430] //doi.org/10.1130/0091-7613%281973%291%3C99%3AWTLATD%3E2.0.CO%3B2

[431] //www.worldcat.org/issn/0091-7613

[432] http://www.mantleplumes.org/WebDocuments/Morgan1968.pdf

[433] http://adsabs.harvard.edu/abs/1968JGR....73.1959M

[434] //doi.org/10.1029/JB073i006p01959

[435] http://adsabs.harvard.edu/abs/1968JGR....73.3661L

[436] //doi.org/10.1029/JB073i012p03661

[437] https://web.archive.org/web/20101221023449/http://science.org.au/fellows/memoirs/carey.html

[438] http://www.science.org.au/fellows/memoirs/carey.html

[439] http://adsabs.harvard.edu/abs/1961GSAB...72.1267R

[440] //doi.org/10.1130/0016-7606%281961%2972%5B1267%3AMSOTWC%5D2.0.CO%3B2

[441] http://adsabs.harvard.edu/abs/1965RSPTA.258....1R

[442] //doi.org/10.1098/rsta.1965.0016

[443] //doi.org/10.4401/ag-4406

[444] http://adsabs.harvard.edu/abs/2006GSAB..118..199S

[445] //doi.org/10.1130/B25734.1

[446] http://www.stephan-mueller-spec-publ-ser.net/2/171/2002/smsps-2-171-2002.pdf

[447] //doi.org/10.5194/smsps-2-171-2002

[448] http://www.es.ucsc.edu/~rcoe/eart290C/Additional%20Papers/Sleep_MartianPlateTectonics_JGR94.pdf

[449] http://adsabs.harvard.edu/abs/1994JGR....99.5639S

[450] //doi.org/10.1029/94JE00216

[451] http://adsabs.harvard.edu/abs/2007P&SS...55.2015S

[452] //doi.org/10.1016/j.pss.2007.04.015

[453] http://szseminar.asu.edu/readings/Rev_Geophys_Spence_1987.pdf

[454] http://adsabs.harvard.edu/abs/1987RvGeo..25...55S

[455] //doi.org/10.1029/RG025i001p00055

[456] https://web.archive.org/web/20070926055021/http://www.tos.org/oceanography/issues/issue_archive/issue_pdfs/16_3/16.3_spiess.pdf

[457] //doi.org/10.5670/oceanog.2003.30

[458] http://www.tos.org/oceanography/issues/issue_archive/issue_pdfs/16_3/16.3_spiess.pdf

[459] //www.ncbi.nlm.nih.gov/pmc/articles/PMC34063

[460] http://adsabs.harvard.edu/abs/2000PNAS...9712409T

[461] //doi.org/10.1073/pnas.210382197

[462] //www.ncbi.nlm.nih.gov/pubmed/11035784

[463] http://www.tandfonline.com/doi/abs/10.1080/14786446308643410

[464] //doi.org/10.1080/14786446308643410

[465] https://web.archive.org/web/20110516165855/http://www.gps.caltech.edu/~gurnis/Papers/2010_Torsvik_etal_EPSL.pdf

[466] http://adsabs.harvard.edu/abs/2010E&PSL.291..106T

[467] //doi.org/10.1016/j.epsl.2009.12.055

[468] http://www.gps.caltech.edu/~gurnis/Papers/2010_Torsvik_etal_EPSL.pdf

[469] //arxiv.org/abs/0710.0699

[470] http://adsabs.harvard.edu/abs/2007ApJ...670L..45V

[471] //doi.org/10.1086/524012

[472] http://adsabs.harvard.edu/abs/1963Natur.199..947V

[473] //doi.org/10.1038/199947a0

[474] https://web.archive.org/web/20100705081509/http://epic.awi.de/Publications/Polarforsch2005_1_3.pdf

[475] http://epic.awi.de/Publications/Polarforsch2005_1_3.pdf

[476] http://adsabs.harvard.edu/abs/1989JGR....94.7685W

[477] //doi.org/10.1029/JB094iB06p07685

[478] http://adsabs.harvard.edu/abs/1963Natur.198..925T

[479] //doi.org/10.1038/198925a0

[480] https://web.archive.org/web/20100806140625/http://www.rpi.edu/~mccafr/plates/reading/wilson_1965.pdf

[481] http://adsabs.harvard.edu/abs/1965Natur.207..343W

[482] //doi.org/10.1038/207343a0

[483] http://www.rpi.edu/~mccafr/plates/reading/wilson_1965.pdf

[484] http://www.geology.cwu.edu/facstaff/huerta/g501/pdf/Wilson1966.pdf

[485] http://adsabs.harvard.edu/abs/1966Natur.211..676W

[486] //doi.org/10.1038/211676a0

[487] https://www.webcitation.org/65EJD0m8X?url=http://hypertextbook.com/facts/ZhenHuang.shtml

[488] http://hypertextbook.com/facts/ZhenHuang.shtml

[489] http://adsabs.harvard.edu/abs/2002ESRv...59..125Z

[490] //doi.org/10.1016/S0012-8252%2802%2900073-9

[491] http://adsabs.harvard.edu/abs/2004ESRv...67...91Z

[492] //doi.org/10.1016/j.earscirev.2004.02.003

[493] http://www-geodyn.mit.edu/mars.deg1.pdf

[494] http://adsabs.harvard.edu/abs/2001E&PSL.189...75Z

[495] //doi.org/10.1016/S0012-821X%2801%2900345-4

[496] http://pubs.usgs.gov/publications/text/understanding.html

[497] http://www.tectonic-forces.org

[498] http://peterbird.name/publications/2003_PB2002/2003_PB2002.htm

[499] http://snobear.colorado.edu/Markw/Mountains/03/week3.html

[500] http://www.geology.wisc.edu/~chuck/MORVEL/

501 http://www.bbc.co.uk/programmes/b008q0sp

502 https://www.youtube.com/watch?v=6EdsBabSZ4g

503 http://www.ucmp.berkeley.edu/geology/tectonics.html

504 http://qz.com/577842/scientists-have-used-groundbreaking-technology-to-figure-out-how-the-earth-looked-a-billion-years-ago/

505 Skinner, B.J. & Porter, S.C.: *Physical Geology*, page 17, chapt. *The Earth: Inside and Out*, 1987, John Wiley & Sons,

506 Daly, R. (1940) *Strength and structure of the Earth*. New York: Prentice-Hall.

507 Donald L. Turcotte, Gerald Schubert, Geodynamics. Cambridge University Press, 25 mar 2002 - 456

508 Nixon, P.H. (1987) *Mantle xenoliths* J. Wiley & Sons, 844 p.

509 http://www.windows.ucar.edu/cgi-bin/tour_def/earth/interior/earths_crust.html

510 http//www.geolsoc.org.uk

511 https://code.google.com/p/open-geomorphometry-project/

512 //tools.wmflabs.org/geohack/geohack.php?pagename=Extreme_points_of_Earth&params=83_40_N_29_50_W_type:landmark&title=northernmost+point+on+land

513 //tools.wmflabs.org/geohack/geohack.php?pagename=Extreme_points_of_Earth&params=83_38_N_32_40_W_type:landmark&title=Cape+Morris+Jesup

514 //tools.wmflabs.org/geohack/geohack.php?pagename=Extreme_points_of_Earth&params=83_S_59_W_

515 //tools.wmflabs.org/geohack/geohack.php?pagename=Extreme_points_of_Earth&params=84_30_S_150_0_W_type:landmark&title=southernmost+point+of+ocean

516 Gould Coast http://geonames.usgs.gov/apex/f?p=gnispq:5:0::NO::P5_ANTAR_ID:5881 US Geographic Survey.

517 A 1995 realignment of the International Date Line http://www.trussel.com/kir/dateline.htm moved all of Kiribati to the Asian side of the Date Line, causing Caroline Island to be the easternmost. However, if the previous Date Line were followed, the easternmost point would be Tafahi Niuatoputapu, in the Tonga Islands chain.

518 The elevation given here was established by a GPS survey in February 2016. The survey was carried out by a team from the French Research Institute for Development, working in cooperation with the Ecuadorian Military Geographic Institute.<ref>

519 (includes description and photos of Aucanquilcha summit road and mine)

520 //tools.wmflabs.org/geohack/geohack.php?pagename=Extreme_points_of_Earth&params=21.214_S_68.475_W_

521 //tools.wmflabs.org/geohack/geohack.php?pagename=Extreme_points_of_Earth&params=32_49_30_N_81_03_45_E_type:waterbody&title=Ating+Ho+%28source%29

522

523 //tools.wmflabs.org/geohack/geohack.php?pagename=Extreme_points_of_Earth&params=28.736667_N_88.386944_W_

524 //tools.wmflabs.org/geohack/geohack.php?pagename=Extreme_points_of_Earth&params=46_17_N_86_40_E_&title=Continental+Pole+of+Inaccessibility

525 //tools.wmflabs.org/geohack/geohack.php?pagename=Extreme_points_of_Earth&params=46_15_N_86_50_E_&title=Suluk

526 //tools.wmflabs.org/geohack/geohack.php?pagename=Extreme_points_of_Earth&params=44_17_N_82_11_E_&title=EPIA1.1

527 //tools.wmflabs.org/geohack/geohack.php?pagename=Extreme_points_of_Earth&params=44_29_N_82_19_E_&title=EPIA1.2

528 //tools.wmflabs.org/geohack/geohack.php?pagename=Extreme_points_of_Earth&params=45_17_N_88_08_E_&title=EPIA2.1

529 //tools.wmflabs.org/geohack/geohack.php?pagename=Extreme_points_of_Earth&params=45_28_N_88_14_E_&title=EPIA2.2

530 //tools.wmflabs.org/geohack/geohack.php?pagename=Extreme_points_of_Earth&params=5.65_N_26.17_E_&title=Continental+Pole+of+Inaccessibility+of+Africa

531 //tools.wmflabs.org/geohack/geohack.php?pagename=Extreme_points_of_Earth&params=23_2_S_132_10_E_&title=Australian+Pole+of+Inaccessibility

[532] Centre of Australia, States and Territories http://www.ga.gov.au/education/facts/dimensions/centre.htm , Geoscience Australia

[533]

[534] //tools.wmflabs.org/geohack/geohack.php?pagename=Extreme_points_of_Earth&params=43.36_N_101.97_W_&title=Pole+of+Inaccessibility+North+America

[535]

[536] //tools.wmflabs.org/geohack/geohack.php?pagename=Extreme_points_of_Earth&params=48_52.6_S_123_23.6_W_type:landmark&title=Point+Nemo

[537] //tools.wmflabs.org/geohack/geohack.php?pagename=Extreme_points_of_Earth&params=54_26_S_3_24_E_type:landmark&title=most+remote+island

[538] Draft Logic – Google Maps Distance Calculator http://www.daftlogic.com/projects-google-maps-distance-calculator.htm, accessed 4 September 2011

[539] //tools.wmflabs.org/geohack/geohack.php?pagename=Extreme_points_of_Earth&params=48_24_53_N_4_47_44_W_

[540] //tools.wmflabs.org/geohack/geohack.php?pagename=Extreme_points_of_Earth&params=48_24_53_N_140_6_3_E_

[541] //tools.wmflabs.org/geohack/geohack.php?pagename=Extreme_points_of_Earth&params=18_39_12_N_110_15_9_E_

[542] //tools.wmflabs.org/geohack/geohack.php?pagename=Extreme_points_of_Earth&params=18_39_12_N_103_42_6_W_

[543] //tools.wmflabs.org/geohack/geohack.php?pagename=Extreme_points_of_Earth&params=76_13_6_N_99_1_30_E_

[544] //tools.wmflabs.org/geohack/geohack.php?pagename=Extreme_points_of_Earth&params=7_53_24_N_99_1_30_E_

[545] //tools.wmflabs.org/geohack/geohack.php?pagename=Extreme_points_of_Earth&params=32_19_0_N_20_12_0_E_

[546] //tools.wmflabs.org/geohack/geohack.php?pagename=Extreme_points_of_Earth&params=34_41_30_S_20_12_0_E_

[547] //tools.wmflabs.org/geohack/geohack.php?pagename=Extreme_points_of_Earth&params=11_30_30_N_70_2_0_W_

[548] //tools.wmflabs.org/geohack/geohack.php?pagename=Extreme_points_of_Earth&params=52_33_30_S_70_2_0_W_

[549] //tools.wmflabs.org/geohack/geohack.php?pagename=Extreme_points_of_Earth&params=68_21_0_N_97_52_30_W_

[550] //tools.wmflabs.org/geohack/geohack.php?pagename=Extreme_points_of_Earth&params=16_1_0_N_97_52_30_W_

[551] //tools.wmflabs.org/geohack/geohack.php?pagename=Extreme_points_of_Earth&params=66_23_45_N_34_45_45_W_

[552] //tools.wmflabs.org/geohack/geohack.php?pagename=Extreme_points_of_Earth&params=77_37_0_S_34_45_45_W_

[553] //tools.wmflabs.org/geohack/geohack.php?pagename=Extreme_points_of_Earth&params=64_45_0_N_172_8_30_W_

[554] //tools.wmflabs.org/geohack/geohack.php?pagename=Extreme_points_of_Earth&params=78_20_0_S_172_8_30_W_

[555] //tools.wmflabs.org/geohack/geohack.php?pagename=Extreme_points_of_Earth&params=5_2_51.59_N_9_7_23.26_W_

[556] //tools.wmflabs.org/geohack/geohack.php?pagename=Extreme_points_of_Earth&params=28_17_7.68_N_121_38_17.31_E_

[557] (Map from gcmap) http//www.gcmap.com

[558] //tools.wmflabs.org/geohack/geohack.php?pagename=Extreme_points_of_Earth&params=25_25_N_66_25_E_

[559] //tools.wmflabs.org/geohack/geohack.php?pagename=Extreme_points_of_Earth&params=59_38_N_163_24_E_

[560] (Map from gcmap) http://www.gcmap.com/mapui?P=25%B025%27N+66%B025%27E+-+59%B038%27S+16%B036%27W+-+25%B025%27S+113%B035%27W+-+59%B038%27N+163%B024%27E%0D%0A

[561] //tools.wmflabs.org/geohack/geohack.php?pagename=Extreme_points_of_Earth&params=25_35_N_58_22_E_

[562] //tools.wmflabs.org/geohack/geohack.php?pagename=Extreme_points_of_Earth&params=17_57_N_101_57_W_

[563] (Map from gcmap) http://www.gcmap.com/mapui?P=25%B035%27N+58%B022%27E+-+17%B057%27S+78%B003%27E+-+25%B035%27S+121%B038%27W+-+17%B057%27N+101%B057%27W

[564] //tools.wmflabs.org/geohack/geohack.php?pagename=Extreme_points_of_Earth&params=46_37_S_168_59_E_

[565] //tools.wmflabs.org/geohack/geohack.php?pagename=Extreme_points_of_Earth&params=52_09_N_6_34_W_

[566] (Map from gcmap) http://www.gcmap.com/mapui?P=46%B037%27S+168%B059%27E+-+52%B009%27S+173%B026%27E+-+46%B037%27N+11%B001%27W+-+52%B009%27N+6%B034%27W%0D%0A

[567] ὕδωρ http://www.perseus.tufts.edu/hopper/text?doc=Perseus%3Atext%3A1999.04.0057%3Aentry%3Du%28%2Fdwr, Henry George Liddell, Robert Scott, *A Greek-English Lexicon*, on Perseus

[568] σφαῖρα http://www.perseus.tufts.edu/hopper/text?doc=Perseus%3Atext%3A1999.04.0057%3Aentry%3Dsfai%3Dra^, Henry George Liddell, Robert Scott, *A Greek-English Lexicon*, on Perseus

[569]

[570] According to planetary geologist, Ronald Greeley, "Water is very common in the outer solar system." Europa holds more water than earth's oceans.

[571] http://capp.water.usgs.gov/GIP/gw_gip/index.html

[572] Lide, David R. *Handbook of Chemistry and Physics*. Boca Raton, FL; CRC, 1996· 14–17

[573] Wallace, John M. and Peter V. Hobbs. *Atmospheric Science: An Introductory Survey* http://cup.aos.wisc.edu/453/2016/readings/Atmospheric_Science-Wallace_Hobbs.pdf. Elsevier. Second Edition, 2006. Chapter 1

[574] Source for figures: Carbon dioxide, NOAA Earth System Research Laboratory http://www.esrl.noaa.gov/gmd/ccgg/trends/#mlo, (updated 2013-03). Methane, IPCC TAR table 6.1 http://www.grida.no/climate/ipcc_tar/wg1/221.htm#tab61 , (updated to 1998). The NASA total was 17 ppmv over 100%, and was increased here by 15 ppmv. To normalize, N_2 should be reduced by about 25 ppmv and O_2 by about 7 ppmv.

[575] Ahrens, C. Donald. Essentials of Meteorology. Published by Thomson Brooks/Cole, 2005.

[576] Geometric altitude vs. temperature, pressure, density, and the speed of sound derived from the 1962 U.S. Standard Atmosphere. http://www.centennialofflight.gov/essay/Theories_of_Flight/atmosphere/TH1G1.htm

[577] Lutgens, Frederick K. and Edward J. Tarbuck (1995) *The Atmosphere*, Prentice Hall, 6th ed., pp. 14–17,

[578] B. Windley: *The Evolving Continents*. Wiley Press, New York 1984

[579] J. Schopf: *Earth's Earliest Biosphere: Its Origin and Evolution*. Princeton University Press, Princeton, N.J., 1983

[580] Christopher R. Scotese, Back to Earth History : Summary Chart for the Precambrian http://www.scotese.com/precamb_chart.htm, Paleomar Project

[581] Peter Ward:http://www.nap.edu/catalog.php?record_id=11630 Out of Thin Air: Dinosaurs, Birds, and Earth's Ancient Atmosphere

[582] Starting from http://www.merriam-webster.com/dictionary/pollution Pollution – Definition from the Merriam-Webster Online Dictionary

[583] https://earth.nullschool.net/

[584] //en.wikipedia.org/w/index.php?title=Template:Weather&action=edit

[585] Merriam-Webster Dictionary. Weather. http://www.merriam-webster.com/dictionary/weather Retrieved on 27 June 2008.

[586] Glossary of Meteorology. Hydrosphere. http://amsglossary.allenpress.com/glossary/search?p=1&query=hydrosphere&submit=Search Retrieved on 27 June 2008.

[587] Glossary of Meteorology. Troposphere. http://amsglossary.allenpress.com/glossary/browse?s=t&p=51 Retrieved on 27 June 2008.

[588] NASA. World Book at NASA: Weather. http://www.nasa.gov/worldbook/weather_worldbook. html Archived copy https://www.webcitation.org/6F17zeqFy?url=http://www.nasa.gov/ worldbook/weather_worldbook.html at WebCite (10 March 2013). Retrieved on 27 June 2008.

[589] John P. Stimac. http://www.ux1.eiu.edu/~cfjps/1400/pressure_wind.htmlAir pressure and wind. Retrieved on 8 May 2008.

[590] Carlyle H. Wash, Stacey H. Heikkinen, Chi-Sann Liou, and Wendell A. Nuss. A Rapid Cyclogenesis Event during GALE IOP 9. http://ams.allenpress.com/perlserv/?request=get-abstract&doi=10.1175%2F1520-0493(1990)118%3C0234%3AARCEDG%3E2.0.CO%3B2 Retrieved on 28 June 2008.

[591] Windows to the Universe. Earth's Tilt Is the Reason for the Seasons! http://www.windows. ucar.edu/tour/link=/earth/climate/cli_seasons.html Retrieved on 28 June 2008.

[592] Milankovitch, Milutin. Canon of Insolation and the Ice Age Problem. Zavod za Udžbenike i Nastavna Sredstva: Belgrade, 1941.

[593] Ron W. Przybylinski. The Concept of Frontogenesis and its Application to Winter Weather Forecasting. http://www.crh.noaa.gov/lsx/science/pdfppt/ron.ppt Retrieved on 28 June 2008.

[594] Michel Moncuquet. Relation between density and temperature. http://www.lesia.obspm.fr/ ~moncuque/theseweb/tempioweb/node6.html Retrieved on 28 June 2008.

[595] Encyclopedia of Earth. Wind. http://www.eoearth.org/article/Wind Retrieved on 28 June 2008.

[596] Spencer Weart. The Discovery of Global Warming. http://www.aip.org/history/climate/chaos. htm Retrieved on 28 June 2008.

[597] NASA. NASA Mission Finds New Clues to Guide Search for Life on Mars. http://www.nasa. gov/mission_pages/odyssey/odyssey-20080320.html Retrieved on 28 June 2008.

[598] West Gulf River Forecast Center. Glossary of Hydrologic Terms: E http://www.srh.noaa.gov/ wgrfc/resources/glossary/e.html Retrieved on 28 June 2008.

[599] James P. Delgado. Relics of the Kamikaze. http://www.archaeology.org/0301/etc/kamikaze. html Retrieved on 28 June 2008.

[600] Mike Strong. Fort Caroline National Memorial. http://www.mikestrong.com/fortcar/ Retrieved on 28 June 2008.

[601] Anthony E. Ladd, John Marszalek, and Duane A. Gill. The Other Diaspora: New Orleans Student Evacuation Impacts and Responses Surrounding Hurricane Katrina. http://www.ssrc. msstate.edu/katrina/publications/katrinastudentsummary.pdf Retrieved on 29 March 2008.

[602] " *Famine in Scotland: The 'Ill Years' of the 1690s https://books.google.com/books?id= RiLjHZdt-sMC&pg=PA21*". Karen J. Cullen (2010). Edinburgh University Press. p.21.

[603] Eric D. Craft. *An Economic History of Weather Forecasting*. http://eh.net/encyclopedia/article/ craft.weather.forcasting.history Retrieved on 15 April 2007.

[604] NASA. Weather Forecasting Through the Ages. http://earthobservatory.nasa.gov/Library/ WxForecasting/wx2.html Retrieved on 25 May 2008.

[605] Weather Doctor. Applying The Barometer To Weather Watching. http://www.islandnet.com/ ~see/weather/eyes/barometer3.htm Retrieved on 25 May 2008.

[606] Mark Moore. Field Forecasting: A Short Summary. http://www.nwac.us/education_resources/ Field_forecasting.pdf Retrieved on 25 May 2008.

[607] Klaus Weickmann, Jeff Whitaker, Andres Roubicek and Catherine Smith. The Use of Ensemble Forecasts to Produce Improved Medium Range (3–15 days) Weather Forecasts. http://www.cdc. noaa.gov/spotlight/12012001/ Retrieved on 16 February 2007.

[608] Todd Kimberlain. Tropical cyclone motion and intensity talk (June 2007). http://www.wpc. ncep.noaa.gov/research/TropicalTalk.ppt Retrieved on 21 July 2007.

[609] Richard J. Pasch, Mike Fiorino, and Chris Landsea. TPC/NHC'S REVIEW OF THE NCEP PRODUCTION SUITE FOR 2006. http://www.emc.ncep.noaa.gov/research/NCEP-EMCModelReview2006/TPC-NCEP2006.ppt Retrieved on 5 May 2008.

[610] National Weather Service. National Weather Service Mission Statement. http://www.weather. gov/mission.shtml Retrieved on 25 May 2008.

[611] National Meteorological Service of Slovenia http//www.meteo.si

[612] Blair Fannin. Dry weather conditions continue for Texas. http://southwestfarmpress.com/news/ 061406-Texas-weather/ Retrieved on 26 May 2008.

[613] Dr. Terry Mader. Drought Corn Silage. http://beef.unl.edu/stories/200004030.shtml Retrieved on 26 May 2008.

[614] Kathryn C. Taylor. Peach Orchard Establishment and Young Tree Care. http://pubs.caes.uga.edu/caespubs/pubcd/C877.htm Retrieved on 26 May 2008.

[615] Associated Press. After Freeze, Counting Losses to Orange Crop. https://query.nytimes.com/gst/fullpage.html?res=9D0CE5DB1E30F937A25752C0A967958260 Retrieved on 26 May 2008.

[616] The New York Times. FUTURES/OPTIONS; Cold Weather Brings Surge In Prices of Heating Fuels. https://query.nytimes.com/gst/fullpage.html?res=9F0CE7D9123AF935A15751C0A965958260 Retrieved on 25 May 2008.

[617] BBC. Heatwave causes electricity surge. http://news.bbc.co.uk/1/hi/uk/5212724.stm Retrieved on 25 May 2008.

[618] Toronto Catholic Schools. The Seven Key Messages of the Energy Drill Program. http://www.tcdsb.org/environment/energydrill/EDSP_KeyMessages_FINAL.pdf Retrieved on 25 May 2008.

[619] American Meteorological Society http://www.ametsoc.org/policy/wxmod98.html

[620] Intergovernmental Panel on Climate Change http://www.grida.no/climate/ipcc/regional/226.htm#extreme

[621] Intergovernmental Panel on Climate Change http://www.grida.no/climate/ipcc/regional/503.htm#overview

[622] Global Measured Extremes of Temperature and Precipitation. http://www.ncdc.noaa.gov/oa/climate/globalextremes.html National Climatic Data Center. Retrieved on 21 June 2007.

[623] Glenn Elert. Hottest Temperature on Earth. http://hypertextbook.com/facts/2000/MichaelLevin.shtml Retrieved on 28 June 2008.

[624] Glenn Elert. Coldest Temperature On Earth. http://hypertextbook.com/facts/2000/YongLiLiang.shtml Retrieved on 28 June 2008.

[625] Canadian Climate Normals 1971–2000 – Eureka http//www.climate.weatheroffice.ec.gc.ca

[626] Jet Propulsion Laboratory. OVERVIEW – Climate: The Spherical Shape of the Earth: Climatic Zones. http://sealevel.jpl.nasa.gov/overview/climate-climatic.html Retrieved on 28 June 2008.

[627] Anne Minard. Jupiter's "Jet Stream" Heated by Surface, Not Sun. http://news.nationalgeographic.com/news/2008/01/080123-jupiter-jets.html Retrieved on 28 June 2008.

[628] ESA: Cassini–Huygens. The jet stream of Titan. http://www.esa.int/esaMI/Cassini-Huygens/SEMQO5SMTWE_0.html Retrieved on 28 June 2008.

[629] Georgia State University. The Environment of Venus. http://hyperphysics.phy-astr.gsu.edu/hbase/Solar/venusenv.html Retrieved on 28 June 2008.

[630] Bill Christensen. Shock to the (Solar) System: Coronal Mass Ejection Tracked to [[Saturn http://www.space.com/businesstechnology/technology/technovel_shock_041105.html].] Retrieved on 28 June 2008.

[631] AlaskaReport. What Causes the Aurora Borealis? http://alaskareport.com/science10043.htm Retrieved on 28 June 2008.

[632] Rodney Viereck. Space Weather: What is it? How Will it Affect You? http://asp.colorado.edu/~reu/summer-2007/presentations/SW_Intro_Viereck.ppt Retrieved on 28 June 2008.

[633] //en.wikipedia.org/w/index.php?title=Template:Atmospheric_sciences&action=edit

[634] //en.wikipedia.org/w/index.php?title=Template:Weather&action=edit

[635] AR4 SYR Synthesis Report Annexes http://www.ipcc.ch/publications_and_data/ar4/syr/en/annexes.html. Ipcc.ch. Retrieved on 2011-06-28.

[636] Intergovernmental Panel on Climate Change. Appendix I: Glossary. http://www.grida.no/climate/ipcc_tar/wg1/518.htm Retrieved on 2007-06-01.

[637] National Weather Service Office Tucson, Arizona. Main page. http://www.wrh.noaa.gov/twc/ Retrieved on 2007-06-01.

[638] Stefan Rahmstorf The Thermohaline Ocean Circulation: A Brief Fact Sheet. http://www.pik-potsdam.de/~stefan/thc_fact_sheet.html Retrieved on 2008-05-02.

[639] Gertjan de Werk and Karel Mulder. Heat Absorption Cooling For Sustainable Air Conditioning of Households. http://www.enhr2007rotterdam.nl/documents/W15_paper_DeWerk_Mulder.pdf Retrieved on 2008-05-02.

[640] United States National Arboretum. USDA Plant Hardiness Zone Map. http://www.usna.usda.gov/Hardzone/ushzmap.html Retrieved on 2008-03-09

[641] Robert E. Davis, L. Sitka, D. M. Hondula, S. Gawtry, D. Knight, T. Lee, and J. Stenger. J1.10 A preliminary back-trajectory and air mass climatology for the Shenandoah Valley (Formerly J3.16 for Applied Climatology). http://ams.confex.com/ams/pdfpapers/118516.pdf Retrieved on 2008-05-21.

[642] http://upload.wikimedia.org/wikipedia/commons/5/5b/BlueMarble_monthlies_SMIL.svg

[643] Susan Woodward. Tropical Broadleaf Evergreen Forest: The Rainforest. http://www.radford.edu/~swoodwar/CLASSES/GEOG235/biomes/rainforest/rainfrst.html Retrieved on 2008-03-14.

[644] International Committee of the Third Workshop on Monsoons. The Global Monsoon System: Research and Forecast. http://caos.iisc.ernet.in/faculty/bng/IWM-III-BNG_overview.pdf Retrieved on 2008-03-16.

[645] Susan Woodward. Tropical Savannas. http://www.radford.edu/~swoodwar/CLASSES/GEOG235/biomes/savanna/savanna.html Retrieved on 2008-03-16.

[646] Michael Ritter. Humid Subtropical Climate. http://www.uwsp.edu/geo/faculty/ritter/geog101/textbook/climate_systems/humid_subtropical.html Retrieved on 2008-03-16.

[647] Climate. Oceanic Climate. http://www.meteorologyclimate.com/Oceanic-climate.htm Retrieved on 2008-04-15.

[648] Michael Ritter. Mediterranean or Dry Summer Subtropical Climate. http://www.uwsp.edu/geo/faculty/ritter/geog101/textbook/climate_systems/mediterranean.html Retrieved on 2008-04-15.

[649] Blue Planet Biomes. Steppe Climate. http://www.blueplanetbiomes.org/steppe_climate_page.htm Retrieved on 2008-04-15.

[650] Michael Ritter. Subarctic Climate. http://www.uwsp.edu/geo/faculty/ritter/geog101/textbook/climate_systems/subarctic.html Retrieved on 2008-04-16.

[651] Susan Woodward. Taiga or Boreal Forest. http://www.radford.edu/~swoodwar/CLASSES/GEOG235/biomes/taiga/taiga.html Retrieved on 2008-06-06.

[652] Michael Ritter. Ice Cap Climate. http://www.uwsp.edu/geo/faculty/ritter/geog101/textbook/climate_systems/icecap.html Retrieved on 2008-03-16.

[653] San Diego State University. Introduction to Arid Regions: A Self-Paced Tutorial. http://www-rohan.sdsu.edu/~batterso/port_arid/formation.html Retrieved on 2008-04-16.

[654] Glossary of Meteorology. Thornthwaite Moisture Index. http://amsglossary.allenpress.com/glossary/search?p=1&query=Thornthwaite&submit=Search Retrieved on 2008-05-21.

[655] Eric Green. Foundations of Expansive Clay Soil. https://web.archive.org/web/20080527223539/http://www.slabongrade.net/DesignSeminars/SeminarDownloads/Science_of_Expansive_Clay.pdf Retrieved on 2008-05-21.

[656] Istituto Agronomico per l'Otremare. 3 Land Resources. http://www.iao.florence.it/training/geomatics/Thies/Senegal_23linkedp6.htm Retrieved on 2008-05-21.

[657]

[658] http://data.giss.nasa.gov/gistemp/

[659] Spencer Weart. The Modern Temperature Trend. http://www.aip.org/history/climate/20ctrend.htm Retrieved on 2007-06-01.

[660] National Oceanic and Atmospheric Administration. NOAA Paleoclimatology. http://www.ncdc.noaa.gov/paleo/paleo.html Retrieved on 2007-06-01.

[661] Arctic Climatology and Meteorology. Climate change. http://nsidc.org/arcticmet/glossary/climate_change.html Retrieved on 2008-05-19.

[662] Illinois State Museum (2002). Ice Ages. http://www.museum.state.il.us/exhibits/ice_ages/ Retrieved on 2007-05-15.

[663] Eric Maisonnave. Climate Variability. http://www.cerfacs.fr/globc/research/variability/ Retrieved on 2008-05-02.

[664] Climateprediction.net. Modelling the climate. http://www.climateprediction.net/science/model-intro.php Retrieved on 2008-05-02.

[665] http//portal.iri.columbia.edu

[666] http://www.americanscientist.org/issues/feature/2012/4/the-study-of-climate-on-alien-worlds

[667] http://img.kb.dk/tidsskriftdk/pdf/gto/gto_0048-PDF/gto_0048_69887.pdf

[668] http://www.climate.gov

[669] http://www.ncdc.noaa.gov/sotc/

[670] https://climate.nasa.gov/

[671] https://web.archive.org/web/20020723014728/http://128.194.106.6/~baum/climate_modeling.html

[672] https://web.archive.org/web/20051125010853/http://climateapps2.oucs.ox.ac.uk/cpdnboinc/

[673] https://web.archive.org/web/20050902220920/http://www.atmosphere.mpg.de/enid/1442

[674] http://www.arctic.noaa.gov/climate.html

[675] http://www.beringclimate.noaa.gov/

[676] http://www.climate-charts.com/index.html

[677] http://www.everyspec.com/MIL-HDBK/MIL-HDBK-0300-0499/MIL_HDBK_310_1851/

[678] http://www.ipcc-data.org

[679] http://historicalclimatology.com

[680] http://www.globalclimatemonitor.org

[681] https://climatecharts.net/

[682] http://www.emdat.be/

[683] http://www.cop21paris.org/

[684] "Wolfram|Alpha Gravity in Kuala Lumpur", Wolfram Alpha, accessed May 2017 https://www.wolframalpha.com/input/?i=gravity+in+kuala+lampur

[685] Resolution of the 3rd CGPM (1901), page 70 (in cm/s^2). BIPM – Resolution of the 3rd CGPM http://www.bipm.org/en/CGPM/db/3/2/

[686] "Curious About Astronomy?" http://curious.astro.cornell.edu/question.php?number=310, Cornell University, retrieved June 2007

[687] "I feel 'lighter' when up a mountain but am I?" http://www.npl.co.uk/reference/faqs/i-feel-'lighter'-when-up-a-mountain-but-am-i-(faq-mass-and-density), National Physical Laboratory FAQ

[688] "The G's in the Machine" https://science.nasa.gov/science-news/science-at-nasa/2003/24jan_micro-g/, NASA, see "Editor's note #2"

[689] Gravitational Fields Widget as of Oct 25th, 2012 http://www.wolframalpha.com/widgets/view.jsp?id=d34e8683df527e3555153d979bcda9cf – WolframAlpha

[690] T.M. Yarwood and F. Castle, Physical and Mathematical Tables, revised edition, Macmillan and Co LTD, London and Basingstoke, Printed in Great Britain by The University Press, Glasgow, 1970, pp 22 & 23.

[691] International Gravity formula http://geophysics.ou.edu/solid_earth/notes/potential/igf.htm

[692] *Department of Defense World Geodetic System 1984 — Its Definition and Relationships with Local Geodetic Systems*,NIMA TR8350.2, 3rd ed., Tbl. 3.4, Eq. 4-1 http://earth-info.nga.mil/GandG/publications/tr8350.2/wgs84fin.pdf

[693] The rate of decrease is calculated by differentiating $g(r)$ with respect to r and evaluating at $r=r_{Earth}$.

[694] http://hyperphysics.phy-astr.gsu.edu/hbase/orbv.html

[695] http://www.csr.utexas.edu/grace/

[696] http://geodesy.curtin.edu.au/research/models/GGMplus/

[697] http://www.universetoday.com/116801/the-potsdam-gravity-potato-shows-earths-gravity-variations/

[698] Structure of the Earth http://scign.jpl.nasa.gov/learn/plate1.htm . Scign.jpl.nasa.gov. Retrieved on 2012-01-27.

[699] , pages 126–141

[700] http://tools.wmflabs.org/timescale/?Ma=3,450

[701] , p. 1.

[702] //en.wikipedia.org/w/index.php?title=Earth%27s_magnetic_field&action=edit

[703] //www.ncbi.nlm.nih.gov/pmc/articles/PMC40105

[704] http://adsabs.harvard.edu/abs/1996PNAS...93..646H

[705] //doi.org/10.1073/pnas.93.2.646

[706] //www.ncbi.nlm.nih.gov/pubmed/11607625

[707] //www.ncbi.nlm.nih.gov/pmc/articles/PMC58687

[708] http://adsabs.harvard.edu/abs/2001PNAS...9811085H

[709] //doi.org/10.1073/pnas.201393998

[710] //www.ncbi.nlm.nih.gov/pubmed/11562483

[711] http://geomag.usgs.gov/downloads/publications/pt_love0208.pdf

[712] http://adsabs.harvard.edu/abs/2008PhT....61b..31H

[713] //doi.org/10.1063/1.2883907

[714] http://adsabs.harvard.edu/abs/1992GeoRL..19.2151L

[715] //doi.org/10.1029/92GL02485

[716] http://www.newton.dep.anl.gov/askasci/gen99/gen99256.htm

[717] http://adsabs.harvard.edu/abs/1984GSAB...95..221T

[718] //doi.org/10.1130/0016-7606%281984%2995%3C221%3ATAGVFA%3E2.0.CO%3B2

[719] http://adsabs.harvard.edu/abs/1954Geop...19..281W

[720] //doi.org/10.1190/1.1437994

[721] http://www.agu.org/sections/geomag/background.html

[722] http://geomag.usgs.gov

[723] http://www.geomag.bgs.ac.uk

[724] https://www.nytimes.com/2004/07/13/science/13magn.html?ex=1247457600&en=
e8f37e14d213ba16&ei=5090&partner=rssuserland

[725] http://news.nationalgeographic.com/news/2004/09/0927_040927_field_flip.html

[726] https://www.pbs.org/wgbh/nova/magnetic/

[727] http://www.psc.edu/science/Glatzmaier/glatzmaier.html

[728] http://www.phy6.org/earthmag/demagint.htm

[729] http://www-spof.gsfc.nasa.gov/Education/wmap.html

[730] https://web.archive.org/web/20080905113925/http://blackandwhiteprogram.com/interview/
dr-dan-lathrop-the-study-of-the-earths-magnetic-field

[731] http://www.ngdc.noaa.gov/IAGA/vmod/igrf.html

[732] http://www.vukcevic.talktalk.net/Global%20Mag%20Anomaly.gif

[733] http://www.ethz.ch/index_EN

[734] http://sciencenewsdigest.com/story/read/132/Earths-magnetic-field-simpler-than-we-thought

[735] http://www.new1.dli.ernet.in/data1/upload/insa/INSA_1/20005b61_51.pdf

[736] (PDF version http://islamsci.mcgill.ca/RASI/BEA/Ibn_Sina_BEA.pdf)

[737] trans in pages 496–500

[738] in

[739] in

[740] Almagestum novum, chapter nine, cited in

[741] See Fallexperimente zum Nachweis der Erdrotation (German Wikipedia article).

[742] When Earth's eccentricity exceeds 0.047 and perihelion is at an appropriate equinox or solstice, only one period with one peak balances another period that has two peaks. UNIQ-ref-0-
467e96e81ac026b0-QINU

[743] Equation of time in red and true solar day in blue http://www.jgiesen.de/planets/img/
EoTGraph.gif

[744] The duration of the true solar day http://www.pierpaoloricci.it/dati/giornosolarevero_eng.htm

[745] http://hpiers.obspm.fr/eoppc/eop/eopc04_05/eopc04.62-now

[746] Physical basis of leap seconds http://iopscience.iop.org/1538-3881/136/5/1906/pdf/1538-
3881_136_5_1906.pdf

[747] Leap seconds http://tycho.usno.navy.mil/leapsec.html

[748] Prediction of Universal Time and LOD Variations http://www.ien.it/luc/cesio/itu/gambis.pdf

[749] R. Hide et al., "Topographic core-mantle coupling and fluctuations in the Earth's rotation"
http://www.gps.caltech.edu/~clay/PDF/Hide1993.pdf 1993.

[750] Leap seconds by USNO http://tycho.usno.navy.mil/leapsec.html

[751] IERS EOP Useful constants http://hpiers.obspm.fr/eop-pc/models/constants.html

[752] Aoki, the ultimate source of these figures, uses the term "seconds of UT1" instead of "seconds
of mean solar time".<ref>Aoki, *et al.*, " The new definition of Universal Time http://adsabs.
harvard.edu/abs/1982A&A...105..359A", *Astronomy and Astrophysics* **105** (1982) 359–361.

[753] *Explanatory Supplement to the Astronomical Almanac*, ed. P. Kenneth Seidelmann, Mill Valley,
Cal., University Science Books, 1992, p.48, .

[754] IERS Excess of the duration of the day to 86,400s ... since 1623 http://hpiers.obspm.fr/eop-
pc/earthor/ut1lod/lod-1623.html Graph at end.

[755] IERS Variations in the duration of the day 1962–2005 https://web.archive.org/web/20070813203913/http://hpiers.obspm.fr/eop-pc/earthor/ut1lod/figure3.html

[756] In astronomy, unlike geometry, 360° means returning to the same point in some cyclical time scale, either one mean solar day or one sidereal day for rotation on Earth's axis, or one sidereal year or one mean tropical year or even one mean Julian year containing exactly for revolution around the Sun.

[757] Arthur N. Cox, ed., *Allen's Astrophysical Quantities* https://books.google.com/books?id=w8PK2XFLLH8C&pg=PA244 p.244.

[758] Michael E. Bakich, *The Cambridge planetary handbook* https://books.google.com/books?id=PE99nOKjbXAC&pg=PA50, p.50.

[759] Sumatran earthquake sped up Earth's rotation http://www.nature.com/news/2004/041229/full/news041229-6.html, Nature, 30 December 2004.

[760] Permanent monitoring http://hpiers.obspm.fr/eop-pc/techniques/techniques.html

[761] http://www.usno.navy.mil/USNO/earth-orientation

[762] https://web.archive.org/web/20130802032453/http://maia.usno.navy.mil/

[763] http://hpiers.obspm.fr/eop-pc/

[764] http://www.iers.org/

[765] http//curious.astro.cornell.edu

[766] Jean Meeus, *Astronomical Algorithms* 2nd ed, (Richmond, VA: Willmann-Bell, 1998) 238. See Ellipse#Circumference. The formula by Ramanujan is accurate enough.

[767] Our planet takes about 365 days to orbit the Sun. A full orbit has 360°. That fact demonstrates that each day, the Earth travels roughly 1° in its orbit. Thus, the Sun will appear to move across the sky relative to the stars by that same amount.

[768] Jerry Brotton, *A History of the World in Twelve Maps*, London: Allen Lane, 2012, p. 262.

[769] Aphelion is 103.4% of the distance to perihelion. See "Orbital characteristics" table. Due to the inverse square law, the radiation at perihelion is about 106.9% of the radiation at aphelion.

[770] For the Earth, the Hill radius is

UNIQ-math-0-467e96e81ac026b0-QINU

where m is the mass of the Earth, a is an astronomical unit, and M is the mass of the Sun. So the radius in AU is about

UNIQ-math-1-467e96e81ac026b0-QINU .

[771] The figure appears in multiple references, and is derived from the VSOP87 elements from section 5.8.3, p. 675 of the following:

[772] Seuss, E. (1875) *Die Entstehung Der Alpen* [*The Origin of the Alps*]. Vienna: W. Braunmuller.

[773] Early edition, published online before print.

[774] National Geographic, 2005 http://news.nationalgeographic.com/news/2005/02/0203_050203_deepest.html

[775] Amri Wandel, On the abundance of extraterrestrial life after the Kepler mission https://www.researchgate.net/publication/269116658_On_the_abundance_of_extraterrestrial_life_after_the_Kepler_mission

[776] http://www.intechopen.com/books/the-biosphere

[777] https://web.archive.org/web/20080623204421/http://www.eoearth.org/article/Biosphere

[778] http://www.globio.info/

[779] http://www.vega.org.uk/video/programme/111

[780] https://web.archive.org/web/20110709103140/http://www.sage.wisc.edu/atlas/

[781] IPCC Special Report on Land Use, Land-Use Change And Forestry, 2.2.1.1 Land Use http://www.grida.no/climate/ipcc/land_use/045.htm

[782] FAO Land and Water Division https://wayback.archive-it.org/all/20090808100750/http://www.fao.org/landandwater/agll/landuse/landusedef.stm retrieved 14 September 2010

[783] JAPA 25:3 http://www.informaworld.com/smpp/content~db=all~content=a787389941~frm=titlelink

[784] Village of Euclid, Ohio v. Ambler Realty Co.

[785] Nolon, John R., Local Land Use Control in New York: An Aging Citadel Under Siege http://papers.ssrn.com/sol3/papers.cfm?abstract_id=1505003 (July/Aug. 1992). New York State Bar Journal, p. 38, July–August 1992.

[786] Zoning

[787] Intergovernmental Panel on Climate Change http://www.ipcc.ch/pdf/assessment-report/ar4/wg1/ar4-wg1-chapter2.pdf

[788] UN Land Degradation and Land Use/Cover Data Sources http://unstats.un.org/unsd/ENVIRONMENT/envpdf/landdatafinal.pdf ret. 26 June 2007

[789] UN Report on Climate Change http://www.ipcc.ch/SPM2feb07.pdf retrieved 25 June 2007 from Web archive https://web.archive.org/web/19960101/http://www.ipcc.ch/SPM2feb07.pdf

[790] Jaeker WG, Plantinga AJ (2007). How have Land-use regulations Affected Property Values in Oregon? http://extension.oregonstate.edu/catalog/pdf/SR/SR1077-E.pdf OSU Extension.

[791] https://web.archive.org/web/20130511205341/http://www.eoearth.org/article/Land-use_and_land-cover_change

[792] http://www.landuselawreport.org/

[793] http://law.wustl.edu/landuselaw/

[794] http://www.publicintegrity.org/investigations/luap/

[795] https://web.archive.org/web/20070715063537/http://web1.msue.msu.edu/wexford/LU/index.html

[796] http://www.landpolicy.msu.edu

[797] http://www.jtlu.org/

[798] https://www.law.cornell.edu/wex/land_use

[799] //en.wikipedia.org/w/index.php?title=Template:History_of_geography_sidebar&action=edit

[800] Simandan, D., 2017. Demonic geographies. Area. 49(4), pp. 503-509. https://doi.org/10.1111/area.12339

[801] http://babel.hathitrust.org/cgi/pt?id=mdp.39015033659353;view=1up;seq=72

[802] ACME journal homepage. http://www.acme-journal.org/Home.html Accessed: May 18, 2015.

[803] http://www.tandfonline.com/toc/rgeo20/current Accessed: July 26, 2017.

[804] Global Environmental Change homepage http://www.journals.elsevier.com/global-environmental-change/

[805] http://www.tandfonline.com/toc/rscg20/current Accessed: July 26, 2017.

[806] In only 200 years, the world's urban population has grown from 2 percent to nearly 50 percent of all people.

[807] https://worldmapper.org/

[808] Terry 2013, p. 226.

[809] Scott, Elaine. *Our Moon: New Discoveries About Earth's Closest Companion.* Houghton Mifflin Harcourt (2016) . page 7.

[810] Collins English Dictionary https://www.collinsdictionary.com/dictionary/english/moon

[811] Oxford Living Dictionaries https://en.oxforddictionaries.com/definition/us/moon#moon

[812] Meaning of "moon" in the English Dictionary https://dictionary.cambridge.org/dictionary/english/moon Cambridge Learner's Dictionary

[813] The American Heritage Dictionary Indo-European Roots Appendix https://www.ahdictionary.com/word/indoeurop.html

[814] *Oxford English Dictionary*, "luna", Oxford University Press (Oxford), 2009.

[815] American Heritage Dictionary https://www.ahdictionary.com/word/search.html?q=selenic

[816] Collins English Dictionary https://www.collinsdictionary.com/dictionary/english/selenic

[817] Oxford Living Dictionaries https://en.oxforddictionaries.com/definition/selenic

[818] NASA: The Moon Once Had an Atmosphere That Faded Away | Time http://time.com/4974580/nasa-moon-had-atmosphere-volcanoes

[819] Global influences of the 18.61 year nodal cycle and 8.85 year cycle of lunar perigee on high tidal levels, U. of Western Australia http//research-repository.uwa.edu.au

[820] There is no strong correlation between the sizes of planets and the sizes of their satellites. Larger planets tend to have more satellites, both large and small, than smaller planets.

[821] Needham 1986, p. 411.

[822] Needham 1986, p. 227.

[823] Needham 1986, p. 413–414.

[824] Needham 1986, p. 415–416.

[825] SpaceX to help Vodafone and Nokia install first 4G signal on the Moon | The Week UK http://www.theweek.co.uk/91979/spacex-to-help-vodafone-and-nokia-install-first-4g-signal-on-the-moon

[826] "Muhammad." Encyclopædia Britannica. 2007. Encyclopædia Britannica Online, p.13

[827] Brooks, A. S. and Smith, C. C. (1987): "Ishango revisited: new age determinations and cultural interpretations", *The African Archaeological Review*, 5 : 65–78.

[828] https://books.google.com/?id=jfQ9E0u4pLAC

[829] https://www.nytimes.com/2014/09/09/science/revisiting-the-moon.html

[830] http://www.bbc.co.uk/worldservice/specials/948_discovery_2008/page4.shtml

[831] http://www.astronomycast.com/astronomy/episode-17-where-does-the-moon-come-from/

[832] http://www.minsocam.org/msa/RIM/Rim60.html

[833] http://adsabs.harvard.edu/abs/2006RvMG...60D...5J

[834] //doi.org/10.2138/rmg.2006.60.0

[835] http://www.hq.nasa.gov/office/pao/History/alsj/

[836] http://www.lpi.usra.edu/expmoon/

[837] http://www.psrd.hawaii.edu/Archive/Archive-Moon.html

[838] http://www.apolloarchive.com/apollo_archive.html

[839] http://ser.sese.asu.edu/GHM/

[840] http://www.lpi.usra.edu/publications/books/rockyMoon/

[841] https://www.nasa.gov/moon

[842] https://www.flickr.com/photos/136797589@N04/albums/72157686992929766/with/35498090194/

[843] https://www.youtube.com/playlist?list=PLBdt9s6ywoD_62doBvzu9pSLDSY1TYtX_

[844] https://www.youtube.com/watch?v=nr5Pj6GQL2o

[845] https://www.youtube.com/watch?v=zNpsy6lBPBw

[846] https://www.google.com/maps/space/moon/@6.1467095,139.2754359,23010541m/data=!3m1!1e3

[847] http://www.lpi.usra.edu/resources/cla/

[848] http://planetarynames.wr.usgs.gov/jsp/FeatureTypes2.jsp?system=Earth&body=Moon&systemID=3&bodyID=11

[849] http://www.cmf.nrl.navy.mil/clementine/clib/

[850] http://moon.google.com

[851] http://www.worldwindcentral.com/wiki/Moon

[852] http://ralphaeschliman.com/id26.htm

[853] https://web.archive.org/web/20120305055023/https://wms.selene.jaxa.jp/index_e.html

[854] http://home.bt.com/techgadgets/technews/explore-the-lunar-north-pole-11363885909226?s_intcid=con_RL_LunarNorthPole

[855] https://web.archive.org/web/20070820075142/http://sunearth.gsfc.nasa.gov/eclipse/SKYCAL/SKYCAL.html

[856] http://sunearth.gsfc.nasa.gov/eclipse/SKYCAL/SKYCAL.html

[857] http://www.timeanddate.com/worldclock/moonrise.html

[858] http://www.crescentmoonwatch.org/nextnewmoon.htm

[859] http//www.esa.int

[860] http://www.esa.int/Our_Activities/Technology/Building_a_lunar_base_with_3D_printing

[861] Science, Evolution, and Creationism http://books.nap.edu/openbook.php?record_id=11876&page=R1 National Academy Press, Washington, DC 2005

[862] ; but see also Cosmas Indicopleustes

[863] An Anthem for the Earth http://www.ekantipur.com/the-kathmandu-post/2013/05/24/related_articles/voices/249135.html Kathmandu Post, May 25, 2013

Article Sources and Contributors

The sources listed for each article provide more detailed licensing information including the copyright status, the copyright owner, and the license conditions.

Earth *Source:* https://en.wikipedia.org/w/index.php?oldid=856136637 *License:* Creative Commons Attribution-Share Alike 3.0 *Contributors:* 1618033golden, A2soup, Abductive, Aeonx, AhmadLX, Anthony Appleyard, Arlo Barnes, Armanschwarz, ArnoldReinhold, Attic Salt, BabbaQ, Bear-rings, BeatlesLedTV, Beland, Bobbie73, Bomb319, Bongwarrior, Cactusframe, Caryncliving, Cgx8253, Charlesdrakew, CheChe, Chiswick Chap, ClueBot NG, CommanderOzEvolved, Computer40, DVdm, Dao1, Dawnseeker2000, Dbachmann, Dbfirs, Dhtwiki, Diverentyaseen, Dlambe3, Dlthewave, Double sharp, Dr.K., DrKay, Drbogdan, Drdpw, Drow, Duncan.Hull, Eevin, Elijah Youngblood, Ermahgerd9, Exoplanetaryscience, Facts707, Firth m, Fmadd, Fotaun, G0mx, Geartooth, Geographyinitiative, Gorthian, Hddty., Headbomb, HopsonRoad, Huntster, I am. furhan., Iboughttoomanygames, Iggy the Swan, Isambard Kingdom, JEH, JJBers, JeanLucMargot, Jeppesn, Joby.CC, Joe Kress, Joeinwiki, Jon Kolbert, Josebarbosa, Just plain Bill, Katolophyromai, Keith-264, Kellner21, Kenwick, Kind Tennis Fan, Kku, Ldweisberg, Lightlowemon, Lopifalko, LuciferZH, Magic9mushroom, Magyar25, Mareklug, Martin Gühmann, MartinZ, Materialscientist, Mean as custard, Mikenorton, Mindbuilder, Miracle Pen, Mojo0306, Mrmw, Mykhal, Narky Blert, Navarre0107, NeatNit, NetrualEditor, Nick.mon, NotARabbit, Nov3rd17, Oliver Goransson, Owain Knight, PaleCloudedWhite, Pbsouthwood, Permstrump, Piledhigheranddeeper, Pinguinn, Pocketthis, Power～enwiki, Praemonitus, Prokaryotes, Pythoncoder, Qzd, Randy Kryn, Ranjitbok, Red Director, Redom115, Rhinopias, Rhombuth, Rich Farmbrough, Rjwilmsi, Robert1947, Rodw, Samantha Ireland, Saurusaurus, Scootalmighty, Scrossuk, Serols, Sidewinder, Sijadthelastpoet, Somedifferentstuff, Someguy432, SophieWhitton, Spidersmilk, Spike-from-NH, Srich32977, SuNaW, SuperTurboChampionshipEdition, Supertali64, TAnthony, The Transhumanist, The snare, TheDragonFire, TheFreeWorld, TheRedCityXFactor, Tomruen, TwoTwoHello, Ubcule, United Massachusetts, Unman Seams, Wayne Elgin, WhatsUpWorld, WikiImprovment78, WolfmanSF, Woodstone, Woscafrench, XeloriaGrand, Zefr, ВоенTex, 七战功成 1

History of Earth *Source:* https://en.wikipedia.org/w/index.php?oldid=856153617 *License:* Creative Commons Attribution-Share Alike 3.0 *Contributors:* A Great Catholic Person, Abelmoschus Esculentus, Alumnum, AntiVan, Apokryltaros, Aquarius-1, Arnav mahani, Bear-rings, Beland, Bender235, Bgwhite, Bobbie73, Byteflush, CLCStudent, CV9933, Caehlla, ClueBot NG, CommonsDelinker, Cryptic, Crystallizedcarbon, Dane, Dawnseeker2000, Dcirovic, Discospinster, Djwiofeoi, Dolotta, Donner60, DrStrauss, Dragonpurl, Drbogdan, Dudley Miles, DuncanHill, Dunkleosteus77, Dw122339, EarthOcean, Edward, Filursiax, G0mx, Gap9551, Gareth Griffith-Jones, GeneralizationsAreBad, GeoWriter, Gilliam, God's Godzilla, Gorthian, GrapefruitSculpin, HMSLavender, Hafeez Depar, Harizotoh9, Headbomb, HiLo48, Hut 8.5, Iacobus, Ilyakor0676, IronGargoyle, Isambard Kingdom, IyrandrarSarhana, IznoRepeat, JamesBWatson, Jarble, Jdaloner, Jerod Franko, Jim1138, Jon Kolbert, Julietdeltalima, Karanblood2525, Kennyz, Khirurg, Kku, KylieTastic, Lappspira, Magyar25, Mikee butt, Minituremeow, My Chemistry romantic, Newone, Nyttend, Osh33m, Oshwah, PaleoNeonate, Pauli133, Peterye2005, PlyrStar93, Quibilia, Red Planet X (Hercolubus), Renamed user 943a06d1c3, Rhinopias, Rjwilmsi, RockMagnetist, Roxy the dog, Ryangosh, SA 13 Bro, Sanfranman59, Seaweed, Semmendinger, Serafart, Serols, Slightsmile, Soul2251974, Specane111, Spike Wilbury, Tajotep, Theroadislong, Thrif, Titiawolflover, UnsungKing123, Vanamonde93, Vsmith, Wgolf, Ynoss, Z0, Zboy Muer, 163 anonymous edits 41

Geological history of Earth *Source:* https://en.wikipedia.org/w/index.php?oldid=843100522 *License:* Creative Commons Attribution-Share Alike 3.0 *Contributors:* 72, Adam9007, Alfie Gandon, Altenmann, Amortias, Amp71, AndrewHowse, Andycjp, Aunt Entropy, Avenue, Awickert, Awsome2212, Azuris, BD2412, BSATwinTowers, Bazonka, Bear-rings, Bejnar, Bender235, Bettymnz4, Bilalnaik, Bongwarrior, Burwellian, Cadiomals, Canthusus, Catalaalatac, Chimesmonster, Chris the speller, Citation bot 1, ClueBot NG, Colonies Chris, CommonsDelinker, Craic Den, CuriousMind01, D.M.N., DSachan, Dawn Bard, Dcirovic, Deadbeef, Deor, Der Golem, DerBorg, Doug Weller, Drbogdan, Dwaipayanc, Eastlaw, ErgoSum88, Excirial, Froth, GeoWriter, Geologyguy, Gilliam, GoingBatty, Greenpac4mp, Hamsterlopithecus, Hardyplants, Harryboyles, Headbomb, Hmains, Hughesyjj, Isambard Kingdom, IznoRepeat, J 1982, J-stan, JLincoln, Jack Greenmaven, Jon Kolbert, JonRichfield, Jpvandijk, KP Botany, Khazar, Kintaro, Kozuch, LeadSongDog, Leovizza, Leszek Jańczuk, LittleOldMe, Lugia2453, MCdance101, Materialscientist, McZusatz, Metiscus, Michael Devore, Micheletb, Mikenorton, Mikeo, Muhends, Mwtoews, NatureA16, NawlinWiki, Neito Nossal, Non-dropframe, NotWith, Novangelis, Onel5969, Onetruepurple, Originalbigj, Oshwah, Parsa, Paul H., Peter M. Brown, Pinethicket, Preciousjfm, R'n'B, RJHall, Rcsprinter123, ReeceStewart, Rjwilmsi, RockMagnetist, Rocket000, Sairjohn～enwiki, Serendipodous, SpaceChimp1992, Spamalot360, Srich32977, Stinger20, Suicidalhamster, Sushant gupta, The Thing That Should Not Be, The Transhumanist, TheCascadian, Tide rolls, Tom.Reding, UBeR, Viriditas, Vsmith, WereSpielChequers, Widr, Yamara, Yngvadottir, 154 anonymous edits 79

Evolutionary history of life *Source:* https://en.wikipedia.org/w/index.php?oldid=856256169 *License:* Creative Commons Attribution-Share Alike 3.0 *Contributors:* AManWithNoPlan, Abdulbasetkurd, Abyssal, Airplaneman, Aknauff01, Alan G. Archer, Animalparty, Anrnusna, Aquarius-1, AstroBio Ben, Azcolvin429, BatteryIncluded, Bear-rings, Bender235, Bernstein0275, Bgwhite, Blainster, C.Fred, Chaya5260, ChrisGualtieri, ClueBot NG, CommonsDelinker, CuriousMind01, Curt99, DMacks, Danno349, Davemck, Dawnseeker2000, Dcirovic, Discospinster, Dpleibovitz, Dratman, Drbogdan, Ebrahim50, Editor2020, Education2015, Edward321, Ellin Beltz, Ettrig, Forward Unto Dawn, Fraggle81, Franciscosp2, Frze, Geo-Science-International, Gvisconty, Harizotoh9, Hberaldi, Headbomb, Henry101cool, IdreamofJeanie, Improvingapoint, J 1982, James brownuyt87467, Jim1138, Jk2005, Jonesey95, Julesd, Jupitus Smart, Just plain Bill, JzG, Keith D, Kevmin, Kfitzgib, Khirurg, L293D, Lappspira, Leptus Froggi, Looie496, Luizpuodzius, MHampton28, Magioladitis, Malerisch, Mama meta modal, Materialscientist, Meredyth, Mindmatrix, MrBill3, Mre env, MusikAnimal, Niceguyedc, OAnick, Pandeist, Pbsouthwood, Peter M. Brown, Philip Trueman, Pitke, Plantdrew, Plantsurfer, Quebec99, RandomEditPro, Reatlas, Redd Foxx 1991, Rhinopias, RichardWeiss, Rjwilmsi, Ryan Shakiba, SCorneliusB, SaRan, SemiHypercube, Skoskav～enwiki, Smartse, Sminthopsis84, Smithsonwick, Somedifferentstuff, Spike Wilbury, Susurrus, Tassedethe, TomS TDotO, Tony 1212, Topbanana, Toyokuni3, Trappist the monk, Ugog Nizdast, Varkman, Venturi Seba, Verbum Veritas, Vsmith, Wavelength, X1\, Xanzzibar, YoursT, 75 anonymous edits 97

Future of Earth *Source:* https://en.wikipedia.org/w/index.php?oldid=855572207 *License:* Creative Commons Attribution-Share Alike 3.0 *Contributors:* 7Sidz, Adam9007, Amishi Srivastava, Arado, Aurora2698, Bender235, Bobbie73, Bryantbarzee, CLCStudent, Canavalia, CedeDancer, Ceoil, Cj3636, ClueBot NG, DVdm, Darylgolden, Dcirovic, DocWatson42, Doug Coldwell, Drbogdan, Dunkleosteus77, EditorFromSpace, EvergreenFir, FFAR, Fixuture, Gaioa, Gap9551, Gilliam, Gob Lofa, Headbomb, Hijuecutivo, IloveRumania, J 1982, Jarble, Jcpag2012, Jim1138, Jleon, Jonesey95, Jusdafax, Justeditingtoday, Kind Tennis Fan, Kku, Knobulose, Ladykiller45, LakesideMiners, Lord Marcellus, Madscribbler, Marcocapelle, MarioProtIV, Materialscientist, MelbourneStar, Mfb, Millennium bug, Neuron1, Newone, NewsAndEventsGuy, Niche-gamer, Nnemo, Noodle poodle, Oshwah, Pdhall99, Person who formerly started with "216", Petrb, Poyekhali, Praemonitus, Quenhitran, Rainald62, RichardWeiss, Rjwilmsi, Rupertslander, Ruslik0, Serpinium, Shellwood, Shrikanthv, Space-Age Meat, Srich32977, Stormchaser89, Surtsicna, Tetra quark, TheHacker1234, Thespaceface, Ticgame, Timpo, Tom.Reding, Transportfan70, Trappist the monk, Trev74, U-95, Vsmith, Wbm1058, Widr, Yamaguchi先生, 魅, 174 anonymous edits 135

Figure of the Earth *Source:* https://en.wikipedia.org/w/index.php?oldid=856073356 *License:* Creative Commons Attribution-Share Alike 3.0 *Contributors:* 000000Soccerman, 3410ankit, Acetotyce, Adam9007, Alanfeynman, Alby, Anaxial, AnonMoos, April J XD, Asdfghjklqwertyuiopo, Auntof6, Ayush Morbar, Beland, Bender235, Bennv3771, Bfinn, Bobbie73, Bobblewik, Bobo192, Boeing720, Brandmeister, Brews ohare, Bryan Derksen, Bubbha, CAPTAIN RAJU, CLCStudent, CarolGray, Cesar1242134759927402, Charles Matthews, Chrismurf, Citynoise, ClueBot NG, Cmglee, Codeman47250, Cubs Fan, Dauto, Dawnseeker2000, Dontdoit, Doug Weller, Eilthireach, Elassint, Feli70, Fgnievinski, Flat-EarthTruthSeeker, Fred Bradstadt, Gaius Cornelius, Gap, Geof, Gerry Ashton, Glafyjk, Glevum, GliderMaven, Gorthian, Hdante, Headbomb, Hello5959us, Heron, Hobartimus, Isambard Kingdom, J 1982, JHarris11, Jagged 85, Jbeyerl, Jc3s5h, Jidanni, Joe Kress, John, JorisvS, Joseph Solis in Australia, Jrvz, Just plain Bill, KH-1, Kaimbridge, Katieh5584, Kostya Wiki～enwiki, Krauss, Kremmen, Kri, Ktotam, Kvwiki1234, Leonard G., Liuhouhan, Lockley, LucasVB, MarkSweep, Martin tamb, Mgiganteus1, Michael Glass, Michael Hardy, Mikael Häggström, Mikenorton, Mindbuilder, Musamies, Nanophosis, NawlinWiki, Ninly, Nk, Obankston, Onel5969, Oshwah, Oxymoron83, Paolo.dL, Patrick, Peak, Peter Mercator, Phi beta, Philip Trueman, Phoenix79, PianoDan, Pmanderson, Quibik, Qwfp, R'n'B, RDBury, RJHall, RP459, Ragesoss, RealGeo, Reatlas, RekishiEJ, Rjwilmsi, RockMagnetist, Rocksandwaves, RoyBoy, Rush1p, S3000, SaliVader, Santifc, Saros136, Scartboy, Schwilgue, Septegram, Serendipodous, Sgv 6618, Shellwood, SilkTork, Ssstttt, St.nerol, Stamptrader, SteveMcCluskey, Steveengelhardt, Steven J. Anderson, Strebe, TakuyaMurata, The Anome, Tim Zukas, Timl, Tom.Reding, Tuxipedia, Twinsday, Tyler's 851424444, UAwiki, Vassasisualii, Vermeer～enwiki, VernoWhitney, Vsmith, Welsh, William Avery, Woohookitty, XJaM, Xmilanz, 132 anonymous edits 157

Abundance of the chemical elements *Source:* https://en.wikipedia.org/w/index.php?oldid=855975282 *License:* Creative Commons Attribution-Share Alike 3.0 *Contributors:* 28bytes, 5ko, ASmartKid, Abitslow, Aedstrom, Alexander Davronov, Anephrazon, Anomalocaris, Arbitrarily0, Art LaPella, BD2412, Bear-rings, Bigjeffmeme, Billinghurst, Brendan Rizzo, CV9933, ChemNerd, Cherkash, Churchgoer251, Citation bot 1, ClueBot NG, Cmglee, Coder Dan, Count Truthstein, Courcelles, DARTH SIDIOUS 2, Dawnseeker2000, DePiep, Denisarona, Dissident93, Doctor C, Donner60, Double sharp, Drbogdan, Drollere, Eteethan, EvergreenFir, Evolution and evolvability, Excirial, Falcon8765, Firetraner, Floozybackloves, Frankie0607, Frietjes, Gap9551, Genewiki1, Glacialfox, GoatGuy, God's Godzilla, Headbomb, Howicus, Hypnosifl, I dream of horses, ILOVETOSCREWSTUFFUP, Ian.thomson, IiKkEe, Iridescent, James Cantor, Jim1138, Jni, Jodosma, Joe2719, Joeinwap, Jonesey95, JorisvS, JukeBoxHero, Junkinbomb, KAP03, Katherine, KlappCK, Korandder, LarryMorseDCOhio, LeyteWolfer, Mabuhelwa, MacHyver, Materialscientist, Michbich, Mikespedia, Mz7, NielsenGW, Nwbeeson, Oliphaunt, Ordinary Person, Orthorhombic, Oshwah, Pbrower2a, Philip Trueman, Pinethicket, QuantumShadow, Quindraco, R0uge, RProgrammer, Reify-tech, Riventree, Rjwilmsi, RockMagnetist, RockMagnetist (DCO visiting scholar), Rodii, Roentgenium111, Rsrikanth05, Runningonbrains, SBarnes, Sbharris, Sboehringer, SchreiberBike, ScottyBerg, Serols, Sleepneeder, Spacepotato, Stepbang, StevenBell, Stupidsaggy, Sun Creator, Swift, Syrthiss, Tamfang, The Thing That Should Not Be, Theoprakt, Tide rolls, Tom.Reding, Trackteur, Trappist the monk, UpdateNerd, Vrenator, Vsmith, Wastednow, Wayne Slam, Wiae, Widr, Wingedsubmariner, Woohookitty, YBG, Ynhockey, Zidonuke, 138 anonymous edits 167

Structure of the Earth *Source:* https://en.wikipedia.org/w/index.php?oldid=854705638 *License:* Creative Commons Attribution-Share Alike 3.0 *Contributors:* 123456789wgs, 72, Ajraddatz, Alamshdir, Alfie Gandon, Anaxial, Aphmausenpaifan, Arjayay, ArnoldReinhold, Beland, Bharat Jatia, Bhavarth shah, Biografer, BobEnyart, Brod.F, CAPTAIN RAJU, CPColin, Cameron11598, CelestialKeystone, Classicwiki, ClueBot NG, Concus Cretus, Daask, Daniel "Danny" Wilson, David.moreno72, DavidLeighEllis, Dbachmann, Dcirovic, DeltaQuad, Donner60, Dragons flight, Eric-Wester, Excirial, FoCuSandLeArN, GeoWriter, Geodyn4, Gigarose, Gilliam, Glane23, God's Godzilla, Guccisnap, Gulumeemee, Hadron137, Hello9879, Home Lander, Hornpipe2, Imminent77, Issyl0, J 1982, JJJOOORRR, Jennica, Jim1138, Jiten D, John Cline, Joshualouie711, Jumpyjackkk, Kacy799, Katlegear, KylieTastic, LimitedTim, Lube Lord, MarnetteD, Materialscientist, Mikenorton, Nick Moyes, Niggnogg, Noyster, Omer164, Oshwah, Ost316, PackMecEng, PedroAlex.20, Perpetualmoss, Plantsurfer, RA0808, Rcsprinter123, RockMagnetist, Rreagan007, Serols, Shazhaibasif73, Shellwood, Sideways713, Simplexity22, Sjö, Smjsjsjs, Sneeuwschaap, SparklingPessimist, Telecineguy, Thegooduser, Tinyds, Tompop888, Vinci2005, Vsmith, Widr, William Avery, Winged Blades of Godric, Winner 42, XXCoolPvPXx, Yamaguchi先生 , YouDunGoofed083, 229 anonymous edits 183

Earth's internal heat budget *Source:* https://en.wikipedia.org/w/index.php?oldid=855917464 *License:* Creative Commons Attribution-Share Alike 3.0 *Contributors:* Afernand74, Beland, Benlisquare, Bkilli1, Br77rino, Chris857, ChrisGualtieri, ClueBot NG, Colonies Chris, Dave souza, Dawnseeker2000, Deacon Vorbis, Ericbarefoot, Fangorn-Y, Giraffedata, Graeme Bartlett, Ground Zero, Jdthood, Jimw338, Julesd, KH-1, Klaas van Aarsen, Mgiganteus1, Mikhail Ryazanov, NameIsRon, Ninjakiller90, No.0ne.057, Offnfopt, Ordlock, Pederony, Physicist5777, Piledhigheranddeeper, Prokaryotes, R8R, Rfassbind, Rpg32, Scootalmighty, Ser Amantio di Nicolao, Sinysee, Tritario, Usp, Vaughan Pratt, Vsmith, Wavelength, Wikishovel, WolfmanSF, 19 anonymous edits 191

Plate tectonics *Source:* https://en.wikipedia.org/w/index.php?oldid=855199820 *License:* Creative Commons Attribution-Share Alike 3.0 *Contributors:* A876, Abyssal, Acquaduct, Ammorgan2, Anrnusna, Asfd666, Atcovi, BIL, Baldy Bill, BeenAroundAWhile, Beland, Bender235, Bgwhite, Bkilli1, Blaxthos, Bluerasberry, CES1596, Capacitor12, CarloMartinelli, Cellodont, Chienlit, Chris55, Ckruschke, Concord113, Corinne, CuriousMind01, Cynulliad, Dawnseeker2000, Dcirovic, Diego Moya, Domdomegg, Download, Drbogdan, DuncanHill, Ego White Tray, Enric Naval, Erikamit, Finetooth, Fomirax, Fredvanner, Geo-Sciences-International, GeoWriter, Geogene, Gilderien, Glevum, Goodtimber, Gorthian, Grammarian3.14159265359, Hairy Dude, Hamiltondaniel, Harizotoh9, Headbomb, Hogyn Lleol, I'm your Grandma., Instagram768, Isambard Kingdom, Izkala, IznoRepeat, J 1982, Jbitz743, John D. Croft, Jonesey95, JorisvS, Josophie, Julesd, Jwratner1, KConWiki, Kelseymh, Kenvandellen, Ketiltrout, LM2000, LightandDark2000, Magioladitis, Mandruss, Marikanessa, Mathwizurd29, Michaelchin78, Migco, Mikenorton, Motivação, MrOllie, Napalatt, Niel Malan, Nightscream, Nihiltres, Nyttend, Pbsouthwood, Phoebe, Phoenix7777, Pragmaticstatistic, Prandr, Proof Pudding, QYYZ, RamblinWreck3, Rcsprinter123, Remotelysensed, Rjwilmsi, Robot psychiatrist, Rost11, Ryoung122, Sanne.cottaar, SciPedian, ScienceDawns, Smyth, Srewter, Srich32977, Sting, Swpb, TAnthony, Takeaway, TheGoodBadWorst, Timothy Robinson12345, Tmangray, Trackteur, Trappist the monk, Tvtonightokc, Tvtvashisth, Twinsday, Veryfaststuff, Vsmith, Ward20, Wen D House, WikiRigaou, Wikiauthor, Wikimag74, Wikirictor, Wzrd1, Xezbeth, Zaslav, Мурад 97 197

Lithosphere *Source:* https://en.wikipedia.org/w/index.php?oldid=855171222 *License:* Creative Commons Attribution-Share Alike 3.0 *Contributors:* Acetotyce, Anasofiapaixao, Anaxial, Andyjsmith, Arbnos, Asukite, Avoided, Axwux, BeenAroundAWhile, Bill7068, Canthusus, CaptainLaptop, Cgingold, ClueBot NG, Dan Koehl, Dcirovic, Deli nk, Discospinster, Donner60, Dpleibovitz, Eleven even, Epicgenius, Erikvinson.enc, Erinbella05, EuroCarGT, Fgnievinski, Flyer22 Reborn, FoCuSandLeArN, Fraggle81, Frosty, GeoWriter, Giraffedata, Gz159, Hgrosser, Hogmonkey22, Ht8338, I dream of horses, Imbrains, Inchiquin, Isambard Kingdom, Jevelynn Bolton, Judahpurwanto55, Jusdafax, KDS4444, Keith D, Kittykat03, Lappspira, Livitup, Lugia2453, MacPoli1, Marek69, Marikanessa, MaskedHero, Materialscientist, Mikenorton, Mtd2006, MusikAnimal, NealeyS, Obama1234, Optakeover, Parluhphone, Pbsouthwood, Philip Trueman, Pinethicket, Potatoassasin, Prokaryotes, Quenhitran, RA0808, Racerx11, Rainbowskull55, Renamed user sdfkjlskdfreu8r98, Robot psychiatrist, Roxy the dog, Sarah Liptay, Satellizer, Seaphoto, Seaver3434, Serols, Sjö, Skamecrazy123, Skizzik, Sosthenes12, Srich32977, Stball2000, Th0rgall, Thegooduser, Thirdright, Tom.Reding, Trut-h-urts man, Vsmith, WiHkibew, Wiae, Widr, Wikipelli, William Avery, 220 anonymous edits 233

Landform *Source:* https://en.wikipedia.org/w/index.php?oldid=854170132 *License:* Creative Commons Attribution-Share Alike 3.0 *Contributors:* A. Scholar (Nabu), AgnosticPreachersKid, Alliecatmeow, AmaryllisGardener, Bender235, Bermicourt, Birutas123, Brambleshire, Burat01, CAPTAIN RAJU, CLCStudent, Cadillac000, Cannolis, Chris troutman, ClueBot NG, Coasterlover1994, Coolkid4578, Crboyer, DanHobley, DavidLeighEllis, Dawnseeker2000, Dcirovic, Delusion23, DemocraticLuntz, Dentren, DerHexer, DiscantX, Discospinster, Docduncan, Donthurtmeiamaniceperson, Drewmutt, El C, Eli1210, EvenGreenerFish, Excirial, Eyesnore, Fakephilosopher, FatiaR123, Favonian, Fgnievinski, Firstdreamding, Flyer22 Reborn, FoCuSandLeArN, Frosty, Funandtrvl, GeneralizationsAreBad, GeoWriter, George8211, Gfoley4, Gilliam, Ginsuloft, Happylulusk8er7, Hdt83, Hector4721, HilaizaPanales, Hmains, Hume42, I dream of horses, Ihaveafaceok, JUSTONY, Jerzy, Jim1138, John F. Lewis, KAP03, Kaobear, Katieh5584, Kejo, Kollo3250, Kubbkubb, KylieTastic, Leitmotiv, Libcub, Lor, LuK3, Lugia2453, MONGO, MRD2014, Materialscientist, Mean as custard, Mercurywoodrose, MnM's are COOL 181920, MusikAnimal, Myasuda, Nick Moyes, Oluwa2Chainz, Onel5969, Oshwah, P. S. F. Freitas, Paul H., Philip Trueman, Pinethicket, Prinsipe Ybarro, ProprioMe OW, Pseudomonas, QueefDonnut, RAJ Shalem, Saksham agarwal1234, Sangdeboeuf, SeoMac, Serols, Shellwood, Simplexity22, Slightsmile, The High Fin Sperm Whale, Vsmith, Wiae, Widr, Wikieonkatrollr, Zawl, 225 anonymous edits 238

Extreme points of Earth *Source:* https://en.wikipedia.org/w/index.php?oldid=855220728 *License:* Creative Commons Attribution-Share Alike 3.0 *Contributors:* Abenporath00, AcidSnow, Ajd, Androl, BIL, BORAGORIDEV, Bandana man95, Banedon, Blahma, Boud, Brycehughes, Buaidh, Chienlit, Chris the speller, ClueBot NG, CrashesToAshes, Cyberbot II, D A R C 12345, DLinth, Dale Arnett, DavidLeighEllis, Dawnseeker2000, DePiep, Delusion23, Deor, DonGlover13, Doremo, Double sharp, Doug1963, EamonnPKeane, Enordlander, EternalNomad, Fioravante Patrone, Florian Blaschke, FreeRangeFrog, Gap9551, Georgestrieby, Giantflightlessbirds, Gilliam, Gob Lofa, Graham87, Harami2000, Interlingua, Ira Leviton, J 1982, JackofOz, JasonAQuest, Jdaloner, Joelbaby72, Johar3482, Johnuniq, Joshua, Juan cosecha, Julietdeltalima, K6ka, Kind Tennis Fan, Kingboyk, Kintetsubuffalo, Kransky, LICA98, LLarson, Lambiam, Leoberacai, Leptictidium, Lerichard, Lightkey, Lotje, Lquilter, LynxTufts, Makeandtoss, Makyen, Maranello10, Marianna251, Markkozlov, Michael Glass, MikeTRose, MjolnirPants, Morriswa, Mountaincirque, Mwtoews, Mx. Granger, Narky Blert, NealeFamily, Nealmcb, Nicksbinvahid, NoahDeKnight, O-Qua-Tangin-Wann 2015, Oiygg, Old Wolf 2, Ori, Oriole4008, Ozverin1, PJsg1011, PaintedCarpet, Parikhbhavin1989, Park3r, Perey, Primefac, Professor Proof, Racerx11, Regulov, Rendesr, Rmhermen, Robert Brukner, Rushabhp, Samee, Scotstout, Scott J MacDonald, Serols, Sethant, Shubh199, Shujianyang, Sriram.iitr, StringRay, Sudhanshu.rat, T3mujin, TU-nor, Takeaway, Td1177, Tjhiggin, Twinsday, Utcursch, Viewfinder, Warofdreams, Weneedwikipedia, Wnt, Xdswiftmeme, YBG, Yowanvista, Zaxius, חרוין, מורג, 雲起林野间 , 207 anonymous edits 241

Hydrosphere *Source:* https://en.wikipedia.org/w/index.php?oldid=856236696 *License:* Creative Commons Attribution-Share Alike 3.0 *Contributors:* Allens, Amaury, Anna Frodesiak, Aqwfyj, Besieged, Bloodyrider, CLCStudent, Cannolis, Ciccc, ClueBot NG, Crow, DASonnenfeld, DVdm, Dawnseeker2000, Dcirovic, Dirkbb, Discospinster, Dketcherside, Donner60, Esszet, Excirial, Flyer22 Reborn, Freebullets, Frigotoni, Funandtrvl, Fylbecatulous, GenuineArt, GokuSSJ820, Gorthian, Graeme Bartlett, HanotLo, Happysailor, Helpsome, Ht8338, ILovePlankton, IronGargoyle, J 1982, Jasmin Ros, Jessicapierce, Jesusmassage, Jevelynn Bolton, Jiten D, Joe1000000, JorisvS, JosueArce, Julietdeltalima, K6ka, KLBot2, Kenvandellen, Keri, Kethrus, Kuzmaabdoo, KylieTastic, Lectonar, Lightemorphous, MBG02, Macedonian, Manray123, Materialscientist, Matthew233551, McSly, Mjs1991, Mogism, Mz7, Navjot1200, NewEnglandYankee, NottNott, Oceanflynn, Omnipaedista, Oxfordwang, Ozric14, Pbsouthwood, Peter M. Brown, Petiatil, Petrb, Philip Trueman, Pinethicket, QPT, R'n'B, RA0808, Racerx11, Rhinopias, Rmashhadi, Rsrikanth05, Runningonbrains, Science girl122405, Scifipete, Sedaashchyan, Smalljim, Steve Quinn, Storkk, Sunmist, SupernovaExplosion, TCN7JM, TYelliot, TexasAndroid, The Unknown1470, There'sNoTime, Thirdright, Ticky111, Tide rolls, Timawesomeness, Titodutta, Tolly4bolly, Transmark25, TwoTwoHello, Tyrannosaurus1337, Umasuthan Sutharson, Velella, Vester97, Vsmith, WadeSimMiser, Wavelength, Webclient101, Wiae, Widr, Yamaguchi先生 , Yamamoto Ichiro, Yourmomspimpdaddy7, 234 anonymous edits 261

Atmosphere of Earth *Source:* https://en.wikipedia.org/w/index.php?oldid=854230106 *License:* Creative Commons Attribution-Share Alike 3.0 *Contributors:* 1a16, 9Questions, Alaney2k, Amaurea, AndrewDressel, ArcadianRefugee, Aristophanes68, Attaboy, Attic Salt, Bear-rings, Belltoes, Benboy00, Bender235, Benwildeboer, Benzx132, Beryllium-9, Bgwhite, BirdValiant, Blackbombchu, BoDeppen, Br77rino, ChrisCarss Former24.108.99.31, ChristopherKingChemist, Ciphers, Clr324, Comfr, Cruithne9, CuriousMind01, DASonnenfeld, Dan Harkless, Daniel.Cardenas, Dave3457, DaveDaytona, Dawnseeker2000, DocWatson42, Double sharp, Dough34, Dr. British12, Drbogdan, ELApro, Edgars2007, Egmonster, Electron9, Eric Kvaalen, Erikmartin, Erikvinson.enc, Eyal Morag, Friecode, Gap9551, George Makepeace, Gggbgggb, Gire 3pich2005, GliderMaven, GoShow, Headbomb, Hessamnia, Huntster, IOLJeff, InedibleHulk, Inks.LWC, Itisifitis, J 1982, Jandalhandler, Jigibb, Jim.henderson, Jkhoury, Joe Kress, John, John Cline, Johnuniq, Jonathan W, JorisvS, Josve05a, Jprg1966, Just granpa, Jxm, KAP03, Kelvin13, Kfitzner, Kolbasz, Kuyi123w, Lentower, LollyBear12, Lotje, LukeSurl, MKar, Magioladitis, Mandruss, Martarius, Martin Hanson, Marvel Guy 1, Master of Time, Meters, Mogism, MrOllie, Mungomba, Munin, NHSavage, Narky Blert, Neelix, Neodinium, Nhak, Nojan, OcarinaOfTime, Omnipaedista, OrenBochman, Originalwana, Pablo Mayrgundter, Parulsingh1478, Pegminer, Petergans, Pine, Pkbwcgs, Plantsurfer, Pragmaticstatistic, Primefac, Prokaryotes, PurpleDiana, RIT RAJARSHI, RPgzLp, Ranze, Raphael.concorde, Reatlas, RedPanda25, Respect.compassion, Rfassbind, Riventree, Rjwilmsi, RockMagnetist, Rwflammang, SS1901, Sankalpdravid, Sbharris, Sjö, SnowMoreMisterIceGuy, Sotkil, Srich32977, Strait, Sweart1, Swpb, Tdadamemd, Th4n3r, The Herald, The PIPE, Thenitingrewal, Thomas.W, TimOsborn, Trappist the monk, Tripodics, Twinsday, Vgy7ujm, Viewmont Viking, Vsmith, Walpurgishacked, Wetman, William M. Connolley, Yerrik, Yitscar, ZackeryTaylor, Zedshort, ZerodEgo 265

Weather *Source:* https://en.wikipedia.org/w/index.php?oldid=844117931 *License:* Creative Commons Attribution-Share Alike 3.0 *Contributors:* 11otsotrenta, 1416domination, 58MUSTANG, A Sad Bear, A. Parrot, Aircorn, Alexius08, AmanLalwani1234, Ampwiki, BD2412, BabylonAS, Beland, Bender235, Bentogoa, Bgwhite, Big universe, Bldkiller7, CAPTAIN RAJU, Calliopejen1, Cellocatiris, Charlesme001, ChrisCarss Former24.108.99.31, Christianvam, Chriswx817, Clarities, ClueBot NG, CommonsDelinker, Condog8, CuriousMind01, Dangebvhdhdhs, DatGuy, Dawn Bard, Dawnseeker2000, Deli nk, DemocraticLuntz, Discospinster, Doctorhawkes, Donner60, Drbogdan, Eurodyne, Excirial, Flyer22 Reborn, Fmadd, Gabriella 062, Gap9551, General Ization, Gflsflflfddfffd, Gilliam, Glane23, Gob Lofa, Greger Henriksson, Gulumeemee, Headbomb, Hellokitty1000, Hiperfelix, I dream of horses, In veritas, IronGargoyle, J 1982, Jd22292, Jess, John Cline, John of Reading, Jpgordon, KAP03, KH-1, Katieh5584, Kintetsubuffalo, KylieTastic, Lambiam, Larry Hockett, Laura riverson, Leaky caldron, LearnMore, Legchild, Legchild2, Legchild3, LizardJr8, MaNeMeBasat, Magioladitis, Master of Time, Materialscientist, Mean as custard, Mojo Hand, Mynameislikejeffsum, Naddy, Napy65, Narky Blert, Nihiltres, Noyster, Nvvchar, Nyttend, Oshwah, Piafcheck, Pinethicket, Primefac, Racerx11, Rivertorch, Rowan Adams, Saif Dalvi, Saipriya123456, Samlow32, Senik123, Serols, Sharonsambu, ShawntheGod, Soggy socks 123456789, Sports Devotee, SpyMagician, Spyglasses, Stemwinders, Supdiop, TerraCodes, TerryAlex, TesLiszt, The Voidwalker, TheFreeWorld,

TheRealWeatherMan, Thomas.W, Tom.Reding, Trailblazer124, Twistedoriginal, Vsmith, Widr, Wikigeek244, Wtmitchell, Xezbeth, Yinf, Yintan, Zackmarino, Zedshort, Андрей Эн, 193 anonymous edits 286

Climate *Source:* https://en.wikipedia.org/w/index.php?oldid=855997369 *License:* Creative Commons Attribution-Share Alike 3.0 *Contributors:* 3 of Diamonds, 72, Acebulf, Adamtt9, Ajsjhdd, Alaney2k, Alfie Gandon, Ammarpad, Anomalocaris, Antandrus, Anthere, Arjayay, Augustjackson, Bgwhite, Bhoot rock, Bojo1498, Bongwarrior, Brandmeister, Bruce1ee, CAPTAIN RAJU, CLCStudent, Cannolis, Chiswick Chap, Chrissymad, CkickenBoy119, ClueBot NG, Cmglee, Colbalt14, Colonel Wilhelm Klink, Crystallizedcarbon, Csigabi, CuriousMind01, Cyberbot II, DHAIRYA, DRAGON BOOSTER, Dan Koehl, Dave souza, Dawn Bard, Dawnseeker2000, Dcirovic, Denisarona, Dinoboyx7, DivineAlpha, DoctorInfo123, Drbogdan, Eagleash, Eleven even, Eurodyne, Faolin42, Felixwie, Frappyjohn, Gap9551, Geogeoron, Gilliam, Gob Lofa, Gulumeemee, HACNY, HJ Mitchell, Hellokitty1000, Hiperfelix, Home Lander, I dream of horses, I.p98, IdreamofJeanie, Inks.LWC, Iridescent, IsraphelMac, J 1982, Jalvarez2, Jeff G., JimVC3, Jodielavery, JorisvS, K6ka, KAP03, KGirlTrucker81, Kahooter999, Kingj12345678m, Kleuske, Laoris, Lappspira, Library Guy, Mandruss, Marcocapelle, Marianna251, Materialscientist, MattRuiz01, McSly, Melcous, Mojo Hand, Mppiyush, Mujtaba!, MusikAnimal, Mx. Granger, NYBrook098, Nakon, NewsAndEventsGuy, Nigelj, Nigos, Noopla2, Onel5969, Oshwah, Packmeister, Petrb, PhantomTech, Philip Trueman, Piguy101, Plantsurfer, Plastikspork, Power~enwiki, Praveen bastia, Prokaryotes, Qwfp, Qzd, Redrose64, Ricky81682, Rrburke, Rubbish computer, Serols, Shellwood, Shendoa, Shubham Dandeva, Shyam274, Sjö, Slightsmile, Smalljim, Srich32977, Stemwinders, TAnthony, Thanhminh2000, The Voidwalker, TheMesquito, Thomas.W, Trackteur, TranquilHope, Varshit1234, Vgenapl, Vsmith, Wiae, Widr, William M. Connolley, Woctaviano, Wrh2, Wtmitchell, Yerpo, Yuka1111, Zabshk, Zppix, 174 anonymous edits 299

Gravity of Earth *Source:* https://en.wikipedia.org/w/index.php?oldid=855734721 *License:* Creative Commons Attribution-Share Alike 3.0 *Contributors:* Abhijangale, Acroterion, Adaviel, Adit.thelu, Adérito Alfredo, Allen McC.~enwiki, Ametrica, ArnoldReinhold, AsSFGhjvchbb, BegbertBiggs, Billhpike, Bongwarrior, ClueBot NG, ColinClark, Con-struct, Count Count, Crystallizedcarbon, CuriousMind01, Cygnus78, DLH, DVdm, Dawn Bard, Dcoetzee, Devil2012, Dewritech, Discospinster, Dmd3e, Domdomegg, Donner60, Double Plus Ungood, Dragons flight, Dspradau, Egsan Bacon, Enpsychopiedia, Entropy1963, Eregli bob, Eric Kvaalen, Eynar, Fgnievinski, Francois-Pier, Frank Klemm, GeoWriter, Gilliam, Glrx, Goalie1998, GregorDS, H7an0f, Hanif Al Husaini, Headbomb, Help me to join Wiki, Imbocile, Inspots, Isambard Kingdom, Izno, J 1982, JKim, Jakec, Jim1138, Jkazanbai, Jl452, JorisvS, Julesd, Kaimbridge, Kendram, Kielvon, KrakatoaNair, KurtHeckman, Longhair, MMAdj, Materialscientist, MathAddict, Meters, Mikenorton, Moe Epsilon, Mogism, MusikAnimal, Nelg, Nilantha.k.herath, Notsofastyou2, Olikrist, Onklo, Oshwah, ParanoidLepton, Patriot1423, Pcmaster, Philippe Colentier, Pinethicket, PoqVaUSA, Prestonmag, Quasar G., Quondum, Racerx11, Reatlas, Reddysiddharth733, Renzmico345, Rjwilmsi, RockMagnetist, Ruslik0, Sameichel123, Samsara, Sarahj2107, Seaphoto, Simplexity22, Spyglasses, Strait, Stultiwikia, Swpb, Tamfang, Teodozjan, The High Fin Sperm Whale, The Voidwalker, Theosch, Tide rolls, Tom.Reding, Tutelary, TwoTwoHello, Velella, Vsmith, WaffleMaster44, Weehugh, Widefox, Widr, Winter-Spw, Wtmitchell, Xindeho, Xinghuei, Zwal11111, 202 anonymous edits 315

Earth's magnetic field *Source:* https://en.wikipedia.org/w/index.php?oldid=856219033 *License:* Creative Commons Attribution-Share Alike 3.0 *Contributors:* 7Sidz, AlternateYou, Anthony Appleyard, AntiCompositeNumber, Babylover74885, Bear-rings, Bongwarrior, CAPTAIN RAJU, CLCStudent, CV9933, Calidum, Cavit2, ChamithN, Chetvorno, ClueBot NG, Congruent-triangles, Dawkeye, Dcirovic, Ddcampayo, Devonson4, DocWatson42, Donner60, Draeath, Drmies, Earthandmoon, Edawg1005, Excirial, Fgnievinski, Frosty, Gab4gab, GeoWriter, Gilliam, Gorthian, Gulumeemee, GünniX, Hansiman33xd, Happy Attack Dog, Headbomb, Hibob222, Hutington, IdreamofJeanie, Ifnord, Isambard Kingdom, Iudshkjhc.ad, JDoe123456789987654321, JGS952, Jackfork, Jaek12020, Jaucafo, Jessicapierce, Jim.henderson, Jim1138, JimVC3, John "Hannibal" Smith, JorisvS, Julia W, Justin15w, Kalmiopsiskid, Kanizak, Kiril1029, Kralizec!, Krishnavedala, KylieTastic, Matt7899, Melonkelon, Mikenorton, MyrddinE, NSH002, Nihiltres, Njarboe, Not remi, Oshwah, Pasinduravimal, Pederony, Praemonitus, ProprioMe OW, RA0808, Rashkeqamar, RileyBugz, Rjwilmsi, RockMagnetist, Rod57, Ryea, SA 13 Bro, Sahil ku, Saryakhran, Saurabmarjara, Serols, Simplexity22, Srich32977, SuperMagicalRiot, TYelliot, TripWire, Vsmith, WMartin74, Wdanwatts, Wikigeek2000, Wishva de Silva, Worldbruce, Yinweichen, مفكر1, 193 anonymous edits 325

Earth's rotation *Source:* https://en.wikipedia.org/w/index.php?oldid=850506687 *License:* Creative Commons Attribution-Share Alike 3.0 *Contributors:* 100zuma, Abedwinger, Antoniosotovega9, Archon 2488, AstroLynx, BU Rob13, Baddullagammanadinuwara, Badkittydemon, Bear-rings, Bencbartlett, Bingston, Bishalajax, BlueGreenYellowRed, Bnicklin, Bobbie73, Bolo king, Brandmeister, Bruno verwimp, CLCStudent, Carlstak, Cattac, Chris the speller, Chris55, ClueBot NG, Cmglee, CommonsDelinker, Crystallizedcarbon, Cyberdog958, DMacks, DatGuy, David.moreno72, DavidLeighEllis, Dawnseeker2000, Dcirovic, Disha Bhattacharjee, Dogbyter, Dondervogel 2, Donner60, Dspark76, EnderLukeD, Excirial, Fgnievinski, Findanotheripaddress, Floquenbeam, Frood, G0mx, Gadget850, Gap9551, Georges T., Gilliam, Gogo Dodo, Harrydbear, Hendrix g, Imkindhearted, InedibleHulk, Isambard Kingdom, J 1982, JJMC89, Jankovoy, Jc3s5h, Jimw338, Jmencisom, Joedeshon, John Sauter, Johnuniq, Jusdafax, Khantopa, Krenair, KylieTastic, Lerf Lerfsson, Lugia2453, MRD2014, Magyar25, Materialscientist, Merelinguists, Mikenorton, Minimac, MusikAnimal, NILKAUSHIKBAHI PATEL, Nealmcb, Niclisp, OlEnglish, Olsensh, Orphan Wiki, Oshwah, Owlsarebae, Paulinho28, Piledhigheranddeeper, Pinethicket, Purple.panda.lover, Qzd, RA0808, RP88, Rfassbind, Riyaaz1, Rjwilmsi, Robertinventor, RockMagnetist, Rod57, Roxy the dog, Rtucker913, Rwflammang, SamuelHaein, Sandvich18, Schnurrbart, ScottieRoadPatriot, Scs, Serols, Smalljim, Tdadamemd dqmb, There'sNoTime, Tom.Reding, TomS TDotO, Tompop888, Twinsday, Vsmith, Wiae, Widr, Wikid77, Wikigcasillas96, William M. Connolley, Wiqi55, Zppix, 192 anonymous edits 349

Earth's orbit *Source:* https://en.wikipedia.org/w/index.php?oldid=855657760 *License:* Creative Commons Attribution-Share Alike 3.0 *Contributors:* 10metreh, 2TonyTony, A guy saved by Jesus, Acroterion, Agtx, Alecneal01, Algocu, Allthingstoallpeople, Am.dh1088, Anbu121, Andreasmperu, Arianewiki1, AsceticRose, Atkinson 291, AtticusX, Bear-rings, BethNaught, BlueEyedChicka, C.Logan, Changingguardsatbuckinghampalace, ChrisGualtieri, Classicwiki, ClueBot NG, Dawnseeker2000, Dbfirs, DemocraticLuntz, Denisarona, Discospinster, Dondervogel 2, Donner60, Dragons flight, Drbogdan, Drewmutt, Dw122339, Ehrenkater, Excirial, Flyer22 Reborn, Frosty, Frozenprakash, Gilliam, Gmporr, Gob Lofa, Hardaker, Hcobb, Hy Brasil, I dream of horses, Incnis Mrsi, Isambard Kingdom, J 1982, Jaswant mohali, Jc3s5h, Jehochman, JoeSperrazza, Johnuniq, JorisvS, K6ka, Katieh5584, Ladygaga241, LjL, MDWeathers, Markshenouda29, Materialscientist, Mikhail Ryazanov, Mild Bill Hiccup, Moe Epsilon, Mpj7, MusikAnimal, MySonLikesTrump, Nabla, NewEnglandYankee, Niceguy149, Onel5969, Oshwah, Pbwelch, Pieeyesquared, Pratyya Ghosh, Psolrzan, Reatlas, Ricardogpn, Roamsdirac, RockMagnetist, Roshan lasrado, Ruslik0, Sachin3468, Saros136, Satellizer, Schmarchist, Scottperry, Seaphoto, Secret agent of Hacin, Shellwood, Slightsmile, Smash1gordon, Snow Blizzard, Solarra, Souljaboy5700, Stroppolo, THetardis123, Tdadamemd, Tetra quark, ThatRusskiiGuy, TheWikiEditor123, Timothy Cooper, ToBeFree, Tom.Reding, Twinsday, Tyzy02, Ubiquity, Vanamonde93, Vaughan Pratt, Widefox, Widr, WikiDan61, Yadsalohcin, Zydecojazz, 194 anonymous edits 360

Biosphere *Source:* https://en.wikipedia.org/w/index.php?oldid=855360518 *License:* Creative Commons Attribution-Share Alike 3.0 *Contributors:* AGH 999, Adarshs018, Alaney2k, Algmwc5, Amortias, Anne drew Andrew and Drew, Arthur Rubin, Babitaarora, Bentogoa, BiologicalMe, Brandmeister, CAPTAIN RAJU, CASSIOPEIA, CambridgeBayWeather, Camyoung54, Carminowe of Hendra, Ccoollaa, ClueBot NG, Coolio385, DASonnenfeld, DVdm, Danielle bb, Dark-World25, Dawnseeker2000, Dcirovic, DemocraticLuntz, Discospinster, DoABarrelRoll.dev, Donner60, Drbogdan, Drmies, Emir of Wikipedia, Es punta, Evilpingouin, Excirial, Fmadd, FoCuSandLeArN, Fraggle81, Frosty, Fugitron, Gilliam, Gnomus, Greyjoy, HalenTaylor02, Headbomb, I dream of horses, IronGargoyle, Isaac Newton 2md, Iseult, J 1982, Jackfork, JinJanJo, Jprg1966, Judahpurwanto55, K6ka, Kevinjudge, Kinyek, Kku, Kunalsukhija, Lakun.patra, Larry Hockett, Leviavery, MarcAnfhoy, Materialscientist, MelbourneStar, Mmyers1976, Mtkumar.27310, NHCLS, Niceguyedc, Nihil novi, North Shoreman, Noyster, OAnick, ODonsky, Omnipaedista, Originalwana, Oshwah, Paleorthid, Pbsouthwood, Pinethicket, Revent, Rivertorch, Rjwilmsi, Robert-Rhys, RockMagnetist, RockMagnetist (DCO visiting scholar), S11565165, SchreiberBike, Serols, Shellwood, Simplexity22, Skamecrazy123, Squidwourde, Stui, Sunrise, TheVVigilantEE, Tom.Reding, Tomdo08, Trappist the monk, Trinidade, VQuakr, Vesuvius Dogg, Vsmith, WadeSimMiser, WarrenSpieker1, Werwolf129, Widr, Wikid77, Xamantha dalas, Zamaster4536, 217 anonymous edits 367

Land use *Source:* https://en.wikipedia.org/w/index.php?oldid=855625760 *License:* Creative Commons Attribution-Share Alike 3.0 *Contributors:* 3 of Diamonds, 7, 83d40m, 90 Auto, Arjayay, Arthur Rubin, AutoGeek, Axlq, Bando26, Beland, Bender235, Bendzh, Bgwhite, Bhny, Binary TSO, Bkonrad, Blarrge, Bobo192, Bruce1ee, Bsadowski1, Caltas, Certes, ClueBot NG, Colonies Chris, Cp111, Crystallizedcarbon, DASonnenfeld, Dave Dial, DavidLeighEllis, DeWikiMan, Deli nk, Denisarona, Der Golem, DocWatson42, Download, Dpakdel, Dragunova, Elekhh, Epbr123, Epipelagic, Escape Orbit, Fastily, Fieldday-sunday, Fugitron, Fyrael, Gaius Cornelius, Ghostalker, Gilliam, Gogo Dodo, Guettarda, Hu12, IKEsofTR, ImperfectlyInformed, Ingolfson, JC Shepard, Jim.henderson, JimVC3, Jterstriep, JueLinLi, KATECS, Kayau, Kentynet, Kern3020, Khushbu Arora, Kku, Kosher Fan, Kurieeto, KylieTastic, Laudosopopo, Lazylaces, Lightmouse, Lugia2453, LukeStewart, Luna Santin, Magioladitis, Mailseth, Marilee cts, Materialscientist, Mattisse, Mayooranathan, Mccapra, Mdicato, Mejor Los Indios, Momoricks, MrBell, Muzikbox, Narayansg, Natg 19, Nathan Johnson, Nirvana2013, Nudecline, Nukeless, Orphan Wiki, Oscarthecat, Oshwah, Paleorthid, People of awesomness, Pinethicket, Podzolman, Pokedigi, Pradeepkoundal, Ps07swt, Pterre, Pwaddell, R45wm800, Randomeditor1000, RayquazaDialgaWeird2210, Rhstafursky, Rivertorch, Rowan Adams, Rrburke, SPACKlick, Satellizer, Serols, Sionus, Sln3412, Some jerk on the Internet, Stwalkerster, Synchronism, TAnthony, Talkov, Tentinator, The wub, Thomasmeeks, Thrisanth rahulan, Tide rolls, Tymon.r, Vegaswikian, Verne Equinox, Vigilius, Vinny Burgoo, Vrenator, Vsmith, W7KyzmJt, Wavelength, Widr, Wikiwinter2016, Wkme20089, Wlchristopher, Ziggyfan23, Zyxw, 219 anonymous edits 375

Human geography *Source:* https://en.wikipedia.org/w/index.php?oldid=855997759 *License:* Creative Commons Attribution-Share Alike 3.0 *Contributors:* 00ff00, 1exec1, ATesarow, Acroterion, Alex Cohn, Allforrous, Altamel, Angela Gh, Antiqueight, Apokryltaros, BD2412, Beyond My Ken, Bgwhite, Bhny, Biogeographist, Bongwarrior, Branstrom, Bronze2018, Byrneryan, CAPTAIN RAJU, Captjons, ChamithN, ClueBot NG, Cooper-42, CsDix, DASonnenfeld, DVdm, Dcirovic, DennouNeko, Devi nandi, Discospinster, Dl2000, DocWatson42, Donner60, Elekhh, Erebus Morgaine, EuroCarGT, Excirial, Eykgamekllr, Fev, Fraggle81, FreddieMartyn, Funandtrvl, Gaia Octavia Agrippa, GeneralizationsAreBad, GeoBenny, Gilliam, Ginsuloft, Gobonobo, Gyalpot, Headbomb, Hobojoename, Hogyn Lleol, Inaki espana, Ipigott, IronGargoyle, IstGebongt, J947, JJMC89, JamiePryor, Jim1138, Jith12, Joefromrandb, Johanna-Hypatia, Jonesey95, Jwisser, Jytdog, K6ka, KBH96, Kelsonyauni, Khazar2, Khunsar, Kirschme, Klilidiplomus, LegoKeoni, Lightningball734, MRD2014, Macedonian, Manoorman, Marcocapelle, MarktheChubRubber, Materialscientist, Mayumashu, McGeddon, Me, Myself, and I are Here, Mean as custard, Menghwani, Mercury McKinnon, Mifter Public, MithrandirAgain, Murky am, MynameisdoofandyoudowhatisayOOHOOH, Narky Blert, Nicole Sharp, Non-dropframe, Nuriatefa, O.Koslowski, Ooand2, Oshwah, Owenstuckey18, Pbsouthwood, Pharaoh of the Wizards, Pinethicket, PlyrStar93, Prisencolin, Quinn Schmidt11, RFWJW, Reatlas, Redolta, RexSueciae, RockMagnetist, Rubicon, Rwtwv, SAUD BIN RAJIBULLAH, SaladPuma, SalineBrain, Savagerpg, Sbutcher1001, Serols, Shellwood, Skdkdk, SkyWarrior, Spicemix, Sprauera, Stwalkerster, Sumi-Jo Zhang, TAnthony,

Image Sources, Licenses and Contributors

The sources listed for each image provide more detailed licensing information including the copyright status, the copyright owner, and the license conditions.

Image *Source:* https://en.wikipedia.org/w/index.php?title=File:Padlock-silver.svg *Contributors:* AzaToth, BotMultichill, BotMultichillT, Gurch, Jarekt, Kallerna, Multichill, Perhelion, Rd232, Riana, Sarang, Siebrand, Steinsplitter, 4 anonymous edits 1

Image *Source:* https://en.wikipedia.org/w/index.php?title=File:Cscr-featured.svg *License:* GNU Lesser General Public License *Contributors:* Anomie 1

Image *Source:* https://en.wikipedia.org/w/index.php?title=File:The_Earth_seen_from_Apollo_17.jpg *License:* Public Domain *Contributors:* NASA/ Apollo 17 crew; taken by either Harrison Schmitt or Ron Evans 1

Figure 1 *Source:* https://en.wikipedia.org/w/index.php?title=File:Beowulf_-_eorthan.jpg *License:* Public Domain *Contributors:* ВоенТех 4

Figure 2 *Source:* https://en.wikipedia.org/w/index.php?title=File:Protoplanetary-disk.jpg *License:* Public Domain *Contributors:* NASA 5

Figure 3 *Source:* https://en.wikipedia.org/w/index.php?title=File:USA_10654_Bryce_Canyon_Luca_Galuzzi_2007.jpg *License:* Creative Commons Attribution-Sharealike 2.5 *Contributors:* Luca Galuzzi (Lucag) 6

Image *Source:* https://en.wikipedia.org/w/index.php?title=File:Interactive_icon.svg *Contributors:* User:Evolution and evolvability 7

Figure 4 *Source:* https://en.wikipedia.org/w/index.php?title=File:PhylogeneticTree,_Woese_1990.svg *License:* Creative Commons Attribution-Sharealike 3.0 *Contributors:* File:PhylogeneticTree, Woese 1990.PNG: Maulucioni derivative work: TilmannR 8

Figure 5 *Source:* https://en.wikipedia.org/w/index.php?title=File:Earth2014shape_SouthAmerica_small.jpg *Contributors:* User:Geodesy2000 10

Image *Source:* https://en.wikipedia.org/w/index.php?title=File:Earth-cutaway-schematic-english.svg *License:* Public Domain *Contributors:* derivative work: Srimadhav Earth_internal_structure.png: USGS 12

Image *Source:* https://en.wikipedia.org/w/index.php?title=File:Tectonic_plates_(empty).svg *License:* Public Domain *Contributors:* User:Ævar Arnfjörð Bjarmason 13

Figure 6 *Source:* https://en.wikipedia.org/w/index.php?title=File:Mount-Everest.jpg *License:* Creative Commons Attribution 2.0 *Contributors:* 98Alex, Esszet, Hedwig in Washington, KurodaSho, LudwigSebastianMicheler, NicoScribe, Till.niermann, 2 anonymous edits 15

Figure 7 *Source:* https://en.wikipedia.org/w/index.php?title=File:AYool_topography_15min.png *License:* Creative Commons Attribution 2.5 *Contributors:* Plumbago 15

Figure 8 *Source:* https://en.wikipedia.org/w/index.php?title=File:Earth_dry_elevation.stl *Contributors:* User:Cmglee 16

Figure 9 *Source:* https://en.wikipedia.org/w/index.php?title=File:Earth_elevation_histogram_2.svg *License:* Public domain *Contributors:* Ciaurlec, MushiHoshiIshi, OgreBot 2, Pitke, Sbelza, Tano4595, باسم, 和平奮鬥救地球 17

Figure 10 *Source:* https://en.wikipedia.org/w/index.php?title=File:MODIS_Map.jpg *License:* Public Domain *Contributors:* NASA 18

Image *Source:* https://en.wikipedia.org/w/index.php?title=File:Felix_from_ISS_03_sept_2007_1138Z.jpg *License:* Public Domain *Contributors:* NASA 19

Image *Source:* https://en.wikipedia.org/w/index.php?title=File:Pressure_ridges_Scott_Base_lrg.jpg *License:* Public Domain *Contributors:* Michael Studinger 19

Image *Source:* https://en.wikipedia.org/w/index.php?title=File:3D-Clouds.jpg *Contributors:* User:Pocketthis 20

Figure 11 *Source:* https://en.wikipedia.org/w/index.php?title=File:Full_moon_partially_obscured_by_atmosphere.jpg *License:* Public Domain *Contributors:* NASA 22

Figure 12 *Source:* https://en.wikipedia.org/w/index.php?title=File:Geoids_sm.jpg *License:* Public Domain *Contributors:* Angrense, B jonas, Ciaurlec, Denniss, Look2See1, LudwigSebastianMicheler, Maddox2, Mapmarks, Mario1952, Minor edit, Rainald62, RedAndr, Stewi101015, 1 anonymous edits 23

Figure 13 *Source:* https://en.wikipedia.org/w/index.php?title=File:Structure_of_the_magnetosphere-en.svg *License:* Public Domain *Contributors:* Original bitmap from NASA. SVG rendering by Aaron Kaase. 24

Figure 14 *Source:* https://en.wikipedia.org/w/index.php?title=File:EpicEarth-Globespin(2016May29).gif *License:* Public Domain *Contributors:* User:Tdadamemd 25

Figure 15 *Source:* https://en.wikipedia.org/w/index.php?title=File:Pale_Blue_Dot.png *License:* Public Domain *Contributors:* Voyager 1 26

Figure 16 *Source:* https://en.wikipedia.org/w/index.php?title=File:AxialTiltObliquity.png *License:* Creative Commons Attribution 3.0 *Contributors:* Dna-webmaster 27

Figure 17 *Source:* https://en.wikipedia.org/w/index.php?title=File:Moraine_Lake_17092005.jpg *Contributors:* Berrucomons, Ernác, Flcelloguy∼commonswiki, Gorgo, Herzi Pinki, Hike395, Ianare, J 1982, JotaCartas, Kalbbes, Kasir, MONGO, Mykola Swarnyk, O (bot), Odysseus1479, Paintman, Para, Raphael17, Shizhao, Sultan11, Takabeg, Thierry Caro, Yuriy75, Überraschungsbilder, باسم, 1 anonymous edits 29

Figure 18 *Source:* https://en.wikipedia.org/w/index.php?title=File:Pavlof2014iss.jpg *License:* Public Domain *Contributors:* NASA 31

Figure 19 *Source:* https://en.wikipedia.org/w/index.php?title=File:Continents_vide_couleurs.png *License:* Creative Commons Attribution-ShareAlike 3.0 Unported *Contributors:* User:Cogito ergo sumo 32

Image *Source:* https://en.wikipedia.org/w/index.php?title=File:FullMoon2010.jpg *License:* Creative Commons Attribution-Sharealike 3.0 *Contributors:* Gregory H. Revera 33

Figure 20 *Source:* https://en.wikipedia.org/w/index.php?title=File:Earth-Moon.svg *License:* Public Domain *Contributors:* Earth-Moon.PNG: Earth-image from NASA; arrangement by brews_ohare derivative work: Cmglee (talk) 34

Figure 21 *Source:* https://en.wikipedia.org/w/index.php?title=File:Tracy_Caldwell_Dyson_in_Cupola_ISS.jpg *License:* Public Domain *Contributors:* NASA/Tracy Caldwell Dyson 35

Figure 22 *Source:* https://en.wikipedia.org/w/index.php?title=File:NASA-Apollo8-Dec24-Earthrise.jpg *License:* Public Domain *Contributors:* AKA MBG, Apalsola, BarnacleKB∼commonswiki, Berrucomons, BoringHistoryGuy, Davepape, Edward, Emijrp, Geitost, George Chernilevsky, Howcheng, J 1982, JalalV, Jcpag2012, Jean-Frédéric, Jimmy Xu, KaragouniS∼commonswiki, Ke4roh, Lx 121, Mario modesto, Medconn, O484∼enwiki, Roberta F., Saperaud∼commonswiki, Schekinov Alexey Victorovich, Sevela.p, Sh1019, Simonizer, Starscream, Surya Prakash.S.A., Takabeg, Tangopaso, TheDJ, Triplecaña, Wieralee, Wstrwald, Марсианский хро, باسم, 5 anonymous edits 36

Image *Source:* https://en.wikipedia.org/w/index.php?title=File:Earth_symbol.svg *Contributors:* OsgoodeLawyer 36

Image *Source:* https://en.wikipedia.org/w/index.php?title=File:Sabaa_Nissan_Militiaman.jpg *License:* Creative Commons Attribution-ShareAlike 3.0 Unported *Contributors:* Christiaan Briggs —Christiaan 21:23, 2 Feb 2005 (UTC) 37

Image *Source:* https://en.wikipedia.org/w/index.php?title=File:Oxalis-pes-caprae0016c.jpg *License:* Creative Commons Attribution-Share Alike *Contributors:* MathKnight and Zachi Evenor אבנור וצחי המתמטיקה אבי"ר 37

Image *Source:* https://en.wikipedia.org/w/index.php?title=File:Oryctolagus_cuniculus_Tasmania_2-mirror.jpg *License:* Creative Commons Attribution-Sharealike 3.0 *Contributors:* JJ Harrison (jjharrison89@facebook.com) 37

Image *Source:* https://en.wikipedia.org/w/index.php?title=File:AlamosaurusDB.jpg *License:* GNU Free Documentation License *Contributors:* Abyssal, BotMultichillT, Dinoguy2, Haplochromis, Kevmin, MGA73bot2, OgreBot 2, 1 anonymous edits 38

Image *Source:* https://en.wikipedia.org/w/index.php?title=File:Clouds_over_the_Atlantic_Ocean.jpg *License:* Creative Commons Attribution-Sharealike 3.0 *Contributors:* Tiago Fioreze 38

Image *Source:* https://en.wikipedia.org/w/index.php?title=File:Halobacteria.jpg *License:* Public Domain *Contributors:* NASA 38

Image *Source:* https://en.wikipedia.org/w/index.php?title=File:Water_drop_001.jpg *License:* Creative Commons Attribution 2.0 *Contributors:* Agung.karjono, Amorymeltzer, Apalsola, Berrucomons, De728631, Dimi z, FlickreviewR, Julia W, Mywood, Sergey kudryavtsev, Tomer T, Tulsi Bhagat, Zaqarbal, باسم, 2 anonymous edits 38

Image *Source:* https://en.wikipedia.org/w/index.php?title=File:Montagem_Sistema_Solar.jpg *License:* Public Domain *Contributors:* ALE!, Lucazdj, Martin H., Ruslik0, WOtP, Zeca pum, 1 anonymous edits 39

Image *Source:* https://en.wikipedia.org/w/index.php?title=File:Rho_Ophiuchi.jpg *License:* Public Domain *Contributors:* NASA/JPL-Caltech/WISE Team 39

Image *Source:* https://en.wikipedia.org/w/index.php?title=File:Milky_Way_Arms_ssc2008-10.svg *License:* Public Domain *Contributors:* Milky_Way_2005.jpg: R. Hurt derivative work: Cmglee (talk) 39

Image *Source:* https://en.wikipedia.org/w/index.php?title=File:Artist's_impression_of_the_Milky_Way_(updated_-_annotated).jpg *License:* Public Domain *Contributors:* Mike Peel, Stas1995, Szczureq, ToBeFree 39

Image *Source:* https://en.wikipedia.org/w/index.php?title=File:Local_Group_and_nearest_galaxies.jpg *Contributors:* User:Cicconorsk 39

Image *Source:* https://en.wikipedia.org/w/index.php?title=File:Local_supercluster-ly.jpg *License:* Public Domain *Contributors:* NASA 39

Image *Source:* https://en.wikipedia.org/w/index.php?title=File:Observable_universe_r2.jpg *Contributors:* User:Azcolvin429, User:JA Galán Baho 39

Image *Source:* https://en.wikipedia.org/w/index.php?title=File:Observable_Universe_with_Measurements_01.png *License:* Creative Commons Attribution-Sharealike 3.0 *Contributors:* Andrew Z. Colvin 40

470

Figure 144 *Source:* https://en.wikipedia.org/w/index.php?title=File:Magnetic_North_Pole_Positions_2015.svg *Contributors:* User:Cavit 330

Figure 145 *Source:* https://en.wikipedia.org/w/index.php?title=File:Magnetosphere_Levels.svg *License:* Public Domain *Contributors:* Magnetosphere_Levels.jpg: Dennis Gallagher derivative work: Frédéric MICHEL 331

Figure 146 *Source:* https://en.wikipedia.org/w/index.php?title=File:Magnetic_Storm_Oct_2003.jpg *License:* Public Domain *Contributors:* Bot-Multichill, DMY, Frank C. Müller, Ghouston, Jochen Burghardt, Kuttappan Chettan, Pmau, ProfessorX, RockMagnetist 333

Figure 147 *Source:* https://en.wikipedia.org/w/index.php?title=File:Earth_Magnetic_Field_Declination_from_1590_to_1990.gif *License:* Public Domain *Contributors:* U.S. Geological Survey (USGS) http://www.usgs.gov/ 334

Figure 148 *Source:* https://en.wikipedia.org/w/index.php?title=File:Geomagnetic_axial_dipole_strength.svg *Contributors:* User:Cavit 334

Figure 149 *Source:* https://en.wikipedia.org/w/index.php?title=File:Geomagnetic_polarity_late_Cenozoic.svg *License:* Public Domain *Contributors:* United States Geological Survey, hand-traced to vector by me (User:Intgr) 335

Figure 150 *Source:* https://en.wikipedia.org/w/index.php?title=File:Brunhes_geomagnetism_western_US.png *License:* Public Domain *Contributors:* Apocheir, DMY, Pieter Kuiper, RockMagnetist, Sarang, 1 anonymous edits 337

Figure 151 *Source:* https://en.wikipedia.org/w/index.php?title=File:Outer_core_convection_rolls.jpg *License:* Public Domain *Contributors:* Ariadacapo, FlickreviewR, Lymantria, RockMagnetist 338

Figure 152 *Source:* https://en.wikipedia.org/w/index.php?title=File:Magnetic_Field_Earth.png *License:* Public Domain *Contributors:* Credit: Terrence Sabaka et al./NASA GSFC 341

Figure 153 *Source:* https://en.wikipedia.org/w/index.php?title=File:Spherical_harmonics_positive_negative.svg *Contributors:* User:Krishnavedala 343

Figure 154 *Source:* https://en.wikipedia.org/w/index.php?title=File:VFPt_four_charges.svg *License:* Creative Commons Attribution-Sharealike 3.0 *Contributors:* Chenspec, Geek3, Sarang 344

Figure 155 *Source:* https://en.wikipedia.org/w/index.php?title=File:Globespin.gif *License:* Creative Commons Attribution-Sharealike 3.0 *Contributors:* Wikiscient 350

Figure 156 *Source:* https://en.wikipedia.org/w/index.php?title=File:Earth_Rotation_(Nepal,_Himalayas).jpg *Contributors:* User:Jankovoy 350

Figure 157 *Source:* https://en.wikipedia.org/w/index.php?title=File:Starry_Spin-up.jpg *Contributors:* Jmencisom 353

Figure 158 *Source:* https://en.wikipedia.org/w/index.php?title=File:Sidereal_day_(prograde).png *License:* GNU Free Documentation License *Contributors:* User:Gdr 355

Figure 159 *Source:* https://en.wikipedia.org/w/index.php?title=File:Earth_rotation_tangential_speed.svg *Contributors:* Cmglee, Waldir 356

Figure 160 *Source:* https://en.wikipedia.org/w/index.php?title=File:AxialTiltObliquity.png *License:* Creative Commons Attribution 3.0 *Contributors:* Dna-webmaster 357

Figure 161 *Source:* https://en.wikipedia.org/w/index.php?title=File:Deviation_of_day_length_from_SI_day.svg *License:* Public Domain *Contributors:* II VII XII 358

Figure 162 *Source:* https://en.wikipedia.org/w/index.php?title=File:Protoplanetary-disk.jpg *License:* Public Domain *Contributors:* NASA 359

Figure 163 *Source:* https://en.wikipedia.org/w/index.php?title=File:North_season.jpg *License:* Creative Commons Zero *Contributors:* Tau'olunga 361

Figure 164 *Source:* https://en.wikipedia.org/w/index.php?title=File:Heliocentric.jpg *License:* Public Domain *Contributors:* Cpt.Muji, Gildemax, Micheletb, Nillerdk, Raymond, Rocket000, Roomba, RuM, Ruslik0, 2 anonymous edits 362

Figure 165 *Source:* https://en.wikipedia.org/w/index.php?title=File:Geoz_wb_en.svg *License:* Creative Commons Attribution-Sharealike 2.5 *Contributors:* Original image by Niko Lang SVG version by User:Booyabazooka 362

Image *Source:* https://en.wikipedia.org/w/index.php?title=File:Seasons1.svg *License:* GNU Free Documentation License *Contributors:* following Duoduoduo's advice, vector image: Gothika. 365

Figure 166 *Source:* https://en.wikipedia.org/w/index.php?title=File:Seawifs_global_biosphere.jpg *Contributors:* Jdx, Julia W, Luis Fernández García, ScotXW, Stewi101015, Tano4595, TheDJ, Túrelio, Yikrazuul, 2 anonymous edits 368

Figure 167 *Source:* https://en.wikipedia.org/w/index.php?title=File:90_mile_beach.jpg *Contributors:* 2000, Basvb, ComputerHotline, Fir0002, MartinHansV, Pierre cb, Saperaud∼commonswiki, Shiftchange, Thuresson 368

Figure 168 *Source:* https://en.wikipedia.org/w/index.php?title=File:Stromatolithe_Paléoarchéen_-_MNHT.PAL.2009.10.1.jpg *Contributors:* Didier Descouens 369

Figure 169 *Source:* https://en.wikipedia.org/w/index.php?title=File:Ruppelsvulture.jpg *License:* Creative Commons Attribution-Sharealike 2.5 *Contributors:* RPS (Rob Schoenmaker) at en.wikipedia 370

Figure 170 *Source:* https://en.wikipedia.org/w/index.php?title=File:XenophyophoreNOAA.jpg *Contributors:* NOAA 371

Image *Source:* https://en.wikipedia.org/w/index.php?title=File:Globe_Spin.gif *License:* Public Domain *Contributors:* Allforrous, Originalwana 372

Image *Source:* https://en.wikipedia.org/w/index.php?title=File:NorthAmericaCycle_Small.gif *License:* Public Domain *Contributors:* Allforrous, Originalwana 372

Image *Source:* https://en.wikipedia.org/w/index.php?title=File:LaNina_Mollweide.gif *License:* Public Domain *Contributors:* Allforrous, Originalwana 372

Image *Source:* https://en.wikipedia.org/w/index.php?title=File:Mollweide_Cycle.gif *License:* Public Domain *Contributors:* Allforrous, Originalwana 373

Figure 171 *Source:* https://en.wikipedia.org/w/index.php?title=File:Biosphere_2_4888964549.jpg *License:* Creative Commons Attribution 2.0 *Contributors:* CGP Grey 373

Figure 172 *Source:* https://en.wikipedia.org/w/index.php?title=File:Indiana_Dunes_Habitat_Fragmentation.jpg *License:* Public Domain *Contributors:* Ciaurlec, Closeapple, Esculapio, Magog the Ogre, Orrling 376

Figure 173 *Source:* https://en.wikipedia.org/w/index.php?title=File:Europe_land_use_map.png *License:* GNU Free Documentation License *Contributors:* Kentynet (talk) 10:12, 30 April 2011 (UTC) 376

Image *Source:* https://en.wikipedia.org/w/index.php?title=File:Kastellet_cph.jpg *License:* Creative Commons Attribution-ShareAlike 3.0 Unported *Contributors:* Berrucomons, Elgaard, Glenn, HBR, MGA73bot2, Mahlum, Orrling, Pauk, Peregrine981, Rl, Thierry Caro, Xenophon, Yuriy75, 2 anonymous edits 378

Figure 174 *Source:* https://en.wikipedia.org/w/index.php?title=File:Snow-cholera-map-1.jpg *License:* Public Domain *Contributors:* John Snow 383

Image *Source:* https://en.wikipedia.org/w/index.php?title=File:OrteliusWorldMap.jpeg *License:* Public Domain *Contributors:* Alexan, AnRo0002, AndreasPraefcke, David Kernow∼commonswiki, Eitan96, Electionworld, Flamarande∼commonswiki, Geagea, Hello world, Itu, Jan Arkesteijn, Mattes, Roke∼commonswiki, Túrelio, Un1c0s bot∼commonswiki, W!B:, ¡0-8-15!, 4 anonymous edits 381

Figure 175 *Source:* https://en.wikipedia.org/w/index.php?title=File:Agriculture_in_Asia.jpg *Contributors:* Achim55, Kirschme, OgreBot 2, P199 385

Figure 176 *Source:* https://en.wikipedia.org/w/index.php?title=File:Shan_Street_Bazaar.JPG *License:* Creative Commons Attribution-Sharealike 3.0 *Contributors:* User:Lionslayer 386

Figure 177 *Source:* https://en.wikipedia.org/w/index.php?title=File:Carl_ritter.jpg *License:* Public Domain *Contributors:* Elya, INS Pirat, Kentin, Kilom691, Matt314, Quedel, Tony Rotondas 388

Figure 178 *Source:* https://en.wikipedia.org/w/index.php?title=File:Lunar_eclipse_October_8_2014_California_Alfredo_Garcia_Jr_mideclipse.JPG *License:* Creative Commons Attribution-Sharealike 2.0 *Contributors:* Aschroet, B dash, ComputerHotline, FlickreviewR 2, Laurent Bélanger, Tomruen, 2 anonymous edits 396

Image *Source:* https://en.wikipedia.org/w/index.php?title=File:LRO_WAC_Nearside_Mosaic.jpg *License:* Public Domain *Contributors:* NASA/GSFC/Arizona State University 397

Image *Source:* https://en.wikipedia.org/w/index.php?title=File:Moon_Farside_LRO.jpg *License:* Public Domain *Contributors:* NASA/GSFC/Arizona State University 397

Image *Source:* https://en.wikipedia.org/w/index.php?title=File:LRO_WAC_North_Pole_Mosaic_(PIA14024).jpg *License:* Public Domain *Contributors:* NASA/GSFC/Arizona State University 397

Image *Source:* https://en.wikipedia.org/w/index.php?title=File:LRO_WAC_South_Pole_Mosaic.jpg *License:* Public Domain *Contributors:* NASA/GSFC/Arizona State University 397

Figure 179 *Source:* https://en.wikipedia.org/w/index.php?title=File:Evolution_of_the_Moon.ogv *License:* Public Domain *Contributors:* NASA/Goddard Space Flight Center 398

Image *Source:* https://en.wikipedia.org/w/index.php?title=File:14-236-LunarGrailMission-OceanusProcellarum-Rifts-Overall-20141001.jpg *License:* Public Domain *Contributors:* Drbogdan, Lotse, Sneeuwschaap, Tulsi Bhagat, 1 anonymous edits 399

Image *Source:* https://en.wikipedia.org/w/index.php?title=File:PIA18822-LunarGrailMission-OceanusProcellarum-Rifts-Overall-20141001.jpg *License:* Public Domain *Contributors:* Drbogdan, Lotse, Sneeuwschaap 399

Image *Source:* https://en.wikipedia.org/w/index.php?title=File:PIA18821-LunarGrailMission-OceanusProcellarum-Rifts-Closeup-20141001.jpg *License:* Public Domain *Contributors:* Badzil, Drbogdan, Lotse, Sneeuwschaap 400

Figure 180 *Source:* https://en.wikipedia.org/w/index.php?title=File:Moon_diagram.svg *License:* Creative Commons Attribution 3.0 *Contributors:* User:Kelvinsong 401

Figure 181 *Source:* https://en.wikipedia.org/w/index.php?title=File:MoonTopoLOLA.png *License:* Creative Commons Attribution 3.0 *Contributors:* Mark A. Wieczorek 402

License

Index

N-body problem, 365
Neanderthal, 73
Near-Earth asteroid, 35
Near-Earth supernova, 135
Near side of the Moon, 395, 397, 404, 413
Nebular hypothesis, 44
Nebular theory, 5
Nectarian, 405
Neil Armstrong, 76, 425, 426
Nematode, 113, 120
Nemertea, 120
Neodymium, 177
Neogene, 63, 91
Neomura, 59
Neon, 168, 268, 395, 409
Neon-20, 170
Neon-22, 170
Neoproterozoic, 8, 83, 117
Nepal, 242, 254, 350
Nepotism, 127
Neptune, 50, 297
Neptunium, 176
Nerthus, 4
Nervous system, 113
Net force, 316
Neumarkt (district), 244
Neuroplasticity, 73
Neutron, 168
Neutron Spectrometer (NS), 407
Neutron star, 169
Nevadan orogeny, 89, 90
Nevado Huascarán, 316
Newark Supergroup, 88
New Brunswick, 212
Newcombs Tables of the Sun, 354
New cultural geography, 384
Newfoundland and Labrador, 212
New International Encyclopedia, 314
New Jersey, 88
New moon, 414, 419
New Scientist, 134
New South Wales, 254
Newton.27s second law, 315
Newtons second law, 323
Newton (unit), 315
New world, 75
New York City, 225, 375
New York Times, 347
New Zealand, 242, 250, 255, 389
Niagara Falls, 142
Nickel, 11, 172, 188
Nickel-56, 169
Nickel-58, 171
Nickel sulfide, 53, 104
Nicolaus Copernicus, 351, 361
Nicole Oresme, 351

Nigel Thrift, 390, 391
Niger, 254
Nigeria, 254
Night, 265
Night sky, 350
Nitrogen, 3, 18, 150, 168, 178, 179, 265, 266, 268, 281, 410
Nitrogen-14, 170
Nitrogen oxide, 295
Nitrous oxide, 19, 139
NOAA, 276, 291, 312
Noble gas, 176
Noctilucent cloud, 271, 284
Nodal line, 342
Nodal precession, 415
Nokia, 429
Nomad, 73
Non-governmental organizations, 375
Non-inertial reference frame, 317
Non-renewable resource, 30
Non-renewable resources, 441
Non-representational theory, 387, 390
Norse mythology, 4, 439
North America, 85, 88, 90–92, 143, 249, 306, 308, 393
North American craton, 49
North American Datum of 1983, 158
North American Plate, 14, 206, 221
North American Vertical Datum of 1988, 158
North American X-15, 276
North Atlantic Treaty Organization, 345
Northern Canada, 21
Northern celestial hemisphere, 350
Northern Europe, 94
Northern Hemisphere, 28, 308, 349, 393, 416
Northern Territory, 249
North geomagnetic pole, 325
North Magnetic Pole, 328, 349
North Pole, 349, 415
Norway, 247, 249
Nostoc, 110
Notochord, 66
Nova Science Publishers, 131
Nova (TV series), 347
NRLMSISE-00, 277
Nuclear binding energy curve, 172
Nuclear fission, 175
Nuclear fusion, 44, 172
Nuclear holocaust, 137
Nuclear weapon, 75
Nucleic acid, 53, 103
Nucleobases, 52
Nucleosynthesis, 171, 172
Nucleotide, 54, 104
Nuclide, 170
Numerical weather prediction, 293

Overgrazing, 31
Oxford, 165
Oxford English Dictionary, 463
Oxford spelling, 4
Oxford University Press, 132, 224
Oxide, 11
Oxidizing, 282
Oxidizing agent, 116
Oxygen, 3, 11, 38, 41, 46, 99, 167–169, 172, 177–179, 265, 266, 268, 270, 410
Oxygen-16, 170
Oxygen catastrophe, 19, 58
Oxygen evolution in nature, 19
Oxygenic photosynthesis, 108
Oxygen saturation, 123
Ozone, 19, 57, 273
Ozone depletion, 283
Ozone–oxygen cycle, 19
Ozone layer, 8, 19, 43, 46, 139, 272, 273, 277, 325

Pachamama, 439
Pacific Ocean, 92, 143, 216, 252
Pacific Plate, 14, 200, 221
Pacific Ring of Fire, 200, 216
Pac-Man, 87
Pakistan, 254
Pale Blue Dot, 26, 440
Paleoarchean, 336
Paleoatmosphere, 281
Paleobiology, 211
Paleocene, 63, 91
Paleoclimatology, 302
Paleogene, 63, 92
Paleogeography, 211
Paleomagnetism, 58, 210, 325
Paleomap Project, 143
Paleontological Society, 130
Paleontology, 34
Paleo-Tethys Ocean, 86, 87
Paleozoic, 63, 83, 87, 90
Palm Springs, California, 247
Palomar College, 134
Pan-African orogeny, 61
Pangaea, 6, 43, 63, 64, 79, 83, 86–88, 90, 142, 221
Pangaea Ultima, 143, 144
Pangea, 83, 209
Pannotia, 6, 43, 61, 79, 83
Panthalassa, 83, 86, 87
Panthéon (Paris), 352
Papunya, 249
Paraceratherium, 70
Paradigm shift, 220
Parallel (latitude), 162
Paranthropus, 72

Parasite, 59, 109
Parasitism, 120
Parmenides, 36
Partial differential equation, 339
Partial melting, 49
Partial pressure, 50
Partial volume, 268
Parts per million, 9, 268, 408
Parts-per notation, 173, 268
Parvancorina, 113
Pascal second, 187
Pascal (unit), 3, 13, 150, 187, 274, 394
Passive continental margin, 87
Pasture, 375, 377
Patrick Moore, 436
Paul Crutzen, 374
Paul Davies, 77, 95
Paul F. Hoffman, 144
Pauli Exclusion Principle, 172
Paul Spudis, 435
Paul Vidal de la Blache, 389
Pavillon de Breteuil, 316
Peak of Eternal Light, 411
Peary (crater), 411
Pedogenesis, 17, 117, 233
Pedosphere, 17, 233
Pelvis, 73
Pelycosaur, 124, 125
Pendulum, 183, 320, 352
Peninsula, 238
Pennsylvanian period, 85
Peptide nucleic acid, 53, 103
Peptide-RNA world, 53
Peptides, 53
Percentage, 268
Per E. Ahlberg, 133
Peridotite, 186, 236
Perihelion, 21, 28, 138
Perihelion and aphelion, 1, 363
Period (geology), 43, 83
Periodic table, 102, 168, 180
Permafrost, 62, 263, 308
Permeability (electromagnetism), 339
Permian, 83, 86, 100
Permian–Triassic extinction event, 63, 100
Perm, Russia, 254
Peroxisomes, 60
Persian Gulf, 143
Perth, 250
Perturbation (astronomy), 44, 135, 411
Peru, 243, 244, 252
Peter Dodson, 133
Peter D. Ward, 148
Peter Haggett, 383
Peter Kropotkin, 389
Petroleum, 30

Post-structural, 384
Poststructuralist, 387
Potassium, 179, 187, 395, 409
Potassium-40, 12, 194
Potassium oxide, 11
Potential, 291
Pound force, 316
Pounds per square inch, 13, 274
Poverty, 75
Power outage, 332
Praseodymium, 177
Preadaptation, 120
Precambrian, 114, 358
Precession, 28, 135, 140, 356, 420
Precession (astronomy), 25, 354, 363
Precious metal, 175
Precipitate, 53
Precipitation, 287, 300, 301
Precipitation (chemistry), 103, 401
Precipitation (meteorology), 288, 294, 309
Predictions, 291
Prehistoric fish, 66
Pressure, 187, 275
Pressure gradient (atmospheric), 291
Pressure ridge (ice), 19
Pressure system, 291
Primary (astronomy), 395
Primary production, 137
Primate, 70
Primitive mantle, 45
Primordial heat, 192
Primordial nuclide, 5
Princeton University, 216
Princeton University Press, 131
Prograde motion, 360
Progress in Human Geography, 391
Project Horizon, 424
Prokaryote, 43, 55, 107, 149, 371
Promethium, 175
Propellant Depot, 429
Prospecting, 341
Prospectors, 320
Protactinium, 176
Protein, 52, 103
Proteins, 53
Protein synthesis, 55
Proterozoic, 43, 82
Protestantism, 440
Protist, 109
Protocell, 104
Proto-Germanic, 4, 396
Proto-Indo-European, 396
Proto-Indo-European language, 434
Proto-Indo-European religion, 431
Proto-mitochondrion, 59
Proton, 168

Proton–proton chain reaction, 146
Protoplanet, 6, 44
Protoplanetary disc, 81
Protoplanetary disk, 43–45, 359
Protorothyrididae, 124
Protostome, 113, 115
Proto-Tethys Ocean, 83
Proxy (climate), 302
Proxy (statistics), 195
Psychoanalysis, 387
Psychogeography, 387
Pterosaur, 124
Ptolemy, 361, 422
Public Broadcasting Service, 347
PubMed Central, 227, 230, 346
PubMed Identifier, 77, 131, 132, 226–228, 230, 346
Puerto Rico Trench, 370
Purple bacteria, 56
Pygostyle, 125
Pyrite, 283
Pyroxene, 16
Pythagoras, 36, 440
Pythagoreanism, 158, 349

Q:Air, 286
Qi, 420
Qinghai–Tibet Railway, 244
Qingshania, 112
Quadrantids, 410
Quadrupole, 342
Qualitative data, 384
Qualitative research, 381
Quality of life, 385
Quantitative data, 293
Quantitative research, 381
Quantitative revolution, 382, 383, 389
Quartz, 16
Quartz (publication), 232
Quasiperiodic motion, 28
Quasi-satellite, 2, 35
Quasi-Zenith Satellite System, 158
Quaternary, 92
Quaternary glaciation, 93, 135, 140
Quaternary ice age, 79, 93
Quebec, 93
Queen Maud Land, 249

Rabi Island, 242
Racism, 382
Radian, 354
Radiation, 278
Radioactive, 339
Radioactive decay, 12, 176, 177, 191, 194, 195, 409
Radioactivity, 211

Sevier orogeny, 90
Sex-determination system, 127
Sexuality and space, 384
Sexual reproduction, 99, 108
Shackleton Energy Company, 429
Shale, 108
Shark Bay, 106
Shear zone, 223
Sheet erosion, 117
Shell theorem, 323
Shen Kuo, 420
Shield volcano, 150, 403
Shi Shen, 420
Shock wave, 44
Shoreline, 238
Shoulder girdle, 123
Shrew, 9, 68
SI, 25, 327, 354
Siachen Glacier, 245
Sial, 199
Siberia, 83, 144, 242, 337, 373
Siberia (continent), 64, 86
Siberian Traps, 69
Sichuan, 244
Sidereal day, 25
Sidereal period, 411
Sidereal time, 354
Sidereal year, 26, 360
Sidereus Nuncius, 422, 423
Siderian, 57
Siderophile element, 45, 189
Silica, 11
Silicate, 186, 187
Silicate mineral, 16, 136, 177
Silicate minerals, 11, 147, 172, 183
Silicon, 11, 102, 168, 172, 177, 179, 199
Silicon-28, 170
Silicon-29, 171
Silicon-30, 171
Silicon dioxide, 400, 406
Silurian, 83, 84, 117, 118
Sima (geology), 199
Simon Newcomb, 354
Simon Winchester, 226
Simoom, 286, 300
Sinauer Associates, 132
Singapore, 316
Single-origin hypothesis, 73
Sinistral, 202
Sinkhole, 31
Sin (mythology), 431
Siple Island, 249
SI prefix multipliers, 13, 44
Sirocco, 286, 300
SK-42 reference system, 158
Skeleton, 66

Slab correction, 319
Slab pull, 205, 206
Slime mold, 111
Slush, 287, 301
Slushball Earth, 59
Small shelly fauna, 114
SMART-1, 427
Smithsonian Institution Press, 436
Smoky Hill Chalk, 90
Snake, 124
Snells law, 186
Snow, 287, 294, 301
Snowball Earth, 8, 43, 55, 68, 83
Snowfall, 306
Snow grains, 287, 301
Snow roller, 287, 301
Snowsquall, 287, 300
Social ecology (theory), 388
Social geography, 384, 388
Social reproduction, 390
Society, 4
Sodium, 179, 186, 395, 409
Sodium-23, 171
Sodium oxide, 11, 400
Soft landing (rocketry), 427
Software engineering, 426
Soil, 4, 17, 117, 371, 377
Soil degradation, 31, 377
Soil depletion, 31
Soil erosion, 377
Soil salinity, 377
Solar calendar, 434
Solar core, 146
Solar day, 360
Solar eclipse, 27, 359, 395, 419
Solar eclipse of August 11, 1999, 419
Solar energy, 262, 288, 290, 374
Solar flare, 332
Solar luminosity, 6, 135, 146, 151
Solar mass, 2
Solar nebula, 5, 44, 79
Solar noon, 353
Solar radiation, 19, 135, 191, 265, 308
Solar radius, 146
Solar System, 3–5, 29, 44, 79, 81, 101, 135, 170, 171, 288, 296, 297, 331, 359, 361, 395, 437
Solar Terrestrial Relations Observatory, 419
Solar time, 26, 355
Solar wind, 6, 21, 45, 82, 145, 152, 269, 289, 297, 298, 325, 407, 409
Solar System, 39
Solid angle, 34
Solnhofen limestone, 89
Solstice, 25, 28, 353, 363
Solvent, 103

Song dynasty, 420
Sounding rocket, 271
South Africa, 57, 247, 249, 252, 254, 308
South America, 73, 90, 92, 143, 211, 249, 252, 306
South American Plate, 14, 221
South Atlantic Anomaly, 327
South Atlantic Ocean, 249
South Australia, 308
Southern Hemisphere, 28
South Magnetic Pole, 325
South Pacific Ocean, 248, 249
South Pole, 28, 143, 349, 415
South Pole–Aitken basin, 402
South Sudan, 249
Sovereignty, 33
Soviet manned lunar programs, 425
Soviet Union, 75, 373, 395, 430
Space debris, 2, 35
Space Race, 423
Spaceship Earth, 441
Space Shuttle, 276, 318
Space Shuttle Challenger, 76
Space Shuttle Endeavour, 273
Space weather, 297, 332
SpaceX, 429
Spain, 160, 254
Spatial analysis, 388
Spatial reference system, 157
Spatial Synoptic Classification system, 302, 304
Species, 3, 42, 52, 97, 120, 137
Spectacled bear, 70
Spectrograph, 430
Speed, 315
Speed of sound, 275, 277
Sperm, 116
Sphere, 158, 159, 315
Spherical Earth, 36, 158, 159, 242, 440
Spherical harmonic, 159
Spherical harmonics, 342
Spherical symmetry, 315
Sphericity, 36
Spheroid, 160, 242, 251
Spinal vertebra, 115
Spiracle, 119
Spirituality, 73
Spirochete, 60
Splitting of the moon, 433
Sponge, 60, 111, 113
Spore, 105
Spores, 266
Spreading ridge, 199
Spriggina, 62, 113
Spring equinox, 28

Springer Science+Business Media, 131, 132, 165
Springer Verlag, 225
Spring (season), 286, 299
Spring tide, 418
Sputnik 1, 76, 163
Sputtering, 145, 409
Squamata, 124
SRID, 158
Stability of the Solar System, 139
Standard gravitational acceleration, 319
Standard gravity, 315
Standard of living, 385
Stanford University Press, 132
Stanisław Masłowski, 434
Stanley Miller, 52
Star, 168, 169
Star and crescent, 433
Starfish, 115
Star trail, 350
State of Palestine, 246
Statistically significant, 138
Statutory law, 375
Steens Mountain, 336
Stellar and sidereal day, 349
Stellar collision, 138
Stellar evolution, 9, 58, 147, 152
Stephen Blair Hedges, 99, 370
Stephen Jay Gould, 132
Steppe, 306
Sterane, 108
Stereoscopic, 240
Stereoscopy, 402
Sterling Publishing Co., 436
Stewardship, 441
STL (file format), 403
Stone tool, 129
Storm, 286, 295, 300, 306
Storm surge, 286, 300
Stramenopiles, 107
Strategic geography, 386
Stratigraphy, 44, 79, 88, 405
Stratocumulus, 290
Stratocumulus cloud, 290
Stratopause, 271
Stratosphere, 21, 149, 273, 277, 283, 288
Stratum, 82, 212, 238
Stratus cloud, 294
Stromatolite, 56, 57, 99, 105, 106, 281, 357
Stromatolites, 42
Strongly magnetic minerals, 336
Structural geology, 212
Structure of the Earth, 45, 164, **183**, 190, 192, 199, 325
Sturtian glaciation, 358
Style guide, 4